Scientific Explanation
and Atomic Physics

Albert Einstein (*left*) and Niels Bohr. Photo by Paul Ehrenfest.
Reprinted by permission of AIP Niels Bohr Library.

Edward M. MacKinnon

Scientific Explanation and Atomic Physics

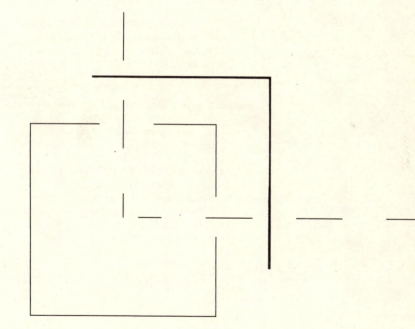

The University of Chicago Press
Chicago and London

Edward M. MacKinnon is a professor in and chairman of the Philosophy Department at California State University, Hayward. He is the author of *Truth and Expression: The 1968 Hecker Lectures* and *The Problem of Scientific Realism*.

The University of Chicago Press, Chicago 60637
The University of Chicago Press, Ltd., London

Library of Congress Cataloging in Publication Data

MacKinnon, Edward Michael, 1928–
Scientific explanation and atomic physics.

Includes bibliographical references and index.
1. Atomic physics—Philosophy. 2. Physics—
Philosophy. 3. Physics—History. 4. Science—Philosophy.
5. Science—History. I. Title.
QC171.2.M23   539   82-2702
ISBN 0-226-50053-5   AACR2

For Barbara

Contents

Preface

This book began its career as an attempt to show how the historical unfolding of one basic scientific problem relates to theories of scientific explanation and accounts of scientific development. By the time the bulk of the first draft was written, I was reluctantly forced to admit that my initial approach was misguided. The philosopher may approach history seeking answers to his own distinctive questions. But he should not attempt to impose rigid patterns or predetermined forms. The difficulty is not merely the ever present danger of distorting history. Of even more concern is the strong likelihood that one who so approaches history will fail to learn much that a more impartial study of history might teach.

The first draft was abandoned, not without reluctance, and I began over, trying to write an account that would be determined by historical sources, not by philosophical theories. My own philosophical values are surely reflected in the way I evaluate the significance, consistency, and adequacy of some of the positions considered. Yet, to the best of my ability, I have avoided imposing my own views on anyone. The study remains a historical one. My philosophical position on the basic issues raised in the present work will be presented later in a separate study.

In the course of the many years that have gone into the writing and rewriting of the present work I have benefited from the assistance, knowledge, and critical feedback of many people and institutions. California State University, Hayward, awarded me a year's leave for research. John Heilbron and the staff of the Office for History of Science and Technology, University of California, Berkeley, were quite helpful in sharing their resources. Some of the ideas developed here, as well as a few of those eventually rejected, were initially presented as talks. I feel grateful for the support and criticism I received from Patrick Suppes and the members of his on-going seminar on philosophical problems of quantum theory; to the members of the Herbert M. Evans History of Science Dinner Club; to the fifth International Congress of Logic, Methodology, and Philosophy of Science; to the West Coast History of Science Society; to the Philosophy of Science Association, the American Philosophical Association, Pacific Division, the Boston Colloquium for the Philosophy of Science; to both the philosophy and physics departments of California State University, Hayward; and to the philosophy departments of the University of San Francisco, San Francisco State University, and the University of Maryland. Preliminary drafts of some chapters were either published or privately circulated. I feel grateful to the editors and referees of *Historical Studies in the Physical Sciences* and the *American Journal of Physics* for inducing me to deepen my research and sharpen my arguments.

Among the philosophers who have commented on various stages of the present work, I owe a special debt to Robert Cohen, Ernan McMullin, Richard Schlegel, and most of all Linda Wessels. I wish to thank Janice Peek, Betty Perry, and Nancy Sadoyama for assistance in typing.

Finally, I must express my special gratitude to my wife and fellow philosopher. Her patience, encouragement, and constructive crtitcism helped me to persevere to the end. To her this book is lovingly dedicated.

Introduction

The proper interrelation of historical and philosophical accounts of scientific development has emerged as one of the most difficult and delicate tasks faced by contemporary philosophers of science. Historians of science are generally, and I believe reasonably, suspicious of any attempt to impose an a priori order on historical developments. An anecdotal approach, which treats the history of science as a quarry from which philosophers may extract illustrations to support their own theories or counterexamples to refute opposing theories, generally produces distorted history. Even philosophers working in the history of science must respect the integrity of the discipline.

The present work is a historical account written from a philosophical perspective. To circumvent possible misinterpretations it may help to comment on both aspects of this claim. This book is not intended as a complete history of atomism as such. It is a historical analysis of the cluster of interpretative problems generated by atomism. Its focus is on the way scientific explanations have been developed and function in atomic theories, particularly in the atomic physics that was developed between 1913 and 1927.

The book is historical; the perspective is philosophical. I am convinced that a searching historical investigation can make a unique and valuable contribution to our understanding of the way explanations function in science. However, I also believe that it is helpful to make a clear distinction between a historical account and a philosophical analysis. The present work is historical rather than philosophical. The philosophical issues it raises will, hopefully, be treated more systematically in a separate study. In the remainder of this Introduction I wish to discuss the bearing this philosophical perspective has on the selection and interpretation of the materials presented here.

This work grew out of a concern with two competing philosophical approaches to the nature of scientific explanation. Logically oriented philosophers tend to treat scientific theories, at least reconstructed scientific theories, as examples of deductive systems. Their concern is with an account of scientific explanation that is general and normative. Historically oriented philosophers, or historians of science functioning as philosophers, often insist that such normative approaches either distort or omit entirely some of the most basic and problematic features in the development of science. The attempt to force scientific explanations into preconceived molds is seen as an obstacle to understanding the way explanations actually develop and function in ongoing scientific systems.

What is now needed is, not further divisiveness, but some basis for interrelating the best work being done in each approach. As a first step in this direction it would help to overcome a shortcoming found in differing developmental approaches to science. One begins with a theory of scientific development and then selects historical examples to fit the theory. Such an approach makes it difficult to understand the underlying continuity that scientific developments manifest. Here I have attempted to trace the development of one basic problem, the role of atomic assumptions in scientific explanations, in a way that is complete enough to bring out both the underlying continuity and the various conceptual and methodological changes involved. Any account of scientific development that cannot accommodate this material must, I believe, be considered unsatisfactory.

The way in which this study might relate to logic-centered approaches to scientific explanation is more indirect and requires a more detailed discussion. As logical positivism gradually died the death of a thousand qualifications, the attempt to cast all theories into essentially the same mold was replaced by more flexible approaches.[1] Instead of attempting to impose normative forms which all theories should fulfill, many philosophers of science have come to stress the primacy of pragmatics. One accepts the theories that work on the grounds of their success. Such functioning physics supplies the intended interpretation which rational reconstruction should clarify. A necessary condition for the success of such an enterprise is that one have an adequate understanding of the way in which functioning physics develops and is interpreted.[2]

In spite of such advances, the recent logic-centered approaches to scientific explanation still tend to share some of the limitations of their predecessors. First, they are technique-oriented. The philosophical problems treated are those that can be accommodated to the type of analysis and reconstruction that philosophers are proficient in. Second, the theories analyzed are still generally treated as conceptually isolated units, rather than as a part of a developing network of theories.

Only gradually, as the historical analysis presented here stretched far beyond the time originally allotted, did it become clear that the most basic philosophical problems this survey generates slip through the nets that logically oriented philosophers impose on science. The two most basic could be labeled 'foundational concepts' and 'coherence'. To bring out the significance of these problems and their relation to the historical material treated here we need a way of looking at scientific theories which fits the practice of science better than the familiar models of scientific theories as interpreted axiomatic systems.

To a crude first approximation the type of scientific theory considered here may be thought of in terms of two components: a physical language and a mathematical formalism. The physical language can, another first-order approximation, be thought of as involving a phenomenological level and a depth level. The phenomenological level is the language used to discuss experiments and report observations. This is basically ordinary language, supplemented by scientific terms and specialized uses. The depth level involves the postulation of entities, structures, properties, or processes, which somehow explains why the phenomenological level appears as it does.

The relation of this physical language to a mathematical formalism inevitably involves a homomorphism of structures embedded in language into mathematical structures. The way in which these are embedded, however, depends on many factors: the type of theory involved, its stage of development, and the way in which mathematics is actually used. For our present purpose, one of supplying a framework for the discussion of foundational concepts and coherence, we may make a rough division of scientific development into three stages: the formative period, deductive unification, and systematic reconstruction. Since the first is our primary concern here, we will begin by contrasting it with the other two.

The formative period may be subdivided into a protoscientific stage, which involves an attempt to organize or reorganize some domain; and a nomic stage, where the emphasis is more on the discovery of laws than on the imposition of categorial schema and classificatory systems.[3] Both are characterized by groping, by a proliferation of tentative hypotheses, by an abundance of postulated entities and processes, and by a very loose fit between the physical language and the mathematical formalism. The emphasis is generally on individual laws, such as Kepler's laws, Boyle's law, Faraday's laws, or Balmer's laws, rather than on mathematical theories.

The stage of deductive unification is characterized by the emergence of theories as explanatory units. This generally leads to a pruning of the luxuriant ontology of the earlier stage, and the replacement of physicalistic by more formalistic reasoning. Thus, instead of making particular adaptions of the Bohr model of the atom, geared to different problems, post-1926 physicists could simply set up and solve the appropriate Schrödinger equation. This pruning transforms some apparently ontological problems into epistemological problems. Thus, where Maxwell and his competitors debated what aspects of reality corresponded to the two electrical and two magnetic vector terms, and to the two potentials, later physicists could focus on the question of what the theory really says about physical reality. Such pruning eventually serves to sharpen the problem of theoretical entities.

Rational reconstruction of scientific theories will not be treated here. The topic is introduced merely to point out some philosophical topics that such reconstructions do *not* treat. A rational reconstruction is characterized by two features. First, a scientific theory becomes an object of study in its own right, rather than a conceptual tool for studying some domain of physical reality. This entails separating the theory from its normal interpretative background. Second, syntactical precision is achieved through semantical impoverishment. Basic concepts must be trimmed and adapted to the structural constraints proper to the logical forms imposed. For these reasons, rational reconstruction, though helpful in treating such problems as consistency, completeness, and independence of axioms, is not the proper tool for clarifying foundational concepts and the type of overall coherence that hinges on concepts which play a basic role in different theories, concepts such as 'wave', 'particle', 'energy', 'mass', 'space', and 'time'. The practicing physicist is likely to be even less adept at clarifying the meaning of such concepts than is the typical philosopher of science. However, if the meaning of a term depends on its normal usage, then it is necessary to attend to the linguistic practices of the fluent native speaker, the functioning scientist.

Within this general framework, which is reflected but not imposed in the text, we may indicate how philosophical problems concerning foundational concepts and explanatory coherence emerge, grow in complexity, and undergo various transformations. Judging by the fragments that survive, Democritus seems to have developed his doctrine of atomism in the form of a quasi dialogue. The mind says, "By convention are sweet and bitter, hot and cold, by convention is color; in truth are atoms and the void." To this the senses rejoined, "Wretched mind, do you, who get your evidence from us, yet try to overthrow us? Our overthrow will be your downfall."[4]

Later forms of atomism, associated with the revival of Lucretius's doctrines, Newtonianism, kinetic theory, nineteenth-century chemistry, and twentieth-century quantum theory, all encountered an underlying clash between an epistemological analysis, which tends to ground knowledge and its expression in the lived world, and an ontological analysis, which sees atoms as foundational and the world of ordinary experience as ontologically derivative. How does one climb from what is basic to us to what is basic in reality—and then dispose of the ladder? There is something of a shared problem in the developments considered. Yet, its formulation grows in sophistication and complexity. Of the various factors that contributed to the advancement and transformation of physics, none was more important than the fusion of physical accounts and mathematical formulations. In the first chapter I have attempted, as a sort of supplementary theme, to indicate the protracted process of conceptual confusion and groping effort that gradually led to a viable synthesis. Even Newton, who brought this synthesis to fruition, was less than clear on the way in which mathematical formalism related to his system of the world.

Later clarifications of this relationship involve a potential source of misunderstanding. A tradition stemming from Cauchy, Dirichelt, and Riemann, and brought to completion by Borel and Lebesgue, has come to define a mathematical function as a mapping of one set of numbers into another. This idea of a function is generally accepted, usually as something nonproblematic, by philosophers concerned with rational reconstructions of scientific theories. This is *not* the idea of a function operative in the mathematical physics treated in the present history.[5] In the classical mathematical tradition, developed especially by Euler and Lagrange, one can have a mathematical formulation of a natural law only when the analytic form of the mathematical expression corresponds to the functional relationship between physical variables. Unless this emphasis on the form of expressions, rather than the sets of numerical values they embody, is appreciated, it is difficult to understand the way physicists argue from physical descriptions to mathematical formulas, and vice versa.

Those concerned with atomism customarily divide physics into classical and quantum, and explain the distinctive features of the latter by contrast with the former. This distinction, valuable as it certainly is, can obscure the difference between the problems generated by atomism as such and those proper to the quantum hypothesis. To bring out this difference we have presented a rather general survey of the attempts to accord atoms a foundational role in prequantum physics. Newton himself had hoped to give his mechanics something of an atomistic foundation. The attempt, he thought, would work only if based on the properties atoms necessarily have. This, in turn, depended on a peculiar sort of inference from the properties of macroscopic bodies to the properties of ultimate corpuscles. If the inference were correct, then atoms, so conceived, should supply a unified foundation for a coherent account of matter and its various manifestations.

Newton never really succeeded in supplying such a foundation. Euler reinterpreted mechanics as essentially a science of macroscopic matter, rather than of ultimate corpuscles. When this was accepted, it transformed the problem of arguing from a macroscopic to a microscopic framework. The new problem was not one of foundations for mechanics, but whether or not one could use the language of Newtonian mechanics to describe even such basic atomic interactions as collisions. No coherent way of doing this was found within the type of descriptive framework Newton had supplied. If atoms are perfectly hard, then the Newtonian account of impact does not apply. If atoms are not perfectly hard and indestructible, then they seem incapable of explaining the stable properties of matter, though this was the basic reason for Newton's insistence on atoms. Eventually this impass was bypassed by relying on a more abstract formulation of mechanics, one stressing conservation principles. In this abstract framework, however, one could not argue from the success of the theory to the properties of atoms.

The nineteenth century witnessed three new sciences concerned with atoms: chemical atomism, kinetic theory, and spectroscopy. Chemical atomists postulated a large number of specifically different atoms. Kinetic theory required only a few structurally different atomic types. Even the more complex types were assumed to have only a small number of internal degrees of freedom. Spectroscopy seemed to require chemically different atoms, each having a large number of internal degrees of freedom. Originally, the dominant tendency was to treat this apparent incoherence as an ontological problem. Since no entity, whether observable or not, should admit of contradictory descriptions, it seemed reasonable to assume that chemists, kinetic theorists, and spectroscopists were not really discussing the same class of entities. Perhaps the chemists and spectroscopists were really discussing molecules, large aggregates of more basic kinetic atoms. When such attempts to achieve coherence failed, there was a tendency to retreat from ontological to epistemological considerations. How, Mach queried, could one reasonably expect to infer something insensible on the basis of sense-observations?

Neither Boltzmann's defense of, nor Mach's attack against, atomism was really cogent. However, when considered as part of an overall development, their conflict brings out the significance implicitly attached to coherence considerations. The operative norm seems to be: any postulation of real entities and processes must ultimately fit into an account of physical reality that is both self-consistent and coherent with the capacities and limitations found in any expression of human knowledge. If the available theories do not meet this requirement, then they must be judged to be incorrect, or at least inadequate.

From the end of the nineteenth century on there was a general tendency in the European physics community to discount the direct ontological significance previously attached to coherence considerations. Following Hertz, Helmholtz, and Planck, and under the indirect influence of Kant, physicists began to insist that physics should supply a coherent picture of reality, but that such pictures need not be interpreted as representations of reality as it exists objectively.

It never proved possible to give a coherent account of real atoms within the general framework loosely labeled classical physics. Both relativity theory and quantum theory represent decisive breaks with that framework. Our concern here is primarily with quantum theory. Though originally developed independently of atomic physics, it soon became the decisive tool for treating atoms. The two-component model of scientific explanation, as applied to the development stage of science, supplies a useful framework for discussing the decisive breakthroughs that transformed atomism. At the stage of deductive unification one may interpret a functioning theory in terms of a relationship between a coherent physical account and a mathematical formalism. However, when the point at issue is a decisive break-through to the beginning of a new theory, the emphasis generally shifts from theories as a whole to the discovery of new laws. Here it often helps to focus on the aspects of the prevailing theories that seem most incoherent.

Almost without exception, the men involved in this revolution still relied on the Lagrangian idea of a function. When a law of nature is given a mathematical expression, then relationships between physical quantities, as conceptualized, should be reflected in relationships between mathematical variables. This idea, implicit in the different breakthroughs considered here, allows for a certain flexibility in proceeding. One may introduce a new concept and then seek its proper mathematical representation. Thus, Einstein introduced the light-quantum hypothesis and Bohr developed the basic features of his atomic model before either sought new mathematical relationships to express these concepts. Or one may begin with new mathematical formulas, interpret these as expressing relationships between physical quantities, and then seek concepts adequate to express these quantities or their relationships. Thus, Planck developed his radiation formula by a simple adaption of Wien's law. We trace in some detail his subsequent struggle to find the proper conceptual expression of the law that proved successful and to see how this fitted into the network of theories he accepted as established. Similarly, Dirac developed both his transformation theory and relativistic quantum mechanics by playing with equations and then seeking the physical significance of the equations that proved successful. More commonly, there is a kind of dialectical interplay between the two components. Thus, Heisenberg used models of the atom which he knew to be unrealistic to suggest new mathematical formulations. Further elaboration and modification of these formulations led to successful formulas, which then required physical interpretation.

The twentieth-century fusion of atomism and quantum theory did succeed in giving functioning physics an atomic foundation, though hardly an ultimate foundation. Many contributed to this development. Only a few shaped the interpretation of the results. Our emphasis is on those few. Even after the technical problems were solved, interpretative problems remained. These came to a climax in the Bohr-Einstein debates. This I consider the most philosophically significant confrontation in the historical development under consideration. Neither position, however, can be adequately understood in terms of the philosophical arguments given in its support. Each position emerged from a lifelong effort to give a coherent account of a world in which man is a part, a knower, and an agent. For this reason I present a detailed development of each man's way of interpreting science before contrasting the two positions in a more systematic way. Even here I have only drawn on other philosophers as a means of rendering some aspect of either position more intelligible, but have tried not to impose my own views. My position will be presented later in a systematic, rather than a historic, manner.

After misunderstandings and technical difficulties were resolved, the ultimate differences between Bohr and Einstein centered on foundational concepts and the way in which they function to make a coherent account of physical reality possible. Einstein thought that quantum mechanics, in spite of its unprecedented success, must be judged unsatisfactory and incomplete. It cannot be used to describe physical reality as it exists objectively. Bohr countered that, since quantum theory is adequate, we must radically reexamine our ideas of how concepts function to make physical descriptions possible.

Though the struggle was public and philosophical, neither man was aided in any significant way by contemporary philosophers. The dominant opinion, among main-line philosophers of science, was that if one could only get logical structures straightened out, then foundational concepts would more or less take care of themselves. There were, to be sure, some dissenting opinions. At the time the new quantum theory was in its gestation period Whitehead declared, "Speculative philosophy is the endeavor to frame a coherent, logical, necessary system of general ideas in terms of which every element of our experience can be interpreted."[6] His *Process and Reality* presented a comprehensive framework in which the new developments in science could be given a definitive interpretation. Such efforts, however, were generally interpreted as attempting to replace, rather than to clarify, the foundational concepts on which the coherence of functioning physics rested. Both Bohr and Einstein were thrown back on their own philosophical resources when it became necessary to analyze concepts and develop doctrines of how concepts come to acquire the meanings they have.

Today there is a renewed interest in the philosophical problems generated by quantum mechanics. The primary emphasis among philosophers working in this area is still on the logical structures needed to systematize quantum theory, rather than on the problems concerning coherence and foundational concepts left unresolved in the aftermath of the Bohr-Einstein debates.[7] This, to be sure, does not mean that such problems have gone unnoticed. Some, who rely on deductive logic, treat overall coherence in terms of a great sphere of knowledge which impinges on reality through sensory stimulation. This approach, however, has not yet been related to the concepts that play a foundational role in functioning physics. At the other extreme, some, who think that inductive logic should assume a more basic role in explaining science, treat coherence in terms of Bayesian confirmation methods. This approach, however, often dissolves theories as explanatory units and supplies an inadequate basis for treating issues concerning the reality of theoretical entities.

It may help to see these aspects of the philosophical problematic in a somewhat simpler setting. Immanuel Kant attempted to make Newtonian physics intelligible by showing how its foundational concepts could be redeveloped. He was not concerned with doing technical physics. He wished rather to explain how a reconstructed physics, based on atomic assumptions, could fit into a coherent account of a world in which man is a part, a knower, and a moral agent. This aspect of Kant's thought is rarely related to the philosophical problems generated by atomism.[8] Yet, from a methodological point of view, it has a more direct bearing on this problem than the bulk of the work being done by contemporary philosophers of science. For this reason, I have included a detailed account of the development of Kant's position on the nature of scientific explanation and the concept of matter. Those not interested in such issues may skip this chapter with no loss in historical continuity.

These are some philosophical problems which the present study suggested to me. Others may interpret it differently. One methodological point, however, seems beyond dispute. Philosophers who approach the history of science merely to seek illustrations for philosophical theses may find what they seek. Yet, one can learn from this history something new about the nature of scientific explanation only by developing a historical mode of explanation. It is for this reason that the present study concentrates on a detailed historical reconstruction, rather than merely focusing on those aspects of the history that have a direct bearing on philosophical problems.

Science, Kant informs us, advances by compelling nature to answer questions of reason's own devising. The form of the question imposed selects and structures the answers allowed. Something similar must be said of historical studies. No historical narrative presents a purely objective account of events as they actually happened. In this Introduction I have attempted to indicate the philosophical perspectives—or prejudices—that shaped the questions I imposed on the history of atomism. This interpretative base delimited the materials selected, the issues treated, and the significance accorded them. The result is a philosopher's history, a somewhat suspect subspecies of the broader species that is the history of science. Yet, within these limitations, the present work is offered as a historical, rather than a philosophical, account. A properly philosophical treatment of the issues raised here requires a radically different manner of presentation, something that will eventually be attempted elsewhere.

1. Atomism in the Newtonian Synthesis

As indicated in the Introduction, I am not attempting to present a history of atomism as such. I intend, rather, to use both historical reflections and philosophical analysis to bring out the interpretative problems generated by the use of atomic assumptions in scientific explanation. Though these interpretative problems grow in complexity and subtlety with the advance of science, there is something of a problematic core shared by classical and quantum physics. In this chapter we will consider some of the elements that contributed to the problematic status of atomism in Newtonian mechanics. Before beginning such historical considerations, it might be helpful to explain the rationale guiding the selection of the elements chosen.

Ancient atomism supplied an account of reality which competed with other philosophical accounts. Implicit in the dialectic of competing positions was a set of ground rules governing the competition. One should not only be able to offer arguments for, and counterobjections against, one's own position; one should also show that the position defended allowed for, or fitted in with, a coherent overall account of reality. With the rise of science and the successive breakaway of the physical, psychological, and social sciences from philosophy, the scope of knowledge vastly expanded and the role of philosophy drastically shrank. It is difficult to take former delusions of adequacy and omnicompetence seriously. Yet, the demand for an overall coherence in one's account of reality was one of the ground rules governing the dialectic of competing positions. Thus, theological and ethical considerations were considered to be pertinent by both the defenders and opponents of atomism.

Atomism, as a doctrine about the nature of physical reality, represents one strand in the problematic core. A second is the rise of mathematical physics. As Duhem, Sarton, and the scholars who followed them have shown, the rise of the new physics was not simply due to the originating genius of such men as Galileo, Kepler, Descartes, and Newton. It had its roots in medieval Aristotelian science. The root that most concerns us is the one involved with the gradual emergence of methods of treating qualities mathematically. However this was to be interpreted, it clearly involved some process of abstraction from physical reality. For this reason, the interpretative problems that seemed most urgent were those concerned with showing that the abstractions used contributed to the understanding of, rather than the distortion of, physical reality. The aspects of this problem that now seem obvious and basic, the idea that attaching numbers to properties requires a unit, a scale, and appropriate dimensions, were not at all obvious in the beginning of this development.

After a long period of incubation involving a gradual transformation in the ways qualities were spoken of and represented, mathematical physics finally emerged, with Newton's *Principia* as its early culmination and crowning success. This very success precipitated a transformation in the way science came to be interpreted by many of its practitioners, a transformation which partially foreshadowed the historicist-logicist split considered in the Introduction. On one side are those for whom science is essentially a way of reasoning about physical reality. Thus, Newton's interpretation of his own achievements inevitably involved a justification of the reasoning that led to these achievements. On the other side are those for whom science is essentially an ordered body of knowledge. In this perspective the justification for mathematical physics does not come from the mélange of epistemological and ontological arguments originally used to justify the introduction of new methods. The real justification of mathematical physics is its success. If a system succeeds better when ontological assumptions and epistemological arguments are reduced to a minimum, then just such a reduction is justified. Thus, the physicists who perfected Newton's mechanics dispensed with the ontological foundations he thought necessary and the epistemological arguments involved in the justification of such assumptions.

A fundamental theme of the present book is that an adequate interpretation of science involves a coherent integration of both the developmental aspects of science and the systematics of scientific explanation. To achieve such an integration it is first necessary, in my opinion, to see how these two phases of science have been related in the historical development of scientific thought. Here, Newton plays a pivotal role in two ways, one rather obvious and one not so obvious. Newton fused together the two strands mentioned, the attempt to explain physical reality in terms of the properties and activities of their ultimate atomic parts, and the mathematical treatment of the properties and activities of observable bodies.

The not so obvious aspect is the fact that Newton's justification of the way he fused together atomic assumptions and mathematical physics failed. His successors clearly showed that classical mechanics does not require the atomic assumptions Newton thought necessary. Yet Newton's failure was more majestic and revealing than the limited success of his followers. He attempted, in accord with the ancient rules of the game, to show how his mechanics, properly interpreted, fits into an overall coherent account of reality. This involved him in some of the epistemological problems later treated by the Copenhagen interpretation. One explains macroscopic properties of bodies in terms of the properties attributed to the atoms constituting these bodies. Yet, one can only speak of the properties of atoms in terms derived from macroscopic bodies. How can the order of explanation reverse the order of meaning on which it depends?

This chapter raises but does not attempt to answer this question. The next chapter will consider the most successful and coherent answer to this question given in the perspective of classical physics. In the present chapter we will first consider the elements that went into Newton's integration of atomic assumptions and mathematical physics and the way in which some of Newton's predecessors interpreted these elements. Next, we will consider Newton's fusion of these two strands and the way in which he sought to justify it. Finally, we will consider how this fusion was dissolved by Newton's successors.

Classical Atomism

Classical atomism, for all its boldness as an innovative hypothesis, had a conceptual simplicity that facilitates a summary account. Early Greek philosophy led from speculative hypotheses about the constitution of matter to a fundamental conceptual problem. Change is, as Heraclitus insisted, an inescapable omnipresent feature of the world we experience. Yet, Parmenides countered, change is unintelligible. Being is; nonbeing is not. If there were change, either being would become nonbeing or nonbeing would become being. Since neither is intelligible, change is impossible.

I believe that Kirk and Raven are correct in seeing the Parmenidean paradox as basic to the development of both atomism and Platonism.[1] Both schools insisted, as had Parmenides, that change is only in appearance. The really real, whether invisible, indestructible, eternal atoms or invisible, timeless, transcendent forms, does not suffer any essential change. Epicurus, who consciously advocated conceptual simplicity, adopted atomism and related it to his ethics of pleasure. In Lucretius's *De Rerum Natura* the historically influential elaboration of atomism is presented with little conceptual subtlety.

The heart of the classical doctrine of atomism is succinctly presented in Democritus's famous dictum:

> By convention are sweet and bitter, hot and cold, by convention is color; in truth are atoms and the void. . . . In reality we apprehend nothing exactly, but only as it changes according to the condition of our body and the things that impinge on or offer resistance to it.[2]

By abstracting the reasons given in support of atomism from Lucretius's meandering dialectic of opposing views, we can indicate the two basic types of arguments Lucretius gave in defense of atomism. First, there are speculative arguments to the effect that the doctrine of an infinite number and variety of atoms moving through a boundless void better explains the change and endless variety observed in the universe than any competing views, such as the doctrine of the four elements. The combination of change in the configuration of things and permanence in the universe as a whole is only intelligible, Lucretius argues, if the ultimate constituents of bodies are permanent and unchanging atoms.

The second type of argument is more empirical, arguing from observable phenomena to the unobservable atoms that explain these phenomena:

> Again, in the course of many revolutions of the sun a ring is worn thin next to the finger with continual rubbing. Dripping water hollows a stone. A curved ploughshare, iron though it is, dwindles imperceptibly in the furrow. We see the cobblestones of the highway worn by the feet of many wayfarers. . . . But to see what particles drop off at any particular time is a power grudged to us by our ungenerous sense of sight.[3]

Other empirically based arguments concern the porousness of apparently solid objects and the invisible particles that must be postulated to explain the action of the wind and the penetration of odors. The most interesting such argument is Lucretius's qualitative anticipation of how Brownian motion, or something like Brownian motion, is to be explained. Dust particles, observed to have a random motion in a beam of sunlight, are under the impact of many invisible blows from all directions.[4]

The eventual success of atomism provides an abiding temptation to overemphasize the philosophical worth of the original doctrine. When judged in relation to competing doctrines, especially Aristotelianism, early atomism labored under many difficulties. Atomistic explanations of observable phenomena ranged from the plausible examples cited above to the arbitrary and even ludicrous. Things pleasant to the taste are made of smooth, round atoms, unpleasant things of rough atoms. Viscous substances have hooked atoms. The slight swerve that ultimately accounts for the coagulation of the primordial moving atoms and the formation of the present universe is a purely ad hoc assumption that admits of no real justification within the system.

The comparative shortcomings of atomism, considered as a competing system, can be grouped under three headings. The *ontological* difficulties concerned atomism's inadequacy in explaining basic features of the lived world. If all observable bodies are simply aggregates of unchangeable atoms, then it is difficult to explain the differences between living and nonliving beings, the natural division of living beings into species, and the reality of emergent nonatomic properties. Here the best that classical atomism offered was, for example, the account of the coherence of bodies through the accidental coming together of atoms with matching hooks. It is tempting to see this as an anticipation of van der Waals forces, but this is a temptation best resisted. In its own historical context the atomic account of these gross features of the lived world simply does not compare with the subtlety, comprehensiveness, and empirical bite of the competing Aristotelian account.

The second type of difficulty is *epistemological*. Epicurus and Lucretius insisted on the primacy of sensation as the source of knowledge. Yet, the doctrine that only atoms and the void are real leads to the conclusion that sensed qualities lack physical reality. Atoms have shape and motion, but no color, taste, or smell. The reasoning supporting atomism seems to contradict its own evidential base. Thus the statement Galen attributed to Democritus: ". . . wretched mind, do you, who get your evidence from us [the senses], yet try to overthrow us? Our overthrow will be your downfall."[5]

In addition to these comparative shortcomings and internal inconsistencies, classical atomism also was constrained by some serious *external* difficulties. A doctrine linked to atheism, hedonism, and the denial of personal immortality was seen as inimical to the established religions of pagan Rome, Islam, and Christianity. It is interesting to note, however, that some church fathers, such as Arnobius, Lactantius, and Isidore of Seville, saw Lucretius's arguments against the pagan gods as supporting their cause and transmitted them to later generations.

The late Renaissance revival of Epicurianism and atomism has been adequately treated elsewhere and need only be indicated here.[6] The novel difficulties due to the juxtaposition of the old atomism and the new physics will be considered later. Fifteenth-century humanists discovered a copy of Lucretius's *De Rerum Natura* and had it published in 1473. Though first studied simply as classical poetry, Lucretius's doctrine was soon seen as contributing to the anti-Aristotelianism of Renaissance humanism. Heron of Alexandria's work on pneumatics, also revived in late Renaissance times, was an independent source of the doctrine that matter is formed of small, hard corpuscles separated by empty space. Two waves of English atomism preceded the work of Newton. The first came from the circle of thinkers assembled around Henry Percy, after he became ninth Earl of Northumberland in 1564. Partially under the influence of Giordano Bruno, this circle accepted atomism and, in Thomas Harriot, produced an atomist who was a significant original thinker. In spite of his brilliance, his influence was slight chiefly because of his dearth of publications.

The second and more influential wave came from the new mechanical philosophy developed in different ways by Hobbes, Descartes, and Gassendi. As Kargon showed, this new mechanism became quite influential in England after 1650 and was widely accepted by the time Newton published his *Principia* in 1687.[7] Before considering the aspects of atomism that proved most problematic in Newton's account, we will consider the second thread we wish to trace.

The Mathematization of Qualities

The mathematical treatment of qualities now seems a simple matter.[8] Most introductory physics texts treat it quickly before moving on to more interesting and challenging topics. It is difficult for anyone with even a modicum of scientific training to realize the centuries-long struggle with conceptual confusion, false starts, and misleading questions that led to the methods that now seem obvious. I believe that this issue had, and still has, an important bearing on the ways scientific theories are interpreted. Yet, it is rarely analyzed since, for the most part, the development hinges on implicit presuppositions, background issues, and ways of speaking, rather than any datable scientific advances. What follows is a tentative interpretation of the stages in the gradual development of the quantitative treatment of qualities.

Aristotle's doctrine of the ten categories, given in slightly different orders in *Metaphysics* 5. 7 (1017a7–1018a20) and in *Categories* 4 (1b25–2a10), derives from his analysis of predication. It is, to be slightly anachronistic, a good example of early ordinary language analysis. 'Red,' for example, cannot be properly predicated of anything unless it is an extended substance. Thus, the quality of being red presupposes the quantity of extension, which in turn presupposes a substance. Substance is not properly predicated of anything. As modern critics have pointed out,[9] Aristotle's treatment of such issues hovers uneasily between explaining what things are and analyzing the ways in which we speak. Regardless of Aristotle's position, or lack of a position, on such issues, medieval Aristotelians gave a strong ontological interpretation to the distinction between a substance and the accidents which inhere in and determine a substance.

St. Thomas Aquinas was, it seems, the first medieval Aristotelian to give a coherent account of the way quantitative determination can be given to qualities in spite of the fact that 'quantity' precedes 'quality' in the ordered list of categories as a matter of conceptual necessity. His account was based on the doctrine of the relation between the three degrees of abstraction and the type of reasoning characterizing the different speculative sciences.[10] The first level, natural philosophy, is based on an abstraction of a whole. Thus, one considers human nature as abstracted from individuating differences. This type of abstraction characterizes Aristotelian physics, or the study of human nature, as subdivisions of the general philosophy of nature.

The second degree of abstraction is an abstraction of form from matter. If, as Aquinas did, one interprets the order of the categories (at least substance, quantity, and quality) as reflecting the intelligibility of being and the abstraction of concepts from the individuals that embody them as a real psychological process, then there is a natural order proper to the process of abstraction. Thus, one can consider quantified matter (or intelligible matter) while prescinding from quality, but one cannot consider any quality of a substance while prescinding from quantity. This gives mathematics a grounding that is both ontological and psychological. Abstraction of a whole, considered simply as a whole, gives the unit which serves as the basis of numbering. Abstraction of a shape gives the intuitive basis for geometry. This supplies a foundation for all the types of mathematics then known. The way in which the third level should be interpreted as supplying a ground for metaphysics is a disputed issue not pertinent to the present discussion.

Arithmetic and geometry are, in this view, sciences of matter, though at a second level of abstraction. By Aquinas's time, however, it was clear that a purely materialistic explanation of mathematics was inadequate to the Christian tradition he sought to systematize. The first four ecumenical councils had declared that there are three persons in one God and, adopting geometrical terminology, that all three are equal. The *Sentences of Peter Lombard*, the compilation of patristic doctrines which served as a basic source book for medieval theologians, taught that rank in heaven depends upon the degree of grace, or charity, that a person has at death. To make sense of such usages, accepted as authoritative, it must be possible to apply quantitative motions to things other than material substances.

Quantity, according to medieval Aristotelianism, is the first determination of any material substance, and quality the second. To speak of the quantity of a quality, St. Thomas distinguished between quantity *per se*, or bulk quantity, and quantity *per accidens*, or virtual quantity.[11] Accidental quantity can have magnitude by reason of the subject in which it inheres, as a bigger wall has more whiteness than a smaller one; or it can have magnitude by reason of the effect of its form. The first effect of a form is a way of existing, e.g., as human or as white. The secondary effect of a form, however, is shown through its action or objects. A comparison of relative effects serves as a measure of virtual quantity. Thus, one with greater strength can lift heavier rocks.

Can such virtual quantities be measured? Here again, it is easy to read modern meanings into detached quotes. Aquinas defined measure as "nothing else but that through which the quantity of a thing is known."[12] This may sound modern; yet, it rests on a reversal of the modern idea of a unit of measure. In medieval thought the Platonic idea that the perfect form is the measure for any being that participates in this form was reinforced by the scriptural statement (Wisdom 9:21) that in creating the world God disposed all things in number, weight, and measure, one of Thomas's favorite citations. Thus, in this perspective, the true measure of grace is not some unit, but the grace of Christ in which man participates.[13]

Subsequent discussions of the quantity of qualities retained the spirit, though not always the doctrines, of Aquinas. Questions concerning the manner in which numerical values are to be assigned were subordinated to questions seen as conceptually more basic. How is quality to be understood as a determination of a substance? How is any change in a quality to be understood in terms of the causes producing change? Two special issues which have a distinct bearing on Newton's treatment of qualities are the questions of the intensification and remission of qualities and the question of how the qualities of components are present in a mixture or a chemical compound.[14]

That qualities can be intensified or remitted is a fact of experience. Heated bodies may become warmer or colder, surfaces brighter or dimmer, students wiser or duller. Late medieval discussions of such intensification and remission were, not surprisingly, cast in ontological terms. What undergoes the change in intensity—the quality itself, the degree of participation? Or does one quality replace another as night replaces day? If the last, rather superficial, answer is rejected, then the central issue is one of understanding how the new quality added merges with the quality already present to become one qualitative determination.

The fourteenth-century nominalists, particularly William of Ockham, attempted to sweep away this interminable ontological speculation. Instead of asking how the intensification or remission of a quality takes place, Ockham sought a criterion by means of which he could decide when a word admits of the adjectives 'strong' and 'weak' and when it can be combined with the terms 'large' and 'small.' When the problem was viewed in this way, changes in local motion, rather than changes in degrees of charity, emerged as the prime example of intensification and remission. If instantaneous velocity is treated as a quality, then any change in velocity, i.e., acceleration or deceleration, is treated as an intensification or remission.

Thanks to the labors of a generation of scholars, late medieval treatments of local motion have become one of the better known segments of the history of science.[15] Here I will only consider these discussions of motion as illustrations of the difficulties involved in developing a mathematical representation of the intensification and remission of qualities. Euclid's theory of proportions, which supplied the techniques and terminology employed, implicitly imposed some conceptual limitation. Thus, Thomas Bradwardine spoke of the ratio between a moving force and a resistance, since both were magnitudes of the same kind. Yet, he did not speak of velocity as a ratio of distance and time or consequently develop any precise notion of inertia. His famous theorem is not, as it is often interpreted, a theorem of uniform acceleration, but a way of relating the quantity of motion in a uniformly variable motion to the quantity of motion in a uniform motion, two quantities of the same kind.

The idea, which now seems obvious, of attaching numbers to qualities relative to some standard taken as a unit measure was not at all obvious at the beginning. The logicians, who then dominated scientific speculation, were not so much concerned with the significance of particular numbers attached to qualities as with the logical consequences of any consistent assignment of numerical values. Thus, Richard Swineshead (or Suiseth), a disciple of Bradwardine who became known as 'The Calculator,' as Aristotle was 'the Philosopher' and Averroes 'the Commentator,' had a procedure of stating a hypothesis as a mathematical relation between qualities, assigning numbers to these qualities on some more or less plausible basis, and then testing the hypothesis by determining the consequences that flowed from the numerical assignments. An interesting example of this technique is his proof of the theorem (generally attributed to Galileo) that a uniformly accelerating body will traverse three times as much space in the second half of a given time as in the first:

> For let a be uniformly accelerated from zero degree to 8. Then if the motion in the first half of the time will correspond to a degree that is less than 2, let that degree be d. Then from the fact that the movement is uniformly accelerated in the second half in the same way as in the first, and the fact that in the middle instant [of the whole time] the degree is 4, it follows that c equally exceeds 4 as d exceeds zero and 6 equally exceeds c as 2 exceeds d. Hence 6 equally exceeds 2 as c exceeds d. Hence since the ratio of 6 to 2 is 3, and c degree is less than 6, it is evident that the ratio of c to d is greater than 3. Consequently, it will traverse more than three times as much in the second half as in the first. But this conclusion is refuted by the initial conclusion. . . . It is evident, therefore, that the motion in the first half will correspond to a degree of 2 in the second half to one of 6.[16]

The most successful medieval attempt to treat qualities mathematically was the graphical method developed by Nicole Oresme in his *Tractatus de configurationibus qualitatum et motuum*. Since this initiates the state space representation, which will later be treated in much more detail, it is of some interest to note how Oresme treats intensities:

> Every measurable thing except numbers is to be imagined in the manner of continuous quantity. Therefore, for the mensuration of such a thing, it is necessary that points, lines, and surfaces, or their properties be imagined. For in them (i.e., the geometrical entities), as the Philosopher has it, measure or ratio is initially found, while in other things it is recognized by similarity as they are being referred by the intellect to them (i.e., to geometrical entities). Although indivisible points, or lines, are non-existent, still it is necessary to feign them mathematically for the measures of things and for the understanding of their ratios. Therefore, every intensity which can be acquired successively ought to be imagined by a straight line perpendicularly erected on some point of the space or subject of the intensible thing, e.g., a quality. For whatever ratio is found to exist between intensity and intensity, in relating intensities of the same kind, a similar ratio is found to exist between line and line, and vice versa.[17]

It is easy to see why Duhem, the modern discoverer of Oresme, thought of him as the real founder of analytic geometry. Later scholars have returned Oresme to his own historical milieu.[18] In modern analytic geometry the abscissa and the ordinate are each assigned numerical values. In Oresme's graphical method, the base line, or *subjectum*, represents the quality in question, while the *altitudo* (ordinate) perpendicular to it represents, by its length, the intensity of the quality at that point, or the *quantitas qualitatis*. The lengths have no absolute significance. A representation is suitable (Oresme's term) if the ratio of the lengths at any two points is the same as the ratio of the intensities of the qualities at those points.

In the application of modern analytic geometry to measurable quantities, the enveloping curve is interpreted as representing a functional relationship, an interpretation which implicitly presupposes the use of some unit of measure. Oresme did not focus on either this curve, or on the area underneath it. His emphasis was on the *configuratio*, or overall shape. Thus, uniform motion has a rectangular configuration and uniformly difform motion, starting from rest, has a triangular configuration. An irregular configuration occurs when an opposite quality repeatedly intrudes on an original, as with the alteration of friendship and enmity, attraction and repulsion. Oresme's discussions of his graphs often read more like an interpretation of René Thom's catastrophe theory than of analytic geometry.

The second question we wish to consider is how the qualities of components are present in a compound. In the classical atomistic account, the qualities of atoms remain forever unchanged, regardless of the combinations these atoms enter into. There is consequently no distinction between a mixture and a true chemical compound. Aristotle, on the other hand, did have such a distinction. Thus, he defined an element functionally: "An element, we take it, is a body into which other bodies may be analysed, present in them potentially or in actuality (which of these is still disputable), and not itself divisible into bodies different in form."[19] In another work he distinguished between an aggregate, in which the parts are unchanged and the whole is not homogeneous, and a true composition, in which the constituents must change to form a homogeneous whole with new properties only potentially present in the parts.[20]

Medieval Aristotelians, especially Averroes, extended this Aristotelian base into a doctrine of *minima naturalia*. These minimal natural units, so conceived, differed from the units of classical atomism in four respects.[21]

First, the *minima naturalia* of different substances were postulated to be qualitatively different. This fitted the importance Aristotelians attached to the concept of natural species. The atoms postulated by classical atomism differ only in size and shape, while macroscopic species have no ultimate significance. Second, in the minimalist doctrine each substance has its own characteristic size units. Third, the shapes attributed to these minimal units play no explanatory role. Fourth, while classical atomism explained chemical composition by configurations of unchanged atoms, the minimalist doctrine taught that minimal parts act upon each other to produce a median quality which disposes the subject for a new qualitative form educed from the potency of the matter.

At the time when classical atomism was revived, the Aristotelian minimalist theory was fairly well known through the writings of Julius Caesar Scaliger (1484–1558), through the Averroistic Aristotelians of the school of Padua in the sixteenth century, and through Kenelm Digby in England (1630–65). In Germany Daniel Sennert (1572–1637) tried to fuse corpuscular doctrines with Aristotelianism as a basis for explaining chemical combinations. These views were popularized by his student Johannes Sperrling in his widely used university textbook, *Institutiones physicae* (1646).[22] *Prima facie*, the minimalist doctrine would seem to have an empirical superiority over classical atomism. It fitted all the arguments for atomism based on observed evidence (slow growth and wearing away of bodies, etc.) and the thought experiments concerning the impossibility of infinite divisibility. It also accepted the observable properties of macroscopic bodies as ontologically real, and it offered a basis for explaining chemical change. This minimalist doctrine relied, to be sure, on the postulation of occult qualities. But so did competing theories. The mid-seventeenth-century acceptance of atomism rather than the competing Averroistic-Aristotelian-minimalist position cannot be attributed to any initial superiority of atomism considered simply as a natural philosophy.

Atomism and Mechanism

The appeal of classical atomism lay, not in its subtlety, but in its simplicity and its potential for furthering a consciously contrived conceptual revolution. Peter Ramus (Pierre de la Ramé, 1515–72), the most influential educational innovator of the sixteenth century, may not have actually begun his career by defending the thesis that everything Aristotle said was wrong, an oft-repeated story. Yet, he certainly mounted a sustained and effective attack against everything Aristotelian.[23] Similarly, though their doctrines differed radically, both Bacon and Galileo insisted that the development of a new science required a new non-Aristotelian methodology. Though these men contributed to the revolution against Aristotelian ideas of science, each rejected atomism as an ontological foundation for the proposed new science. In his early period (1603–12), Bacon had defended atomism as a necessity plainly inevitable in natural philosophy. Yet, as his ideas on scientific methodology developed, he came to realize that the induction he stressed as fundamental relied on matter as observed, rather than a priori but unobservable concepts.[24] Galileo was familiar with atomism and with the minimalist doctrine of the Paduan school. Yet, his work on dynamics depended, not on a priori concepts of matter, but on matter as measurable. Clavelin concludes his study of Galileo with the observation, "It is impossible to discover the slightest link between Atomistic ideas and the geometrization of the motion of heavy bodies."[25] The underlying difficulty is one that haunts the history of atomism. One may attribute a foundational role to observation and empirical testing or to unobservable atoms, but it is hard to hold both.

If observation is considered basic to scientific method, then an atomistic concept of matter seems nonfunctional. If, however, one is willing to admit a priori concepts, then atomism may have a different status. Newton's atomism is justified by a peculiar mixture of a priori considerations and stress on the primacy of observation. This mixture reflects two sources which Newton is known to have studied carefully in his student days, Descartes's *Principles of Philosophy* and Boyle's *Origins of Forms and Qualities*.[26] Here we will consider briefly some ideas on the concept of matter and on scientific methodology which these sources treated in a way that influenced Newton.

The well-known argument in the *Meditations* concludes that extension is the only ontologically real property of material bodies. Descartes's *Princi-*

ples of Philosophy used this concept of matter as a basis for explaining the system of the universe. His vortex theory of the solar system involved three types of particles: the original rough particles; the bits of dust that are ground off these particles through collisions and which form the ether; and the smooth round particles that remain. This atomism, Descartes argued,[27] differs from the atomism of Democritus in three key respects. First, Descartes rejected the idea of a void as unintelligible.[28] Since extension is the essence of a material substance, extension without something extended is a contradictory concept. Space empty of ordinary matter is filled by ether. Second, Descartes rejected Democritus's idea of indivisible atoms. Extension implies divisibility. Though there may be limits to man's power of dividing matter, there is no limit to God's power.[29] Third, Descartes rejected the idea of gravity as an intrinsic property of matter; only extension is an essential property of the matter.[30]

Though Newton eventually became highly critical of Descartes, he was strongly influenced by the precedent Descartes set in explaining the system of the world in terms of matter and motion as ultimate constituents. Two aspects of this are of particular concern here. The first is Descartes's argument that the properties ultimate particles must have are the properties essential to being a material body. Though Newton, in his Third Rule of Reasoning, admitted other basic properties beyond extension, he used the same basic argument. Second, Descartes's concept of matter grounded a mathematical treatment of qualities, though it was a rather nonarithmetical mathematics. Boyer argues that Descartes's *la géometrie* is more properly interpreted as a reduction of algebra to geometry than of geometry to algebra.[31] Since extension and shape, as intuited, are the ultimate ingredients in Descartes's geometrical exposition, he did not have to concern himself with the conceptual problems involved in attaching measurable values to observable qualities.[32]

Where Descartes relied on a priori concepts and deductive arguments, Robert Boyle tried to show how a doctrine of atomism supplied a unified basis for explaining the observable properties of bodies. His atomism, however, was not the austere deductivism of Descartes, but an éclectic doctrine proper to the spirit of the post-Cartesian age. Pierre Gassendi had revived traditional atomism and showed how it could be purged of atheism and reconciled with Christianity. Huygens implemented Descartes's general program, but argued that the activity of ultimate particles cannot be explained if size, shape, and motion are the only properties with which these particles are endowed. A mechanistic explanation of the propagation of light requires that these ultimate particles have perfect hardness and springiness. Though this may, Huygens allows, admit of a deeper explanation in terms of the ether particles filling the interstices of coarse matter, such an account need not be considered.[33]

Robert Boyle was an atomist who was strongly influenced by Descartes and Gassendi. Yet, Boyle was even more strongly influenced by Bacon's stress on observation and induction.[34] The famous experiments which so impressed Boyle's contemporaries seem today not only inconclusive but often ludicrous as proofs of atomism and mechanism. Boyle's work, however, is more intelligible when seen as part of the overall dialectic Boyle confronted in attempting to explain the qualitative properties of material bodies. Though influenced by Descartes's atomism and mechanism, Boyle was too much an observation-centered scientist to accept Descartes's stark distinction between qualities like extension and shape which are ontologically real properties of bodies, and qualities like color which really reside in the consciousness of sensitive organisms. For one who took the ontological status of such qualities seriously, the real choice was between the scholastic doctrine of substantial and accidental forms, which explains the qualities of mixtures as emergent forms conditioned by the disposition of the matter of the component parts, and a mechanistic account, which explains the properties of mixtures in terms of the invariant mechanical properties of ultimate particles.

Mid-seventeenth-century mechanism, far from being uniform, embraced a variety of related doctrines. Boyle's distinctive contribution was a mechanistic account of the ontologically real dispositions in bodies which produce secondary qualities in sensitive organisms. These dispositions, in turn, should be explained through the mechanical properties of ultimate particles: size, shape, motion, and contact action. Boyle's concern here was not the contemporary concern with a quantitative account of measurable properties, nor even with a Cartesian-style deductive account of properties. His arguments and experiments were intended to show that a mechanistic atomism supplies a better basis for explaining the properties of bodies and the ontological dispositions that explain secondary qualities than does the competing scholastic account. Most of Boyle's experiments were designed to show that certain sensible qualities of macroscopic bodies could be introduced, altered, or removed by mechanical means. Thus, to introduce magnetism in a metal one moves it into contact with a loadstone. One can vary colors by heating bodies—Boyle thought of heat as a form of motion—or by whipping fluids to a froth. Pulverizing a glass, a purely mechanical action, destroys its transparency. Some of the physical and chemical properties Boyle attempted to explain in this way are the mechanical origin of heat, cohesion, fluidity and firmness, corrosiveness, chemical precipitation, magnetism, electricity, and color. Only gravity seemed to resist such a mechanical explanation. In each case the goal was to explain observable properties of macroscopic objects in terms of the universal properties, or to use Boyle's quaint term, the catholic affections, of the minute, insensible, undivided, ultimate parts of matter. Boyle's methods or argumentation had some severe limitations. As he said, he could not distinguish experimentally between the atoms of Descartes and Gassendi. Yet, he convinced himself, and many of his influential contemporaries, that mechanistic atomism offered a better program for explaining observable properties than the scholastic doctrine involving substantial forms and occult qualities.

The second distinctive feature in Boyle's position was his fusion of nominalistic metaphysics and voluntarist theology. Though the distinction between voluntarist and intellectualist theologies now seems totally irrelevant to science, it was as vital a concern to Boyle, Newton, and Leibniz as it had been to the medieval scholastics. For all the participants in this protracted debate, God as Creator was considered the ultimate source of whatever intelligibility the universe has. The intellectualist position, developed in different ways by Aquinas and Leibniz, stressed the idea that the plan of creation, something derived from God's contemplation of his own nature as imitable, is intrinsic to the universe considered as a product of the act of creation. The universe, accordingly, must be intelligible in terms of what it is. One need have recourse to God as an explanatory source only for ultimate issues, such as the existence of a universe of contingent beings, a first source of motion or the overall plan responsible for either natural laws or preestablished harmony. One may also invoke God to explain something transcending nature, such as the Incarnation, grace, or miracles. But one should never invoke God for a proximate, or scientific, account of the ordinary working of nature. Thus, this tradition is intellectualist in the special sense of holding, as an interpretative principle and explanatory goal, that the universe must be explainable in terms of its constitutive parts and intrinsic laws.

The voluntarist tradition, developed by medieval nominalists, became a popular issue in the Reformation with Luther's rallying cry, "Let God be God." God in this view is absolutely free to do whatever he wills without regard for the limitations of human understanding or even the laws of logic.[35] The creation of the universe manifests God's sovereignty, his absolute power. Since God is free to do with the universe whatever he wills, the universe cannot, as a matter of theological principle, be intelligible exclusively in terms of its constitutive parts and intrinsic laws. Atomism, as Boyle interpreted it, fits a voluntarist theology far better than the scholastic doctrine of natural laws.[36] An atomistic universe is a collection of nonrelated particulars. Physical laws are not expressions of any rational plan intrinsic to the fabric of creation. The only universally valid causally efficacious agency in nature is God's will. The laws of science in this view are not transcriptions of laws of nature. Rather, they depend on categories imposed upon nature by the human mind in the light of the observed regularities of experience, or the regularities produced or uncovered by the virtuosi—to use the then current title for scientists.

Newton's Atomism

The development and proper interpretation of Newton's thought is a topic that has fascinated and frustrated scores of scholars, and one that continues to supply a virtually inexhaustible source of specialized studies. Here, I am simply skating across the surface of Newtonian scholarship. My primary concern is not Newton's own development, though this must be given some consideration, but the doctrines presented in the *Principia*. In the later editions the atomistic explanation of universal properties is presented as the foundation of all philosophy and the ultimate ground of Newton's mathematical method. In spite of such strong claims, atomism plays an incidental and ultimately dispensable role in the *Principia*. What I wish to focus on, accordingly, is the set of problems Newton encountered in his attempt to use atomic assumptions as the justification of his mathematical treatment of qualities.

Newton was a convinced atomist in the tradition of Robert Boyle. On this point there are no serious grounds for doubt.[37] He rejected Descartes's idea of the plenum and in his early years seemed content with a doctrine of atoms and the void. Though he speculated about ether, he did not discuss it publicly before the second edition of the *Principia* in 1713. His doctrine differed from classical atomism in two key respects. First, his most distinctive contribution, Newton assigned a basic role to *forces*, instead of the contact action of the earlier mechanists. This presented notorious difficulties and led to many unsolved problems. Yet, difficulties notwithstanding, Newton clearly seems to have held for a hierarchy of forces from the gravitational force binding the solar system together to the unknown forces joining ultimate particles together to form composites.[38]

Second, instead of the simple classical atomistic dichotomy of atoms and aggregates, Newton held for a great chain of being. Unlike Boyle, Newton was strongly influenced by neo-Platonic doctrines and alchemical speculations. He was willing, especially in his later years, to consider vital forces and various forms of spirits, without a sharp matter-form distinction.[39] Though Newton's theology was, in terms of the then accepted standards, unorthodox, it was strongly in the voluntarist tradition. The universe is not, in Newton's view, completely intelligible in terms of its matter, motion, and natural laws. It requires God as first mover, as controlling power, and, when necessary, as one who intervenes to ensure stability. Whatever intrinsic intelligibility the universe has is based on the immutable properties of ultimate particles.

In spite of his strong commitment to atomism as the ultimate source of matter's intelligibility, Newton accorded atomic assumptions an apparently peripheral role in the first edition of the *Principia*. Here I follow McMullin[40] and Westfall[41] in attributing this paradoxical gap to the way Newton's ideas on scientific methodology developed. Newton inherited 'force' as a concept with overtones of occult qualities. In the early speculations contained in the *Waste Book*,[42] a collection of notes Newton wrote for his own benefit around 1666, Newton tried to develop a doctrine of force that differed from Galileo's revision of impetus theory and Descartes's concept of the force of a body's motion. The ontological analysis Newton presented involved unresolved ambiguities. He soon began to see the advantage of getting away from a causal analysis of force and concentrating instead on the *effect* of force. The only quantity which he thought could serve as a measure of the effect of force was change in motion.

After a gap in his physical research which still perplexes historians, Newton was stimulated to further investigations by the question Hooke proposed to him in 1679: How can one explain elliptical planetary orbits by supposing a central attractive power operating from the focus of an ellipse according to an inverse square law? Five years later occurred Halley's famous visit to Newton and the writing of Newton's *De motu corporum*, treating chiefly the mathematics of central-field motion with some attendant speculations on the ontology of forces. This was in effect the first draft of the treatment of motion given in the *Principia*. A revision and gradual enlargement followed with the definitions and laws making their appearance about a year later.

In composing his *Principia* Newton seems to have had an idea of scientific explanation as involving three steps: mathematics, physics, and philosophy.[43] By 'mathematics' Newton meant neither Descartes's *mathesis universalis* nor pure mathematics in a modern sense. His special use of the term in the *Principia* was based on his idea that, though force considered as a cause requires an ontological analysis, force considered in its effects may be studied by a mathematical analysis of the consequences of different force laws. Hence, books 1 and 2 of the *Principia* supply a descriptive account of motion following the method of mathematical deduction.

In the draft preface of the first edition, Newton wrote that he suspected that all phenomena depend on the attractive and repulsive forces by which the particles of bodies are reciprocally moved.[44] In the draft conclusion, Newton argued that the attractive and repulsive forces of particles, which can be known by analogy, should explain the observable motions.[45] Both were suppressed in the published version. This vacillation on introducing atomism into the *Principia* continued in later editions. As the Halls summarized it:

Newton was genuinely anxious to discuss his theory of matter
in detail and publicly. Three times he tried to find a place for
it in the *Principia*: in the *Conclusio*, in the *Preface* and in the
General Scholium; and three times he rejected it. Somewhere,
he seems to have felt, some notice should be given of the microscopic architecture of nature side by side with the majestic
system of celestial motion unfolded by mathematical analysis.
But all that at length emerged after painful reflection was a
cautious hint in the printed version of the Preface to the first
edition, to be followed years later by the oracular but confusing conclusion to the General Scholium in the second edition.[46]

Newton's motives for this vacillation and suppression are complex and
debatable. Yet, the basic tension he felt seems clear. The overall intelligibility of the universe requires God as first mover and controlling power, and
the invariant properties of indestructible atoms as the ultimate intrinsic
source of the phenomena treated in the *Principia*. But the mathematical
positivism of the *Principia* seemed able to dispense with both God and
atoms. The peculiar methodological problem Newton faced here was that
atomistic assumptions played a basic role in much of the physical reasoning
behind the *Principia*, yet proved dispensible in the final mathematical
formulation. This tension can be clarified by considering the two components separately.

In spite of Newton's claim that he solved the problem of gravitational
attraction between the earth and the moon in 1666 it seems clear that he did
not have an adequate mathematical solution until he had learned how to
apply his method of inverse fluxions, or integration. His original idea was
that the attractive force exerted by the earth on the moon is the vector sum
of the attractive force between each atom in the earth and each atom in the
moon. The atomistic assumption grounding his reasoning is clearly brought
in in the summary-review of the *Principia* written by Edmund Halley,
Newton's amanuensis, secretary, and factotum for the first edition:

> This done our Author with his usual Acuteness proceeds
> to examine into the Causes of this Tendency or centripetal
> Force, which from undoubted Arguments is shown to be in all
> the great Bodies of the Universe. Here he finds that if a
> Sphere be composed of an infinity of Atoms, each of which
> have a *Conatus accendi ad invicem*, which decreases in dupli-
> cate Proportion of the Distance between them; then the whole
> *Congeries* shall have the like tendency towards its Center, de-
> creasing in Spaces without it, in duplicate Proportion of the
> Distances from the Center.[47]

The role of atomism in Newton's reasoning is clear. Yet, the problems Newton encountered in his treatment of gravity made it prudent for him to minimize any overt reliance on atomic assumptions. He publicly professed not to know the cause of gravity nor to be able to explain how gravitational forces acted at a distance. Hence, he developed the *Principia* in a way that did not overtly involve any dependence on such assumptions. The first two books treat the mathematical consequences of different force laws. The third book, treating the system of the world, begins with "Phaenomena," i.e., data that the scientist can observe but cannot manipulate. The rotation of the planets and their moons, the ebb and flow of the tides, the periods of pendulums—all fit the mathematical consequences of an inverse square law of attraction and so serve to establish its general validity. Accordingly, when Newton treated the gravitational attraction of heavenly bodies as the vector sum of the gravitational attraction of their constituent particles, he was not obliged to identify these constituent particles with physical atoms.[48]

Atomism was also implicit, though again submerged, in book 2, where gases and fluids were treated as collections of particles.[49] Newton relied on a particulate theory of fluids to derive the Boyle-Townley law.[50] Yet, these corpuscular assumptions were not reflected in his mathematics, which required continuous, rather than discrete, quantities. Since the present analysis is primarily concerned with Newton's attempt to fuse an atomic theory of matter with a mathematical treatment of qualities, we will consider the mathematical methods Newton employed in the *Principia*. Following what seems to have been Newton's terminology, we will use 'quality' when speaking about the properties of bodies and 'quantity' when speaking about the same properties as measurable. There are two related questions to be considered. How are numbers attached to quantities? and How do the resulting numbers enter into a mathematical system?

Newton was extremely concerned with obtaining and using the most accurate measurements available for the phenomena that confirmed his mathematical principles of natural philosophy. As Westfall has shown,[51] Newton was not above manipulating fudge factors to enhance the appearance of precision. Yet, in spite of this appearance of modernity, Newton's method of attaching numbers to quantities is still within the Euclidean tradition of proportions. He never manifests any clear idea that the assignment of numbers to quantities requires a unit and a scale. In the *Principia*, for example, Newton employs tables of distances, times, and velocities, and clearly obtains the numbers attached to velocities by dividing distances by time intervals.[52] Thus, he says that a sound will go forward about 1,088 feet in one second of time.[53] Yet, he does not express velocity in units of distance/time. Rather, he follows the Euclidean tradition of relying on proportions of like quantities. Thus, discussing the problem of relative velocity on a moving ship, he has: "As if that part of the earth where the ship is, was truly moved towards the east with a velocity of 10010 parts; while the ship itself, with a fresh gale, and full sails, is carried towards the west, with a velocity expressed by 10 of those parts; but a sailor walks in the ship towards the east, with 1 part of said velocity."[54]

The technical and conceptual difficulties attendant upon attempts to attach numbers to quantities are clearly manifested in Newton's attempt to develop a temperature scale. His text, "Scala Graduum Caloris," lists various natural phenomena related to heat and relates these to two columns of numbers: the left giving numbers in arithmetical proportion, the right in geometrical proportion.[55] Some sample points are:

0. .   0 The heat of the air in winter, when the water
              begins to freeze; and it is discovered exactly by
              placing the thermometer in compressed snow,
              when it begins to thaw
0,1,2 . .   0 The heat of air in winter
2,3,4 . .   0 The same in spring and autumn
4,5,6 . .   0 The same in summer
. . . .
17 . . $1\frac{1}{2}$ Greatest degree of heat of a bath, which a man
              can bear for some time without stirring his hand
              in it.

In composing this table Newton used some unspecified kind of thermometer[56] and a red hot iron, on the assumption that the rate of cooling of the iron gives the heat loss. He has no distinction between heat and temperature, nor any idea of the numerical values given as multiples of some unit.

Difficulties of mensuration notwithstanding, Newton did have a mathematical physics based on the assignment, whether actual or merely possible, of numbers to quantities. In the *Principia* he treated these by the method of the first and last ratios of quantities. As he himself explained it,[57] the demonstrations involved are shorter by the method of indivisibles, but since this seems harsh he follows the geometrical method of ratios. On the grounds that the modern reader is much more likely to find the method of ratios harsh and quite unfamiliar I will attempt to explain the key points in terms of Newton's theory of indivisibles.[58]

Quantities which vary continuously are called 'fluents,' while their rates of change are called 'fluxions.' In this terminology the basic problem of the differential calculus is, given the ratio of two fluents, to find the ratio of their fluxions, while the related problem of integral calculus (or quadrature) is one of going from the ratios of fluxions to the ratios of fluents. Thus, in this method integration is essentially antidifferentiation, while differentiation is, in the terminology of the *Principia*, a matter of determining the first and last ratios of quantities.[59] This methodology is best illustrated in terms of an example developed by Newton.[60] Consider $y = x^n$. To get the fluxion, let x by flowing become $x + 0$ (the notation Newton used where modern notation, following Leibniz and Euler, would use $x + dx$). Then y flows into $(x + 0)^n$. To treat this Newton employed the binomial expansion he had discovered. Then,

$$\frac{\text{(increase of } y)}{\text{(increase of } x)} = \frac{(x^n + n0x^{n-1} + \ldots) - x^n}{(x + 0) - x}$$

$$= nx^{n-1} + (1/2)(n^2 - n)0x^{n-2} + \ldots.$$

To get the ratio of fluxions (i.e., dy/dx) let the increments vanish. Then the only term left, the last ratio of the quantities, is the first term in the expansion:

$$\frac{\text{the fluxion of } y}{\text{the fluxion of } x} = \frac{nx^{n-1}}{1}.$$

The method of quadrature (integration) as presented in the *Principia* also depends on the limits of the ratios. Thus, the ratio between the area under a curve and a set of parallelograms which cover the curve approaches one as the breadth of the parallellograms diminishes and their number increases to infinity. Neither Newton nor Leibniz, cofounders of the calculus, had a satisfactory treatment of the limiting process. Thus, Newton wrote, "And in like manner, by the ultimate ratio of evanescent quantities is to be understood the ratio of the quantities not before they vanish, nor afterwards, but with which they vanish."[61]

The idea that the ratio of two vanishing quantities could be finite and definite only made sense if the quantities in question were conceived as continuous rather than discrete. On this point Newton was altogether clear: "Therefore if hereafter I should happen to consider quantities as made up of particles, or should use little curved lines for right ones, I would not be understood to mean indivisibles, but evanescent divisible quantities."[62] Even in the problem areas previously considered, where Newton clearly thought the problems through in terms of the activities of particles, his mathematical method required continuity rather than discreteness. Thus, he treated the gravitational attraction between a corpuscle (a deliberately vague term) and a sphere by the now familiar method of determining the attraction between the corpuscle and a surface (or shell) and then integrating over shells. He then comments:

> By the surfaces of which I imagine the solids composed, I
> do not mean surfaces purely mathematical, but orbs so ex-
> tremely thin, that their thickness is as nothing; that is, the
> evanescent orbs of which the sphere will at last consist, when
> the number of the orbs is increased, and their thickness dimin-
> ished without end.[63]

Similarly, though he manifestly thought of gases and fluids as collections
of particles, he realized that his mathematical treatment of gases and fluids
did not establish their particulate nature:

> But whether elastic fluids do really consist of particles so re-
> pelling each other, is a physical question. We have here
> demonstrated mathematically the property of fluids consisting
> of this kind, that hence philosophy may take occasion to dis-
> cuss that question.[64]

The universe, in Newton's opinion, is fully intelligible only in terms of
God as first cause and controlling director of all activity and of the invariant
properties of indestructible atoms as the ultimate intrinsic source of matter's
phenomenal properties. Yet, the first edition of the *Principia* (1687) explained the system of the world with no essential reliance on any doctrine of
atomism and with only one incidental reference to God. "Therefore God
placed the planets at different distances from the sun so that each might
receive heat from the sun in accord with its degree of density."[65] In the years
following this publication Newton struggled to develop the third level of
explanation, the philosophical account of matter's motions, properties, and
diverse forms, requisite to a truly intelligible account of the universe. He
never completed this task to his own satisfaction. It is only now when
scholars are exploring his unpublished manuscripts that the herculean effort
Newton expended to achieve such an intelligible account can be properly
appreciated. He resented the criticism that the *Principia*'s treatment of the
system of the world was a mathematical tour de force lacking intelligibility.[66]
In the second (1713) edition of the *Principia* he attempted to answer such
criticisms, chiefly through a revision of the "Rules of Reasoning" at the
beginning of book 3 and through a "General Scholium" appended to book
3. Though this "Scholium" is the best known and most quoted of all
Newton's writings, it does not even profess to clarify the methodology of the
Principia, merely to give supplementary considerations, chiefly concerning
the relative roles of God and atoms in an overall account of the universe.
There remain the "Rules of Reasoning," particularly Rule III, as the
proposed link between Newton's atomism and the methodology of the
Principia. Fortunately, it is now possible to understand the basic develop-
ment leading to the formulation of Rule III. In place of the "Rules of

Reasoning" at the beginning of book 3, the first edition of the *Principia* had three hypotheses. The first two were essentially the first two rules of reasoning. Hypothesis III, however, read, "Any body can be transformed into any other type of body whatsoever, and can successively put on all intermediate grades of qualities."[67] In his own copy of the first edition Newton rewrote Hypothesis III as "Proprietates . . .," crossed out "Proprietates" and inserted "Qualitates" to get, "The qualities of bodies that cannot be intended and remitted, and that belong to all bodies in which one can set up experiments, are the qualities of bodies universally."[68] This text represents a peculiar and rather unsatisfactory fusion of the two traditions we have been tracing. The doctrine that any body can be transformed into any other type of body seems quite contrary to ordinary experience. It is, however, a consequence of the doctrine that all bodies are composed of similar atoms. What qualities should be attributed to these atoms? Here Newton fell back on the old tradition of the intension and remission of qualities. This, however, did not quite succeed in explaining what particular qualities atoms must have and how any such attribution related to the methodology of the *Principia*.

In another interleaved edition Newton changed "Hypotheses" to "Regulae Philosophandi" and on the interleaf wrote out his new Rule III with a long justification which was printed in the second (1713) and third (1726) editions of the *Principia*. Because of the importance of this text in the present analysis, it is helpful to have a literal translation (rather than the Motte-Cajori interpretative translation) of the pertinent parts of Rule III:

> *The qualities of bodies which cannot be intensified or remitted, and which belong to all bodies in which it is possible to institute experiments, are to be taken for the qualities of all bodies.*
>
> Since the qualities of bodies do not become known except through experiment, therefore those are to be considered general which square with experiments generally; and which can not be diminished, can not be taken away. Certainly dreams are not to be rashly devised contrary to the course of experiments, nor should one recede from the analogy of nature, which is accustomed to be simple and always consonant with itself. The extension of bodies does not become known except through the senses, nor is it sensed in all bodies: but because this fits all sensible bodies, it is affirmed of all. We experience the fact that many bodies are hard. The hardness of the whole, however, arises from the hardness of the parts. Hence we rightly conclude to the hardness of the indivisible particles, not only of the bodies which we sense, but of all others as well. That all bodies are impenetrable we gather not

from reason, but from sense. Those which we handle are
found to be impenetrable, and hence we conclude impenetrability to be a property of all bodies. That all bodies are
mobile, and endowed with certain powers (which we call forces
of inertia) of persevering in motion or rest is something we
gather from the properties of the bodies seen. The extension,
hardness, impenetrability, mobility and force of inertia of the
whole arise from the extension, hardness, impenetrability,
mobility, and inertial force of the parts; and hence we conclude the minimal parts of all bodies to be extended, and hard,
and impenetrable, and mobile, and endowed with forces of inertia. And this is the foundation of the whole of philosophy.[69]

The remainder of the explanation affirms that the experience of even
one ultimate particle being divided would be a basis for inferring, in accord
with the stated rule, to the infinite divisibility of all particles. The same rule,
coupled to observation, leads to the conclusion that all bodies mutually
gravitate toward each other. This need not imply, however, that gravity is an
essential property of bodies.

The unique and puzzling justification given for Rule III is best understood as the tip of an iceberg, rather than as an argument valid in its own
right. The first edition of the *Principia* manifested the first two stages which
Newton thought proper to scientific explanation: the mathematical analysis
of the consequences of different force laws, and the application of this
mathematical system to the phenomena observed or inferred in nature. In
the twenty-six years between the first and second editions Newton labored
intermittently at developing concepts of matter and spirit, of force and
space, and eventually even of the ether he had rejected in his earlier years,
all in the hope of giving an adequate philosophical account of how force is
transmitted and matter moved. Most of these speculations remained unpublished, reflecting Newton's lack of satisfaction with his results. However, he
found that the "Queries" appended to his *Opticks* provided a convenient
and suggestive means of indicating the nature of his speculations and the
problems he considered to be significant and outstanding. In the second
(1706) edition his new Queries 20–23 make explicit the particulate nature of
matter within the general framework of attractive and repulsive forces and
suggest that this atomic doctrine should serve to explain the reflection and
emission of light, volatility and evaporation, fermentation and putrefaction,
elasticity, disjunction, and, perhaps, the force of electrical attraction, which
had just been revealed through the experiments performed by Francis
Hauksbee. Arguing from such phenomena and the simplicity and uniformity of nature, Newton made his most explicit public commitment to
atomism:

All these things being consider'd, it seems probable to me, that God in the Beginning form'd matter in solid, massy, hard, impenetrable, moveable Particles, of such Sizes and Figures, and with such other Properties, and in such Proportion to Space, as most conduced to the End for which he form'd them; and that these primitive particles being Solids, are incomparably harder than any porous Bodies compounded of them; even so very hard, as never to wear or break in pieces, no ordinary Power being able to divide what God himself made one in the first Creation.[70]

When Newton began to revise the *Principia* he once again had to face the question, What is the relation between the atomic theory of matter and the mathematical treatment of matter in motion? Here, however, in place of the informal style proper to the queries, he had to consider the *two* methodologies proper to the *Principia*. The first two books followed the mathematical method and gave the deductive consequences of different force laws. As already indicated this presupposed, more as an idealization than as a definite doctrine, a continuous theory of matter. In the third book, the natural philosophy contribution, Newton intended to follow the method of analysis and synthesis. This ancient Greek method had been revised and transformed by Descartes.[71] In Newton's further adaption, analysis began with phenomena and proceeded inductively to general laws:

By this way of Analysis we may proceed from Compounds to Ingredients, and from Motions to the Forces producing them; and in general from Effects to their Causes, and from particular Causes to more general ones, till the Argument end in the most general. This is the Method of Analysis; and the Synthesis consists in assuming the Causes discover'd and established as Principles, and by them explaining the Phenomena proceeding from them, and proving the Explanation.[72]

The third book of the *Principia* began with a list of phenomena. If the ultimate causal explanation of these phenomena is to be found in the properties and activities of atoms, then the method of analysis should ideally reveal this. Yet, the argument supporting the third rule is not analysis in Newton's sense. It is, in McGuire's modification of Mandelbaum's term, *transduction*.[73] The problem Newton faced is basically the problem that bedeviled the doctrine of atomism from its inception, the polarization between the methods of explanation proper to an ontological doctrine of atomism and that proper to an epistemological doctrine of empiricism. If atoms are ontologically basic, then one should be able to explain the observed properties of macroscopic bodies in terms of the properties and activities of atoms. If, on the other hand, all knowledge begins with sense experience, a doctrine Newton professed, then the properties of atoms are only known by inference from the properties of observed macroscopic bodies. Could one use this as an inferential basis in establishing an ontology of atomism and then, on the basis of this very ontology, argue that these observed properties are not ontologically real?

This perennial problem grew in subtlety and complexity as science and reflections on the nature of scientific explanation developed. In Newton's case the operative epistemology was not any simplistic epistemological reduction of all knowledge to sensation. It was a consequence of the success of the *Principia*, and the methodology that grounded it, in expressing the most complete and precise knowledge of the material universe that man had yet obtained. This suggested some principles which supplemented the method of analysis, albeit implicitly, and seemed to mitigate the polar opposition between epistemological and ontological reductionism. The first presupposition is the uniformity of nature, something explicitly affirmed in Rule II (same effects—same causes) and Rule III (nature simple and consonant to itself). Yet, a point McGuire develops in detail, the supposed simplicity and uniformity of nature do not suffice as a basis for arguing from the macroscopic to the microscopic unless one presupposes a chain of being from ultimate atoms to complex living beings, and one in which each stage of the chain manifests a strong similarity in ontologically basic qualities. As his unpublished writings manifest, Newton speculated extensively on the types of forces, active powers, and vital principles that might make such a gradation intelligible. Though dissatisfied with his own hypotheses and acutely aware of a lack of experimental evidence, he never abandoned the chain-of-being concept.

Against the background of the method of analysis and synthesis, as the method characterizing natural philosophy, and the ontological presuppositions of the analogy of nature and the great chain of being, Newton's argument proceeds by utilizing elements derived from both the Cartesian and the British empiricist tradition. Ostensibly, Newton is quite anti-Cartesian. The extension of bodies, he insists, is known only because it is experienced, not because it is intuited. Yet, in his search for a criterion to decide which properties should be considered the universal properties of all bodies Newton implicitly follows Descartes more than Boyle. The criterion is conceptual rather than sensual. The properties that do not admit of intensification and remission are to be considered the properties of material bodies as such. As McMullin has noted,[74] this intension-remission criterion, once introduced, quickly vanishes. It plays no operative role in the rest of book 3. Newton already knew which properties he wished to treat, the extensional properties that his mathematical method could handle. Since

late medieval times the extension-remission criterion had supplied a justification for the mathematical treatment of physical qualities.

What is the net result of this protracted effort to supply a mathematical
treatment of mechanics with an atomistic foundation? The answer, I believe, hinges on one's interpretive perspective. Newtonian mechanics, interpreted in a technical sense as a mathematical treatment of the motions
induced in material bodies by forces, does not require or even utilize an
atomic theory of matter. It relies on the idealization of continuity, rather
than on any theory of discreteness. Yet, an intelligible account of the
universe of which man is a part seems to require a doctrine of indestructible
atoms with invariant properties as well as such other doctrines as animating
spirits, vital forces, and God as controller and first mover. Newton effectively tried to link the two realms by invoking atomism, not as a technical
part of the *Principia*, but as a presupposition necessary to make the subject
matter of the *Principia* intelligible, material bodies with the properties and
activities his methodology presupposes.

The Newtonians

Newton's successors divided rather neatly into two distinct groups. The
first, predominantly Englishmen, focused on the subject matter treated in
the *Principia*, and even more on the *Opticks*, and extended the sort of
qualitative physicalistic reasoning natural philosophers had traditionally
employed in attempting to make this subject matter intelligible. These
people, who will be considered later, generally retained a doctrine of
atomism and, in fact, frequently extended it to include various other types of
unobservable entities. The second group, chiefly Continental Newtonians,
extended and revised the mathematical formalism of the *Principia*. This
division to be sure is a bit oversimplified. People like Colin Maclaurin and
the Dutch Newtonians cannot be confined exclusively to either camp. Yet, it
is adequate for our purpose of exploring the way in which the Continental
Newtonians were able to extend Newton's mechanics while dispensing with
the atomistic foundations that Newton had attempted to supply.

The key figure in the transition from an atomistic to a nonatomistic
interpretation of Newtonian mechanics was Leonhard Euler (1707–83).
Because of his pivotal role in this transition and also because the basic
surveys of the development of mechanics do not adequately treat Euler's
doctrines, I will consider his views in more detail than those of the French
Newtonians. Euler's first public presentation, after receiving an M.A. in
philosophy, was a comparison of the philosophical positions of Descartes
and Newton. This set the tone for much of his subsequent work. As Yehudi
Elkana summarized it,[75] Euler was Cartesian in his metaphysics, Newtonian
in his scientific doctrine, and Leibnizian in his ideas of scientific explanation.

Euler's first important scientific work was his treatise, *Mechanica, sive Motus scientia analytice exposita* (1736), in which he tried to make the laws of motion a matter of logical necessity. In the rationalist tradition discussions of logical necessity concerned conceptual entailment more than formal logic. After defining motion as a change of place and explaining the difference between absolute and relative motion in terms of a doctrine of absolute space, Euler faced the question; What moves through space? If the answer is to fit into a network of logical necessity in Euler's sense, it must flow from the definitions of space and motion. Euler concludes, "These laws of motion, which bodies obey that are left either at rest or in continuing motion, strictly concern infinitely small bodies, which may be considered points."[76] Can such points be said to have mass? Again, Euler proceeds by definition. One can define force in terms of its effect, a change in the speed or direction of motion. Then mass can be introduced rather operationally by postulating that the number of points in a body is proportional to its mass at this time.[77]

At this period in his development Euler was not denying the reality of atoms. He was simply substituting the idealized concept of point-masses for the Newtonian concept of corpuscles. Though point-masses involve infinitesimals, their properties are not interpreted as the properties of real particles. Instead, these point-masses were simply attributed the properties which upon integration (the 'analytice' of the title refers to Euler's use of differential equations) yield the gross properties mechanics treats. They have in other words no more ontological significance than the *dx, dy,* and *dm* employed in contemporary mechanics texts.

It is easy but quite misleading to give this a positivist interpretation. But Euler, far from being an antimetaphysician, was attempting to ground Newtonian physics in Cartesian metaphysics. His distinctive way of developing Cartesian metaphysics is most clearly revealed in two works written some eight years after the treatise on mechanics. The first, an address to the influential Berlin Academy of Science, couples a doctrine of the molecular composition of matter to a Cartesian explanation of the properties of matter.[78] In this address Euler postulated an ether pervading both the whole of space and the interstices of bodies. Though he rejected action at a distance, he did not return to the simple mechanistic idea of contact (or more properly collision) forces. In its place he substituted the more subtle idea of ether pressure proportional to the surface of a molecule as the agency responsible for transmitting gravitational force. Since the gravitational force exerted on a molecule by ether pressure is proportional to the molecule's surface, it follows that extension is the basic property involved. Dense and rare bodies, he argues, differ in the number of molecules per unit volume, not in the mass of their molecules. Euler suggested that these molecules are probably not the elements of traditional atomism, but some sort of homogeneous nonporous matter.

The second work, probably written in 1644 but not published till much later, is an attempt to determine the essence of bodies.[79] For Euler, this meant a specification of the properties which all bodies have and which are not possessed by such noncorporeal entities as spirits, space, and time. He concludes that these properties are, in order, extension, mobility, constancy (*standhaftigkeit*), and impenetrability. Though impenetrability is presented fourth, Euler argues that it implies the other properties.[80] Though Euler did not know it, Newton had reached the same conclusion in an unpublished study.[81] Elkana has argued that these essentialistic deductions reflect a rather Leibnizian idea of scientific explanation. Where Newtonian empiricists would stress precise correspondence with observational data, and Cartesians deduction from clear and distinct ideas, intelligibility manifested through conceptual consistency is the most characteristic stress of the scientific spirit stemming from Leibniz.

The atomism that had been a rather speculative issue in Euler's younger days became a much more polemical issue after the early 1740s when Euler and Maupertuis led a fight against the monadology of Christian Wolff and his followers. The central point of disagreement in physics was the Wolffian doctrine that macroscopic bodies are composed of ultimate simple elements with active powers of changing themselves. The disagreements were not confined to physics. They also extended to metaphysics (Cartesian versus Leibnizian) and to religion (Euler's acceptance of revelation in opposition to the religion of reason championed by Leibniz and Wolff). These doctrinal disputes were exacerbated by the struggle for dominance in the two academies where Euler spent his professional career, Berlin and St. Petersburg. This new spirit of antiatomism carried over to Euler's technical works.

Euler's 1755 paper, which transformed the science of hydrodynamics, begins in typical Eulerian style, with a conceptual clarification of the nature of fluids.[82] The atomic hypothesis, it declares, may be true, but is absolutely sterile. The starting point in studying fluids should be an isolation and adequate mathematical representation of the property distinguishing fluids from solids. This property is the balance of forces necessary for equilibrium. In other respects, a large collection of small solid spheres freely sliding over each other might be an adequate model of a fluid. Yet, Euler argued, it cannot supply the stability necessary for equilibrium. For this reason Euler replaced the molecular model of fluids with a model of homogeneous, continuous material. Truesdell, the editor of Euler's collected works on hydrodynamics, summarized the significance of this: "In §§ 5–8 is Euler's final and rather disgusted rejection of corpuscles. Henceforth, the *principles of mechanics themselves* are to be applied directly to the bodies of physical experience, and 'particle' is to mean only a mathematical point in a continuum model of matter." As Truesdell's historical account shows, a very successful science of continuous mechanics was developed by following Euler's novel method of representing continuous properties by differential equations.[83]

In 1768 Euler published his immensely popular interpretative survey of physics and philosophy, *Letters to a German Princess*. These two volumes of letters were printed, translated, and reprinted in twelve French, ten English, six German, four Russian, two Dutch, two Swedish, one Italian, one Danish, and one Spanish editions.[84] Together with the French *Encyclopédie*, these letters taught the reading public of the Enlightenment era what they needed to know of the new science, how it was to be interpreted, and how these novel advances related to traditional Christian beliefs. In these letters Euler popularized the account of matter we have been considering. Extension, mobility, and impenetrability are the essential characteristics of bodies, with impenetrability the most basic.[85] In his detailed refutation of monadology, Euler also attacked the doctrine that the process of dividing bodies would lead to ultimate simples.[86] This was, of course, the oldest and most fundamental conceptual argument for an atomic theory of matter.

The older accounts of the French Enlightenment depict a triumph of Newtonian science, beginning with the popularizations of Fontenelle and Voltaire and the intellectual influence of Mme du Chatelet, and finding its culmination in the theoretical mechanics of Clairault, d'Alembert, Lagrange, and Laplace. Though this interpretation stems from the scientists themselves, it is a bit one-sided. Newton's concept of force was eventually accepted and the content of his physics redeveloped and extended. Yet, the explanatory framework in which this new mechanics functioned owed at least as much to Descartes as it did to Newton.[87]

The new explanatory ideals are most clearly seen in the writings of d'Alembert, the ablest philosopher in the group. His epistemological orientation was summarized in his famous preliminary discourse.[88] He claimed to hold, with Locke, that all knowledge comes through the senses. However, he argues, the first thing our senses recognize is our own existence, grounding our first reflex idea about the thinking being that constitutes our nature. Among external objects we know first our own bodies. In addition to immediate sensory awareness, we can also abstract universal properties common to all bodies. Some universal properties, such as the ability to move, remain at rest, or communicate motion, are grounded in *impenetrability*, the principal property by which bodies are distinguished from pure space.

This doctrine of abstraction serves to ground three basic sciences. The most abstract possible consideration of bodies prescinds from specifiable properties and simply considers bodies as existing units. This supplies a basis for numbering and, through numbers, for the science of algebra. Adding the property of extension yields the science of geometry. The addition of the further universal property of impenetrability supplies the ground for the science of mechanics. Mechanics, d'Alembert insists,[89] is properly developed as a science when mechanical conclusions are deduced from the clearest notions and most general principles. The necessary notions are extension, impenetrability, space, and time. None need be explained metaphysically. All that is required is a clear and distinct way of representing their interrelation in the trajectories and interactions of bodies. The principles involved can be reduced to three: the force of inertia, the law of the composition of velocities according to the parallelogram of forces, and the principle of equilibrium under constraint.

D'Alembert's deductions from such principles proceeded smoothly only when he was treating statics. He became more of an operationalist when treating collisions. If this were interpreted as a technical limitation rather than a deficiency in the basic method, then d'Alembert might seem to have completed the task that Descartes instituted some one hundred years earlier, deducing the laws of mechanics from matter and motion alone. Yet, this impression is a bit misleading. D'Alembert was not so much a metaphysical rationalist as a formal deductivist. Where Descartes was intent on determining the essence of all material bodies, d'Alembert was concerned with a systematic ordering of human knowledge and found the attribution of properties a convenient basis for the division of science.[90] Extension and impenetrability were presented as the properties of bodies basic to the science of mechanics, rather than as the metaphysical essence of bodies as such.

This trend toward a formal deductive reconstruction of mechanics reached its culmination in Lagrange's *Mécanique analytique*.[91] As he explained in his preface, those who love analysis will find pleasure in seeing mechanics as a new branch of analysis. What Lagrange presents is, in contemporary terms, a pure hypothetical-deductive system. The basic hypotheses are the two principles of statics, the lever and the composition of forces, and the four principles of dynamics: the conservation of living forces, the conservation of the motion of the center of gravity, the conservation of moments, and the principle of the least quantity of action.[92] The only justification offered for these principles is a historical sketch of their discoveries. The implicit metaphysics is minimal and functional, attributing to bodies only those properties needed to ground and interpret mechanics.

Lagrange is quite explicit about using the idealized point-masses required by the calculus, rather than relying on any consideration of real atoms:

> I note that in place of considering the given mass as an assemblage of an infinite number of contiguous points, it suffices, following the spirit of the infinitesimal calculus, to consider it as if composed of infinitely small elements which should be of the same order of dimension as the entire mass.[93]

Considerations of real atoms no longer seemed necessary to supply a foundation for mechanics. Concomitant with this reinterpretation of mechanics there was a gradual reinterpretation of the epistemological and mathematical considerations that, in Newton's view, had supported the foundational role attributed to the invariant properties of indestructible atoms. Thus, in the writings of Galileo, Boyle, Newton, and Locke, the distinction between primary and secondary qualities was closely connected with the doctrine of atomism. Primary qualities are those inherent in matter as such. These accordingly are the properties that atoms must have. Berkeley and Hume severed the relation between this primary-secondary distinction and atomism and rested the distinction on epistemological rather than ontological considerations. From this time on the operative consideration was the reduction of all knowledge to sense impressions and the ideas such impressions produce, rather than the reduction of matter to atoms and their invariant properties.[94]

Calculus had not yet been reinterpreted in terms of mathematical analysis and a theory of limits. This was achieved by Bolzano, Cauchy, and others in the early nineteenth century. Yet, the difficulties which made such a reinterpretation necessary had undercut the plausibility of the argument from the use of calculus to the extensive properties all bodies must have as a basis for such successful application. In spite of the Euclidean mathematical ideal of rigorous deduction from clear principles, the early development of calculus was based chiefly on induction from particular examples, intuitive reasoning, loose geometrical evidence, and plausible physical arguments. This was especially true of the generally sterile and stunted sequels to Newton's fluxions. Thanks chiefly to the Bernoullis and Euler, the successful development of calculus proceeded along Leibnizian, rather than Newtonian, lines and was less dependent on geometric reasoning. Yet, its justification was still entangled with the idea that functions are meaningful only if they can express extensive magnitudes. This correspondence between the analytic form of mathematical expressions and the possible relations between measurable quantities received its simplest and most abstract expression in Lagrange's treatment of functions:

> We call a function of one or several variables any expression
> of calculus in which these quantities enter in in any manner
> whatsoever, whether or not mixed with other quantities which
> one treats as having given and invariable values, provided that
> these quantities of the function can receive all possible values.
> Thus, in the functions one considers only the quantities which
> one assumes to be variable, without any regard for the con-
> stants which could be mixed with them.[95]

Lagrange's variable quantities, which can take on all possible values,
represent the final stage in the development of the late medieval doctrine
that the qualities that can be represented mathematically are those that
admit of intensification and remission. By severing the proposed Newtonian
relation between such variable quantities and a doctrine of atomism and
relating to them, rather, to the measurable properties of macroscopic
observable bodies, the Euler-Lagrange tradition sought to preserve the
plausibility of the interpretation of mathematical variables as expressions of
qualities that admit of intensification and remission. But this plausibility
foundered on the question of the physical significance to be accorded to the
limiting process basic to differential calculus. In *The Analyst* (1734) Berkeley pointed out the ambiguities, inconsistencies, and outright contradictions
involved in the idea of zero increments, vanishing quantities, and division by
zero. The ultimate source of these errors in his influential view was the
physicalistic interpretation accorded the calculus:

> But the velocity of the velocities, the second, third, fourth,
> and fifth velocities, &c, exceed, if I mistake not, all human
> understanding. The further the mind analyseth and pursueth
> these fugitive ideas the more it is lost and bewildered; the ob-
> jects, at first fleeting and minute, soon vanishing out of sight.
> Certainly, in any sense, a second or third fluxion seems an
> obscure mystery.[96]

Newton had built his mechanics on conceptual foundations which included a doctrine of extensive magnitudes as universal properties of all
material bodies; an atomistic theory of matter, which justified the applicability of mathematical principles to physical properties; and an interpretation of calculus, which made it a natural vehicle for giving mathematical
expression to these universal physical properties. Within the hundred years
following the first edition of the *Principia* each of these supporting pillars
suffered serious erosion. Yet, the edifice of Newtonian mechanics grew in
size, strength, and stability, and continued to supply convincing proofs of its
essential soundness. Where Newton had thought that mechanics required
metaphysical foundations to render it truly intelligible, mechanics itself now

seemed to supply a foundation for the rest of physical science. Even the uncommon incomprehensibility of forces acting at a distance became, Mach's apt phrase, a common incomprehensibility.

With the exception of Euler in his later years, most of the scientists involved in the extension of Newtonian mechanics were not antiatomists. The thrust of their collective effort was to liberate mechanics, not merely from a doctrine of atomism, but from an explicit dependence on any metaphysical assumptions. Yet, even a metaphysically liberated mechanics implicitly relies for its interpretation on some presuppositions concerning both matter and its knowability, and man's relation to the universe of which he is a part. Perhaps the most sustained and penetrating effort ever made to render such presuppositions explicit, subject them to critical analysis, and fit this analysis into a coherent and inclusive view of knowledge and reality was the effort that constituted the life work of Immanuel Kant. This will be considered in the next chapter.

2. Kant's Concept of Matter

Few, if any, men have ever labored as long and as hard to achieve an overall coherence of thought as Immanuel Kant. His struggles with particular issues tended to take on a characteristic pattern. Kant might begin by defending some side of a disputed issue only to defend the opposite side at a later stage of his own development. Eventually, however, he would try to find a higher perspective which would allow him to present the opposed positions as complementary rather than contradictory.

This dialectical development presents any expositor of Kant's thought with a challenge and a choice. To present Kant's views adequately one should first supply the requisite historical background, explaining not only such opposed positions as rationalism–empiricism; relative time and space–absolute time and space; mechanistic explanation–teleological explanations; metaphysics as the highest science–metaphysics as the highest delusion, but also the ways in which these traditional positions were transformed in Kant's thought. Next, since the higher perspective inevitably involves Kant's critical synthesis, one should explain enough of Kant's whole philosophical system to make this higher perspective intelligible.

The choice is whether to present such background material, which would involve either a very long account or a rather superficial account, and perhaps both; or to presuppose that the reader is, or can become, sufficiently familiar with Kant's general background and the main lines of his thought, especially during his most creative period. After attempting the former, I have, not without reluctance, opted for the latter choice. I intend to focus on only one point, the development of Kant's concept of matter. As I hope to show, this began as a speculative attempt to explain what matter is. Gradually, this was transformed into an attempt to develop a concept of matter which could play a foundational role in a rational reconstruction of physics. The type of reconstruction attempted reveals Kant's views on what physics should be and reflects his views on the way in which physics, so reconstructed, fits into a coherent account of the universe in which man is both a knower and an agent. His work supplies the initial prototype for the present effort.

45

The exposition that follows will focus on the interaction between Kant's concept of matter, which changed somewhat, and his concept of scientific explanation, which changed considerably in the course of his development. Kant's theory of knowledge and criticism of metaphysics will be brought in only when their introduction seems indispensable for clarifying Kant's physics or theory of science.[1]

Kant's Early Views on Physics

Kant's earliest published work, *Thoughts on the True Estimation of Living Forces*, was an attempt to resolve the *vis viva* controversy by characteristically attempting to synthesize the best elements in both the Leibnizian and Cartesian positions.[2] Kant was not yet a full-fledged Leibnizian dynamist.[3] What he took from Leibniz was basically the idea that a substance must contain the complete source of all its determinations. However, he rejected the Leibnizian idea of windowless monads simultaneously ticking in preestablished harmony. Instead, he insisted that substances must have forces which explain both their own extension and their interaction with other bodies.[4] Where the Aristotelian-Scholastic tradition had thought of the concept of body as having the ordered set of notes substance–quantity–quality–and so on, and the Cartesian conceptualization involved the order substance–extension, Kant is now presenting a concept of body involving the ordered set of notes substance–living force–extension. At that time, he thought that, though living force may be transmitted to a body, it resides in the body and grounds its extension.

Cartesianism comes in only when one distinguishes between the physical and mathematical concept of a body. Measurement, and therefore mathematics, is concerned with velocity, mass, and duration. None of these determine living force, which Kant explained in more metaphorical than mathematical terms. Mathematization of the concept of body requires two things. First, there is the Cartesian law grounding the applicability of mathematics: arbitrarily (of infinitely) small magnitudes must have the same properties as the gross magnitude.[5] Velocity, mass, and duration meet this requirement; living forces do not. Second, the applicability of mathematics requires conserved quantities. Dead, or impact, forces are conserved and accordingly can be treated mathematically. Living forces are due to an inner striving. Hence, they are not conserved and cannot be represented mathematically.[6]

The resolution of the *vis viva* controversy eventually accepted by scientists stemmed from the work of d'Alembert rather than Kant.[7] What interests us is not so much his proposed solution as his initial presentation of themes which perdure as abiding concerns in the subsequent writings we will consider. These include a concept of body in which inner forces are more basic than and somehow ground extension, a distinction between physical and mathematical concepts, and the attempt to explain why mathematics should be applicable to physical quantities.

The label 'precritical' applied to Kant's early works places them in an interpretative perspective in which one concentrates on elements and issues leading up to the first *Critique*. The young Kant, however, thought of himself as a practicing scientist as well as a philosopher. His growing concern with the foundations of scientific method should not be interpreted simply as an analysis of the physics of Newton and Euler. It also reflects the method followed in the scientific work Kant himself did. For this reason we will consider Kant's scientific work and the methods he employed before considering his first attempts to revise his early doctrine of scientific method.

Kant's early scientific works reflect two methodological principles that he later tried to integrate in a coherent way. The first, which will be treated in discussing Kant's monadology, is his attempt to endow Newtonian physics with Leibnizian intelligibility. The second is the method of analysis and synthesis as *the* method through which scientific advance is secured. Recent historical studies of this method have concentrated on the roles of analysis and synthesis in the search for proofs in Greek geometry, especially on the summary account of Pappus.[8] Kant seems to have been unfamiliar with these texts. For him, analysis and synthesis was the method Newton used to discover the fundamental laws of mechanics and to build a science of mechanics. Analysis proceeds from effects to causes or from wholes to their constituent parts, and often relies on speculative hypotheses. Synthesis proceeds from cause to effect or from parts to whole. Ultimately synthesis aims at a coherent account of the universe. This, of course, is oversimplified, but deliberately oversimplified. Kant's early views on this method are revealed chiefly through what he did in his scientific works. It was only after his Copernican revolution that he began critical reflection on the presuppositions underlying this method. We shall follow his order of development.

Kant's first public attempt to make a contribution to physics came in his treatment of the question of whether the earth is slowing down. He was the first to argue from Newton's law of gravity to tidal friction as a cause producing a gradual decrease in the earth's rate of rotation.[9] On the basis of one of his very rare attempts to do, rather than simply discuss, mathematics Kant concluded that in a period of two thousand years the length of a year should decrease by about eight and a half hours. Present calculations, also based on Newtonian physics, lead to a decrease of .032 seconds in a two-thousand-year period. Kant's calculations were in error by a factor of one million.[10]

Kant's *Universal Natural History and Theory of the Heavens* is not only explicitly Newtonian in its subtitle, *An Essay on the Constitution and Mechanical Origin of the Whole Universe Treated according to Newton's Principles*, it also contains an admission that Kant's idea of cosmic evolution resembles those proposed by the classical atomists.[11] Here, however, Kant was not espousing the classical doctrine of hard extended atoms, but the idea that cosmic evolution (or the Kant-Laplace hypothesis) should be explained exclusively in terms of scientific laws of attraction and repulsion rather than through any divine intervention. The analogy of nature, which Newton had presented as the basis for explaining the applicability of macroscopic concepts to atoms, was used by Kant to discuss resemblances between men and possible inhabitants of other planets.[12]

In other works Kant attempted to explain the phenomenon of fire in terms of an atomic composition of solids and fluids, with ether supplying both the cohesive force of bodies and the material that appears as fire.[13] He also attempted a causal analysis of prevailing wind patterns.[14] If the atmosphere is considered as a sea of air in basic equilibrium, then any excess local heating or cooling would cause a wind flow. Similarly, the prevailing patterns, e.g., the gradually increasing westerly trends of winds flowing from the equator toward the North Pole, is explained as an effect of the earth's rotation. After the Lisbon earthquake Kant wrote a paper explaining earthquakes as effects which could be produced by fluid pressures in huge, wide-ranging subterranean cavities.[15]

Kant was attempting to do science, not by working out the deductive implications of axioms, but by using the method of analysis and synthesis to argue from phenomena, interpreted as effects, to a cause capable of producing such effects; from a whole to the parts that constitute it; or from parts to the universe as an organized whole. His initial attempt to accommodate such inferences was, not surprisingly, an adaption of the Leibniz-Wolff metaphysics which still dominated the German universities.[16]

According to the Leibnizian account, which the young Kant accepted,[17] God considered all possible worlds and then created the most perfect. In Wolff's systematization, philosophy begins with ontology, the study of possible beings. For this, one principle suffices as a deductive basis, the principle of noncontradiction. To go from the possible to the actual one needs a further principle, the principle of sufficient reason. The fact that ours is the most perfect of all possible universes is a sufficient reason for God's choosing it rather than any other. These two principles supply a deductive basis for all philosophical explanations. If the sufficient reason for some phenomenon cannot be found, as in Newton's failure to give a causal account of gravity, then one does not yet have a philosophical understanding of this phenomenon. Wolff was quite explicit on this point:

> Unless he [a Newtonian] can distinctly explain how circular
> and especially elliptical motion arises from an impressed force
> and gravity towards the center of the body about which a rev-
> olution takes place, and thus can demonstrate that the planets
> are moved by an impressed force of gravity, he does not pos-
> sess philosophical knowledge of celestial motions.[18]

This idea of scientific explanation as logical deduction from self-evident axioms was Kant's point of departure. He accepted the rationalistic criterion that a proposition is true when the notion of the subject involves or includes the notion of the predicate. At this stage of his development, Kant accepted existence as a predicate. His concern in the *Nova Dilucidatio* was with the ultimate grounds of the possibility of truth. For this, Kant argued, one needs two rather than one first principle, identity as well as noncontradiction. These two principles, however, do not suffice to determine the truth of contingent propositions, propositions whose truth cannot be determined by a conceptual unpacking of the subject and predicate terms. Such proposi-tions require a determining reason, whether antecedent or consequent. Kant's examples illustrate his concern with inferences from one existent to another existent. The ground of the possibility of change in a body is the existence of real relations to other bodies. The ultimate ground of the possibility of a universe of coexisting, really related bodies is an intelligible plan of creation in God's mind. This modification of Leibnizian theology allows physical action and reaction as the cause of change, rather than a preestablished harmony.

Consequent determining reason is more problematic. The reason deter-mining our knowledge of the truth may not be the reason determining the truth. Thus (Kant's example), from the retardation of the eclipses of Jupi-ter's moons we know the velocity of light. But this retardation is not the cause of light's velocity. In a recent study of the *Nova Dilucidatio* Reuscher has argued that Kant's treatment of consequent determining reason is probably a later reworking interpolated into the text when Kant realized the inadequacy of his original account.[19] Even the revised account, however, does not give the argument from phenomena, interpreted as effects, to a cause responsible for the phenomena, the type of intelligibility which the Leibniz-Wolff system considered basic to philosophical understanding.

Of all Kant's early writings on metaphysics, the *Nova Dilucidatio* is the most obviously precritical. Yet, two aspects of this work deserve comment. First, Kant's attempt to implement the Leibnizian notion of intelligibility took the form of trying to determine the grounds of the possibility of the knowledge we already have. The second is Kant's contention that the fundamental object of our knowledge is, though only in a confused way, the universe as a whole. To advance in understanding is to clarify this obscure background knowledge:

> Without doubt the infinite perception of the whole universe, which is always present to mind, although extremely obscure, already contains within itself whatever reality must be in the cognitions afterwards to be sufficed with more light, and the mind merely by turning attention afterwards to certain of these, while withdrawing an equal degree from others, and illuminating the former with intenser light, acquires greater knowledge day by day.[20]

Kant's *Monadologia Physica*, his most successful early attempt to endow Newtonian physics with a Leibnizian intelligibility, was published the same year (1755) as the *Nova Dilucidatio*.[21] The treatise begins with an apparent conflict between metaphysical and geometrical analysis. Metaphysical analysis leads to the conclusion that gross bodies consist of ultimate parts which occupy space. Geometrical analysis, on the other hand, treats space as infinitely divisible and, in Newtonian physics, uses the metaphysically repugnant notion of gravitational attraction as an internal intrinsic force acting at a distance. Kant's solution to this conflict is a dynamic doctrine of atomism, physical monads rather than hard extended particles. A physical monad, in Kant's account, is an inextended point surrounded by two force fields. The force by which it occupies space, or is impenetrable, is a repulsive force. There is also a long-range attractive force obeying an inverse square law. If the repulsive force is an inverse cube force, then the radius at which the two forces balance defines the sphere of impenetrability and should be the same for all atoms.[22] Since the repulsive force becomes infinite at zero, physical monads have perfect elasticity.

In the *Monadologia Physica* Kant subordinated Newton's geometrical method to metaphysics. After some delay, undoubtedly due to the new duties of his post as *Privatdozent*, Kant returned to the methodological problems and began to focus more on differences between the methods of mathematics and metaphysics than on their subordination. Since Descartes's introduction of *mathesis universalis*, philosophers in the rationalist tradition had thought of the mathematical method of logical deduction from axioms as the method proper to philosophy. Kant's treatise *The False Subtlety of the Four Syllogistic Figures* (1762) is more concerned with determining the limits of applicability of logical reasoning than with making a contribution to logic as a science. Kant's immediate focus is on judgments and on the way in which the notes or characteristics (*nota, Merkmal*) of the predicate are included in the subject. Demonstration, guided by logic alone, can only analyze what is given through concepts. It does this by using, as a medium of reasoning, the notion of a characteristic of a characteristic. The principle presented as the first rule of reasoning is: "The characteristic of a characteristic is the characteristic of the thing itself."[23]

The difference between logical and real connection was further developed in Kant's treatise "The Attempt to Introduce the Concept of Negative Quantities into Science" (1763).[24] Metaphysicians reject the idea of negative quantities chiefly, Kant claims, because they do not appreciate the difference between the misguided attempt to imitate mathematical *method* in philosophy and the very successful attempt to apply mathematical propositions to philosophical subjects. In the latter regard the reknowned Euler has shown how to use negative quantities. Negative and positive are relative to a standard, which can be arbitrary. It was in this context that Kant first introduced the distinction between analytic and synthetic judgments, a topic which will be considered much later.

Kant's *Prize Essay* of 1764, though ostensibly a treatise on the principles of natural theology and ethics, is in fact his most developed precritical work on scientific method.[25] By this time Kant had concluded that all previous attempts to develop a science of metaphysics must be dismissed as failures. The basic reason for this failure is the fact that metaphysicians have attempted to follow the mathematical method of logical deduction from self-evident axioms rather than the method that Newton used to develop physics as a science.

To understand the position Kant is developing it is essential to make a sharp distinction between two different uses of the term 'mathematical'. The first sense concerns the use of mathematics in the treatment of physical quantities. This Kant strongly defends as an invaluable contribution to natural philosophy. The second sense is derived from Descartes's doctrine that *mathesis universalis* is *the* method of philosophy. In the Wolffian systematization of rationalistic philosophy, a philosophical explanation is a logical deduction from self-evident, or analytically true, principles. Kant now insists that nothing is more harmful to philosophy than the attempt to follow this mathematical method.[26]

Kant then developed proper philosophical method by contrast with mathematical method. In Kant's constructivist interpretation mathematics achieves its definitions *synthetically*. Thus, one defines a trapezoid by constructing, whether on paper or in imagination, a diagram which represents the universal concept. Proper philosophical method, on the other hand, relies on ordinary language usage, rather than on constructed concepts. It proceeds *analytically* by analyzing the confused concepts that we already have. Analysis in this context means segregating characteristics and testing them in different cases. This method of analysis will be considered in more detail when we consider Kant's lectures on logic. What is noteworthy is the relation Kant saw, in 1764, between the type of analysis he was proposing for analyzing concepts and the method Newton had introduced into physics:

> The genuine method of metaphysics is, in fundamentals, identical with that which was introduced into natural science by Newton and which had such useful consequences there. It says that, by means of certain experiences and always with the aid of geometry, a search should be conducted for the rules according to which particular appearances of nature occur.[27]

This is obviously misleading when considered as a historical interpretation. However, it is particularly helpful in bringing out the change in the significance of 'analysis' from Kant's precritical to his postcritical writings. In the *Prize Essay* Kant gave two rules for the method of attaining philosophical certainty. The negative rule, effectively replacing the principle of noncontradiction in the *Nova Dilucidatio*, is that philosophy should not begin with definitions.

> The second rule is that one should particularly note immediate judgments of the object with respect to that which is first found with certainty in it; and after one is certain that one such judgment is not included in the others, they should, like the axioms of geometry, be made the basis of all inferences.[28]

Here, Kant is clearly holding that our immediate judgments of objects are determined by the objects themselves. Though these judgments are immediate, they can serve in mediate inference, since mediate inference depends on the characteristic of a characteristic and, according to Kant's first rule of reasoning, the characteristic of a characteristic is a characteristic of the thing itself. Given such a method, the key to scientific advancement comes from learning to pick out the right characteristics and then knowing how to exploit them. Kant prefaced his essay with a revealing excerpt from Lucretius: "But for an acute intelligence these small clues should suffice to enable you to know the rest safely." Newton had discovered the type of clues that supply an inferential basis for natural science, the quantitative aspects of physical reality that admit of systematization through general laws from which both new and old phenomena can be deduced.

Kant's example of how this method should be applied in metaphysics is one more redevelopment of his doctrine of atomism.[29] One begins with the confused concept of body. Without establishing what a body is, or defining 'body', one can conclude that a body must consist of parts which would exist even if they were not compounded. Space, however, does not consist of simple parts. From this follows the now familiar conclusion that simple bodies occupy space through their impenetrability; that impenetrability is to be explained through force; and that absolutely simple bodies, point-centers, do not fill space.

The stress on analysis into simples given in the *Prize Essay* is complemented by the role of synthesis explained in Kant's *Inaugural Dissertation* (1770):

> In a substantial composite, just as analysis does not come to an end until a part is reached which is not a whole, that is to say a *simple*, so likewise synthesis does not come to an end until we reach a whole which is not a part, that is to say *a world*.[30]

In this work Kant manifests an awareness of some inherent limitations in the method of analysis and synthesis. It does not fit the *continuous*, where there can be no regression from whole to parts, or the *infinite*, where there is no progression from part to whole. The provisional solution to the antinomies generated by the method of analysis and synthesis given in the *Inaugural Dissertation* need not detain us, for we will consider the more definitive solutions given in the first *Critique*. Of more interest is a further distinction which Kant gave between the place of methodological rules in science and metaphysics. In any science whose principles are given intuitively, i.e., natural science and mathematics, it is the use which gives the method. In metaphysics, however, where the intellect must rely on real rather than merely logical principles, method must come before all science.[31]

Physics in the "Critique of Pure Reason"

The birth of the critical method is a matter of continuing debate among Kantian scholars. Kant himself has given its origin different interpretations.[32] Regardless of how precisely or exclusively Hume influenced him, it seems clear that Kant felt that some particular criticisms given by David Hume in his *Inquiry* raised serious difficulties in the development Kant was pursuing. What I wish to present here is not one more attempt at a historical reconstruction, but the bearing of the particular Humean criticism, that Kant claimed influenced him, on the issues we have been considering.

Hume did not doubt the practical indispensability of the concept of cause. Nor did he question the use of this concept made by Newton and others to explain observable phenomena as effects of postulated unobservable causes:

> It is confessed that the utmost effect of human reason is to reduce the principles productive of natural phenomena to a greater simplicity, and to resolve the many particular effects into a few general causes, by means of reasoning from analo-

gy, experience, and observation. But as to the causes of these
general causes, we should in vain attempt their discovery, nor
shall we ever be able to satisfy ourselves by any particular explication of them. These ultimate springs and principles are
totally shut up from human curiosity. Elasticity, gravity, cohesion of parts, communication of motion by impulse—these are
probably the ultimate causes and principles which we shall ever
discover in nature: and we may esteem ourselves sufficiently
happy if, by accurate inquiry and reasoning, we can trace up
the particular phenomena to, or near to, these general principles. The most perfect philosophy of the natural kind only
staves off our ignorance a little longer, as perhaps the most
perfect philosophy of the moral or metaphysical kind serves
only to discover larger portions of it. Thus the observation of
human blindness and weakness is the result of all philosophy,
and meets us, at every turn, in spite of our endeavors to elude
or avoid it.[33]

The dogmatic slumber which Kant claimed such criticisms interrupted[34]
is generally interpreted as a relatively uncritical acceptance of, if not the
doctrines, at least the basic orientation of the Leibniz-Wolff (or dogmatic, in
Kant's terms) metaphysics. In spite of its widespread acceptance, this interpretation seems to me to be quite untenable. In his slightly earlier work,
Dreams of a Visionary Explained by Dreams of Metaphysics, in his *Prize
Essay*, and in his private correspondence, Kant had repeatedly insisted that
a science of metaphysics simply did not exist. Thus, when Moses Mendelssohn, whom Kant respected very highly, criticized the negative attitude
toward metaphysics displayed by Kant, Kant replied:

As to my expressed opinion of the value of metaphysics in
general, perhaps here and again my words were not sufficiently
careful and qualified. But I cannot conceal my repugnance,
and even a certain hatred, toward the inflated arrogance of
whole volumes full of what are passed off nowaways as insights; for I am fully convinced that the path that has been
selected is completely wrong, that the methods now in vogue
must infinitely increase the amount of folly and error in the
world, and that even the total extermination of all these
chimerical insights would be less harmful than the dream
science itself, with its confounded contagion.[35]

There is simply no way in which a total rejection of all previous
metaphysics as worthless can be interpreted as an uncritical acceptance of
metaphysics. Though Kant rejected the metaphysics of the past, he insisted
repeatedly that there did exist a *method* for the development of metaphysics
as a science, namely, his own adaption of the method Newton had used in

physics. Kant explained how this method was to be adapted in his *Prize Essay* and then attempted to implement this new method in his *Inaugural Dissertation*. Yet, he had never submitted this method to the critical probing accorded metaphysics. Kant himself explained why. In science practice precedes method. To discover the right method one must see what methods actually worked. In metaphysics, however, method must precede practice. One cannot learn the correct method by studying the metaphysical successes of the past, for there are none.

The method which Kant accepted as the basis of Newton's success and which Kant had attempted to adapt into a new method for metaphysics was the method of analysis and synthesis. This method, Kant had come to insist, differs from the method of logical deduction in that it depends on *real* rather than merely logical connections. Logical analysis in Kant's account merely makes explicit what is implicit in the concepts that function as subject or predicate in some argument. The analysis used in physics goes from one real being to another real being, from an effect to its cause, from a whole to the ultimate simples that constitute it. Synthesis goes from parts to a whole, ultimately to a coherent world view. Both the argument from effect to cause and the argument from part to whole hinge on necessary connections between really existent things.[36] Even in his most antimetaphysical writing, *Dreams of a Visionary*, Kant had accepted experience as a basis for causality: "Questions like 'How something can be a cause, or possess power,' can never be decided by reason; but these relations must be taken from experience alone."[37]

In his *Inaugural Dissertation*, where he attempted to begin implementing this new method for developing metaphysics, Kant explained the causal relations obtaining among intelligible (or in his later terminology, noumenal) beings.[38] God, as first cause, is responsible for the existence of the universe considered as an ordered whole. The interconnection of alterations in the substances which constitute the created world should be understood, not in terms of preestablished harmony, but through the physical influx of one being on another. Even logic, as Kant interpreted it, seemed to support this reliance on real connections. The characteristic of a characteristic which serves as a medium of inference must, in accord with Kant's first rule of reasoning, also be interpreted as a characteristic of the thing itself.

If Hume's criticism is correct, then the method on which Kant had hoped to base a viable metaphysics had no theoretical justification. To determine the possibility of metaphysics as a science it was now necessary to make a prior determination of the intelligibility, i.e., the conditions of the possibility, of the method that worked in physics. It is this aspect of the first *Critique*, the submerged reinterpretation of physics as a science, that will be the focus of our concern. We will consider three points related to the first

Critique: the role of physics in the overall argument, the status of synthetic a priori judgments, and the presuppositions that ground physicalistic reasoning.

In spite of the complexity of the architectonic structure and the intricate interrelation of nested arguments, the basic argument of the *Critique* is quite simple. At issue is one question, whether a science of metaphysics is possible. Previously Kant had answered that metaphysics could be developed as a science by following the method of advancement that had worked in physics. Now he introduces a novel transcendental argument as a new means of answering this basic question. We shall outline the structure of this transcendental argument only to the degree necessary to bring out the pivotal role that Kant's theory of physics played in his appraisal of metaphysics and to indicate the interpretative problems this involves.

Mathematics and physics have succeeded as sciences. This point is not argued: their success is simply accepted as established. They succeeded, Kant insists, when scientists learned how to compel nature to answer questions of reason's own formulation. This, in turn, implies that successful sciences have an a priori aspect which is supplied by the mind itself. Hence, Kant's Copernican revolution. Instead of asking how our knowledge conforms to objects, we must inquire how objects of human knowledge conform to our way of knowing.

Empirical knowledge is expressed in a synthetic a priori judgment. An examination of the a priori aspects of such knowledge must accordingly concern itself with synthetic a priori judgments. The problems this doctrine poses will be considered in the next section. The crucial contention for Kant's overall argument is that the axioms of any empirical science must be judgments that are both synthetic—the predicate is not contained in the subject—and a priori—the aspect that is brought to, rather than derived from, experience.

This key assumption anchors Kant's transcendental argument in the stream of developing science. Since any science with empirical content must have synthetic a priori judgments as axioms, an analysis of the conditions of the possibility of such judgments supplies a basis for determining the limits to which human knowledge can stretch. An analysis of the a priori contributions imposed upon the empirically given by sensibility, imagination, understanding, and reason supplies a basis for clarifying the hierarchy of syntheses involved in knowing an object: the synthesis of apprehension in intuition, the synthesis of reproduction in imagination, the synthesis of recognition in a concept, and the ultimate synthesis in the transcendental unity of apperception. Without the unification of all these elements, no knowledge of an object is possible. Pure reason cannot transcend these limits, for, Kant argues, reason is regulative rather than constitutive. Hence, he concludes, there can be no science of metaphysics in the traditional sense of a science concerned with objects such as God, the world as a whole, or the soul as immortal which are not objects of possible experience.

Though physics is not the proper subject of Kant's concern in the first *Critique*, it serves both as the prototype of an empirical science and as a testing ground for Kant's transcendental argumentation. Unless Kant's transcendental method can show how physics is possible, the analysis itself must be judged invalid. The interpretation he gives to physics serves, as will be shown, to supply a framework in which a concept of matter can be developed as a foundational concept for a science of matter.

Synthetic A Priori Judgments

Kant's doctrine that the axioms of science must be synthetic a priori judgments has provoked controversy ever since its original formulation. The earliest criticism came either from the Wolffians, who wished to preserve the rationalist ideal of science as logical deduction from self-evident premises, or from Kant's idealistically oriented followers, who wished to streamline the critical philosophy by dropping the idea of the given.[39] Contemporary phenomenologists are inclined to criticize Kant's doctrine for not going far enough in the direction of a priorism,[40] while analytically oriented philosophers object to the same doctrine on the grounds that it misinterprets the way in which we instantiate general concepts.[41] Even those who defend Kant's doctrine in the wake of Quine's denial of the analytic-synthetic distinction generally strip the doctrine of its psychologism and defend synthetic a priori propositions as propositions which are logically synthetic, but true *ex vi terminorum*.[42]

For our present, rather limited, purpose we need only bring out two points. First, when Kant's distinction is understood in the light of his other writings on the same topic, it is more flexible and less psychologistic than is generally supposed. Second, Kant's use of this distinction in the reconstruction of a concept of matter does not suffer from the shortcomings which may be present in the general doctrine. The second point will be treated when we consider Kant's *Foundations of Natural Science*.

Kant first introduced the analytic-synthetic distinction in his treatise on negative quantities in a context where there was no concern with psychology. At that time Kant was beginning to see the importance of the distinction between *logical* consequence—what follows from the meaning of a concept—and *real* consequence—how one thing follows from another thing not related to it by logical principles:

> I understand quite well, how a consequence will follow from a
> ground in accord with the rule of identity, because it will be

found through the dissection of the concepts contained in it.
Thus necessity is a ground for unchangeableness, composition
a ground for divisibility. . . . But how one thing can alter
another thing, which does not flow from it by the principle of
identity, that is something which I would be happy to have
made clear to me.[43]

After his Copernican revolution Kant ceased looking to things-in-
themselves for the justification of physical inference and began to seek a
justification in the way we think objects. This made it necessary to speak
about different mental processes and the proximate sources of their differences. The first *Critique* undoubtedly manifests a reliance on faculty
psychology, speaking of mental operations in terms of such faculties as
sensibility, imagination, understanding, reason, and will. It seems clear,
however, that Kant did not intend to accord a foundational role to any
ontological doctrine of the soul as the source or possessor of such faculties.
This had been the foundation for faculty psychology in both the Scholastic
and Wolffian traditions. Kant's treatment of the Paralogisms of Pure
Reason clearly shows that he did not hold this as the foundation or justification for the concepts of inner faculties used in the *Critique*.

Kantian scholars seem to agree that the faculty psychology Kant used in
the *Critique* was strongly influenced by Johann Tetens's work, *Philosophische Versuche über die menschliche Natur und ihre Entwicklung*, a two-
volume treatise published in 1777.[44] Tetens had attempted to develop an
analytical, rather than a metaphysical, psychology. He insisted that we have
no intuitive knowledge of the soul as an object, nor of the causes of our
internal experiences. The division of faculties, even the doctrine of introspection itself, is based on the effects mental processes have in our experience. This is not quite the logical behaviorism developed in Ryle's *Concept
of Mind*: When we describe people as exercising qualities of mind, we are
not referring to occult episodes of which their overt acts and utterances are
effects; we are referring to those overt acts and utterances themselves. Yet,
it is closer to Ryle than to traditional faculty psychology. The distinction of
the faculties is based on a distinction of different types of behavior, not on
any doctrine of the inner nature of the soul. It seems clear that in revising the
Critique Kant tried to minimize even this dependence on psychologistic
language.[45]

Like many of the other doctrines of the first *Critique*, the analytic-
synthetic distinction is expressed in psychologistic terms. Must we think the
predicate when we think the subject of some judgment? Yet, Kant developed the distinction prior to his study of the psychological doctrines
reflected in the *Critique*. It emerged from his distinction between logical and
real inference and his reflections on the role the latter played in the method

Kant's Concept of Matter

of analysis and synthesis. By the time he finished the first edition of the *Critique* in 1781, Kant's initial distinction had evolved into a general doctrine interrelating analytic and synthetic methods, concepts, and judgments in subtle and complex ways. The developed doctrine is most clearly seen in Kant's 1782 lectures on logic[46] and in the treatment of the Discipline of Pure Reason in the methodological part of the first *Critique*.[47]

Within general, as opposed to transcendental, logic the method of analysis and synthesis has a restricted use. Analysis serves to make concepts distinct, synthesis to make distinct concepts. A concept, for Kant, is a predicate of a possible judgment. When one is concerned with the concept of a thing, one is treating *logical essences*. For, Kant insists, we do not know real essences because we do not know things in themselves.[48] The analytic method proper to logic clarifies this logical essence. The general concept of a thing is the concept of all the other concepts which, as predicates, can refer to that thing. To clarify a thing-concept which is given either a priori, e.g., substance, or a posteriori, e.g., metal, one begins on a descriptive level by collecting representative judgments in which other concepts, functioning as predicates, are attributed to the thing as subject. Thus: metals are hard, metals are shiny; substances have qualities, . . . No *given* thing-concept can ever be strictly defined, for one can never exhaust the concepts that are possible predicates attributable to that thing. However, one can approximate a definition or logical essence by picking out the predicates that are conceptually most basic in that other predicates presuppose them, but not vice versa. Thus, in works considered earlier, Kant argued that 'impenetrability' is a characteristic of 'material body' from which other characteristics may be derived.

In addition to given concepts, i.e., concepts whose meanings are determined by the cluster of judgments in which they function, there are also synthetic, or constructed, concepts. These do admit of precise definition, as in the geometrician's definition of 'triangle'. Any judgment which simply explicates a defined concept is necessarily analytic. Thus, "A triangle has three sides." With a *given* concept, however, both analytic and synthetic judgments can serve as phases in the process of concept clarification. Here it is important to distinguish between the expositional and the definitional level of analysis. On an expositional level, a judgment attributing a concept to a thing may be considered analytic, in the loose sense proper to the expositional level, if the characteristic represented by the predicate is part of the actual concept represented by the subject. Kant contrasts such analytic characteristics with synthetic characteristics, which are parts of a merely possible whole concept.[49]

While the philosopher is concerned with making given concepts distinct, the mathematician and the physicist are concerned with making distinct concepts.[50] This means going beyond ordinary language usage in the interests of greater precision. In the first *Critique* Kant is more concerned with analyzing a priori aspects of axioms than with explaining how to form them. This formative procedure is best illustrated in the *Foundations of Natural Science*, which will be considered shortly. After analysis has clarified the characteristics basic to the logical essence of some thing and given an approximative definition, one can then construct a concept which embodies, but may go beyond, this logical essence. An axiom stating what is known by means of such a concept is synthetic, because of its experiential basis, and also a priori, because of its manner of construction.[51] It is the use of such concepts in a systematic way that distinguishes science from ordinary knowledge.[52] As Lambert, whose views on science strongly influenced Kant, pointed out, geometrical optics became a science when the empirically simple, a ray of light, was identified with the synthetically simple, a geometric line.[53]

If Kant's doctrine is interpreted in this way, then what the first *Critique* gives is an ideal of scientific explanation. Functioning science is related to this ideal in much the same way that an approximative analytical definition of a given concept is related to the synthetic or defined concept that may replace it for scientific purposes. The *Metaphysical Foundations* and *Opus Postumum* were written as part of an attempt to close the gap between functioning and ideal science. The basic method of closing this gap, rational reconstruction of an existing science in accord with the a priori ideals given by philosophy, often involves transforming empirical generalizations or putative laws of nature into synthetic a priori judgments.

Presuppositions of Physicalistic Reasoning

Before his Copernican revolution, Kant thought that the methods of physics were so securely established that metaphysics could be developed by following the same methods. In the first *Critique* he still accepts the same methods as valid *within* physics, but no longer accords these methods an unrestricted applicability. This internal-external distinction is most clearly presented in the first edition treatment of the "Fourth Paralogism of Pure Reason," where Kant argues that his doctrine of transcendental idealism, rather than transcendental realism, supports a doctrine of empirical realism:

> The transcendental idealist is, therefore, an empirical realist and allows to matter, as appearance, a reality which does not permit of being inferred, but is immediately perceived. Transcendental realism, on the other hand, inevitably falls into difficulties, and finds itself obliged to give way to empirical idealism, in that it regards the objects of outer sense as something distinct from the senses themselves, treating mere appearances as self-subsistent beings existing outside us.[54]

This is quite similar to Carnap's distinction between internal and external questions. Within the discipline of physics, the physicist should act as an empirical realist. He is concerned with studying real bodies having an objective existence independent of the observer, bodies which exist in space and perdure through time. Time and space are also considered as having an objective reality independent of the observer. This empirical realism can extend from perceived to inferred or postulated entities. Thus, to illustrate the postulate of empirical thought that the actual must be connected with sense perception, Kant instances the inference from the attraction of iron filings to a magnetic material pervading all matter. This too should be accepted as actual, for the grossness of our senses is not the criterion of reality.[55]

The external questions arise only when one steps out of the framework of physics as a science and inquires, in a critical way, into what it means for anything to be an object of knowledge. Here, Kant is distinguished not only from his predecessors, but also from some of his would-be followers in his insistence that our knowledge not only of objects 'outside us', but also our knowledge of ourselves is a knowledge of representations, of phenomena, rather than of things in themselves. Only on this critical external level does the question of the a priori aspects of scientific knowledge become a topic of critical inquiry. We will return to a discussion of this level after listing the presumptions that Kant thought operative within physics.

In addition to the presumption that one is dealing with real physical bodies having an objective existence in a space and time that are also real, physics relies on further presuppositions about objects, their properties and interrelations. On the presumption that many might accept these as the operative presuppositions of classical physics and yet question or reject Kant's justification for these presuppositions, I will first list the presuppositions Kant thought basic to physics and then outline how these fit within Kant's system. These presuppositions are:

1. Both the properties and activities of physical bodies can be expressed through mathematical laws which are necessary and universal;

2. Whatever is basic in physical reality is conserved;

3. Any alteration in the state of a physical system can be explained in terms of antecedent causes; and

4. The universe is an ordered totality of mutually interacting objects.

Roughly speaking, presuppositions (1) and (2) ground the mathematical expression of physical laws while suppositions (3) and (4) ground physicalistic inference, or inference from the existence, state, or activity of one object to the existence, state, or activities of other objects. These are, one and all, presuppositions about objects. They cannot be justified or even critically examined within a conceptual framework which simply accepts objects as given, as having a reality independent of the subject. They can only be examined critically and, to the degree possible, justified within a framework in which one explains what it means for *anything* to be an object of knowledge. Part of this examination and justification is supplied in the *Critique*, with a promise of completion in later works. Here, we will indicate how the first *Critique* justifies the general features, though not the particular details, of these presuppositions.

On the presumption that the basic doctrines of the first *Critique* and the *Prolegomena* are generally familiar, we can skim over the question, How can there be a science of nature? The science of nature is concerned, not with *noumena*, things-in-themselves, but with *phenomena*, things as represented by us. The a priori elements supplied by the mind ground the necessity and universality of scientific laws.[56]

A more difficult question is, Why do the objects of possible experience conform to mathematical laws? This has a submerged complexity in that Kant felt constrained to explain the type of mathematical laws that have proved successful. This requires, in his account, a step-by-step clarification of the different a priori contributions that enter into the formation of the *synthetic a priori* judgments which serve as scientific axioms. The first step is the imposition of space and time as a priori forms of sensibility on the raw given of sensation. This in Kant's view is the ultimate basis for the treatment of extension and duration as extensive magnitudes.

Kant's famous twelve categories are interpreted as pure a priori categories of understanding. However, the subsumption of appearances under these categories requires the mediating influence of *schemata*. A schema is essentially a rule for the construction of images which serve in the application of concepts. Each type of category supplies its own pure schema. The schema for quantity is number; for quality it is degree. Relation supplies three schemata—causality, community, and substance—while modality supplies the schemata of possibility, actuality, and necessity. Though products of the imagination, these schemata are neither images, which are particular, nor concepts, which are strictly universal. They are, rather, implicit rules for the synthesis of representations in accord with concepts. Their use is manifested, for example, in the ability to construct a variety of images which accord with a general concept, such as 'triangle' or 'dog'. Since the synthetic a priori propositions proper to mathematics relate to the *construction* of concepts, pure schemata play an essential role in the intuitive constructions involved.[57]

This schematization of the categories leads to principles or rules for the objective use of categories, i.e., their use in determining objects. The first two principles constitute the basis of Kant's explanation of the reason why mathematics applies to the objects of experience. From the category of quantity Kant derives the axioms of intuition expressed in the principle, "All intuitions are extensive magnitudes."[58]

Rather than focus, as many commentators do, on the difficulties associated with Kant's doctrine of intuition and his constructive interpretation of mathematics, I will simply consider the significance this principle has in Kant's interpretation of mathematical physics. A magnitude is extensive when it has the property of additivity. Given two rods of lengths a and b, their combined length, when joined in a straight line, is the sum $a + b$. If this additivity carries over to infinitesimal parts, it justifies the use of integration in treating any extensive magnitudes. The way this fits into Kant's system has two aspects that should be noted. First, the extensivity of intuitive magnitudes is, in Kant's logical terminology, a characteristic of a characteristic. It can accordingly supply a basis for inferential reasoning. Second, not only the extension of bodies, but spatial extension and temporal duration in general are extensive magnitudes because of a priori constraints. In this case the constraint comes from the implicit rule by which the category of quantity is schematized to apply to the products of outer sense, giving extension, and inner sense, giving duration.

From the category of quality, Kant derives the "Anticipations of Perception" expressed in the principle, "In all appearances, the real that is an object of sensation has intensive magnitude, that is, a degree."[59] In the corresponding section of the *Prolegomena* Kant relates this to the traditional terminology treated in the last chapter: "A degree is the quantity of a quality."[60] Here again, Kant's psychology is debatable and ultimately dispensable, while his physics is workable. The psychology hinges on the idea that degrees are quantities of *sensation* rather than of intuition. No one had yet developed a clear idea that a measurement of the degrees of a quality requires a unit and a scale. Kant's example, degrees of pain, admits of imprecise measurement in that one pain may be graded as more intense than another. But this is not a matter of intuition in Kant's sense, for degrees of pain are not perceived as extended.

Kant claims[61] that the first two principles allow of *intuitive* certainty, because they are concerned with mathematical principles. Here again, it is helpful for our purpose to focus more on the significance of his doctrine for science than on the psychology he is using as a prop. To take a very simple example, if a box has length a, width b, and height c, then it *must* have volume $a \times b \times c$. If a direct measurement of the volume yielded some other result, the discrepancy would be interpreted as a failure in measurement rather than a failure in the applicability of mathematical reasoning. In drawing such a conclusion, one is guided, at least implicitly, by the a priori requirement that mathematical reasoning must be valid for measurable quantities.

The next two sets of principles are "The Analogies of Experience," derived from the category of relation; and "The Postulates of Empirical Thought in General," derived from the category of modality. Kant interprets these as dynamical rather than intuitive principles, and attributes to them discursive rather than mathematical certainty.[62] Here again, we will concentrate more on the significance of these principles for physics than on their justification within Kant's system. In physics the basic task of these principles is to supply the presuppositional background needed to validate *real*, i.e., from one existent to another existent, as opposed to merely logical inference.

The basic principle underlying the three analogies of experience is, in effect, Kant's reply to Hume.[63] Hume had argued that perception is incapable of yielding a necessary connection between the things perceived. Kant answered, "Experience is possible only through the representation of a necessary connection of perceptions." This necessity is not given *by* experience. It is supplied a priori by the schemata or implicit rules governing the time-ordering of the manifold of experience. Our experience is not a rhapsody of sensations, William James's blooming buzzing confusion. It is the experience of a world of interrelated objects.

In the words cited earlier, Kant had argued that we only perceive individual objects or events against a horizon, a world which is dimly perceived as a unified whole. While analysis resolves a whole into constituent parts, the ultimate task of a synthesis is to construct a unified world. In the critical synthesis Kant has now found a justification for this doctrine as an a priori requirement of human experience.

The general doctrine that there must be necessary connections in perception does not suffice to determine the nature of these connections, except for the general requirement that these principles must supply a basis for the *unity* of empirical knowledge. The three analogies, all based on temporal ordering of the manifold of experience through schemata, give the details of this a priori unification.

The terminology of the first analogy contains a concealed ambiguity: "In all changes of appearances substance is permanent; its quantum in nature is neither increased nor diminished."[64] The term 'substance' can have different meanings. In the *philosophia perennis* it is used to refer to the substratum of properties. There is also a vaguer, more indeterminate sense of 'substance' as stuff. The first edition version definitely seems to reflect the idea of substance as substratum.[65] Before writing the second edition, Kant wrote the *Foundations of Natural Science*, where he included spatial intuition and clearly held for 'substance' as indeterminate stuff. This is the meaning reflected in the second edition.

The problems concerning substance in this sense center around the doctrine sometimes argued for (more often simply postulated or presumed) that the intelligibility of a perduring world requires the conservation of what is most basic to that world. For the atomists, it was the number of atoms; for

Descartes, it was motion; for Maupertuis, it was action. Newton denied such conservation laws. Here, however, the exception supports the principle. The intelligibility gap generated by this denial was filled by a postulated divine intervention. Kant's protracted attempt to endow Newtonian physics with a Leibnizian intelligibility involved the introduction of conservation laws. His earliest work, "On Living Forces," linked conservation laws to quantities which admit of mathematical representation.

Any postulate concerning the perduring existence of things in themselves is clearly a metaphysical doctrine in the *verboten* sense. On the other hand, simple sense experience does not justify a doctrine of perduring substances, a point Hume developed in detail. Kant's a priori justification did not simply sail between the Scylla of empiricism and the Charybdis of rationalism; it also restored the older intelligibility arguments in a metamorphosed form. The objects of experience are things as they appear to us. This appearance involves an empirically given plus the imposition of forms of sensibility, the schematism of imagination, and categories of understanding. Object *qua* object is the understanding's contribution. The basic point of the transcendental deduction is to establish the indispensability of the categories in thinking objects. The schemata are schematized categories. The principles derived from the temporal aspects of this schematization concern the way in which we experience objects. Alterations, and a world experienced as changing, are unintelligible unless something perdures. However, what it is that perdures is to be determined on the basis of understanding rather than perception. Thus (Kant's example), wood is transformed by fire into smoke and ashes. In this change of one object, wood, into other objects, something must perdure. This is an a priori requirement. Whatever it is that perdures is to be called 'substance.' But what should count as substance in this sense is to be determined in terms of what is understood, i.e., through the development of the laws of physics, as basic to objects.

The second analogy states, "All alterations take place in conformity with the law of connection of cause and effect."[66] As with the first analogy, the precise meaning to be attached to the key terms, 'cause' and 'effect', should be determined by analyzing what the proof establishes. The basic question underlying the first two antinomies is, What must be present if alteration is to be understood as alteration? The first analogy covered one aspect. Change is unintelligible as change, rather than creation and destruction, unless something perdures. The second analogy covers a complementary aspect. Change requires that something is actually altered. Alteration, in turn, is only intelligible as alteration if two features obtain. First, the present must differ from the past. Second, there must be some intelligible, or rule-governed, relationship between the present and the preceding state. Without this, one cannot distinguish alteration from simple replacement.

This supplies an initial clarification of the meaning of the term 'alteration'. Something more is needed and intended. What can be said about the rule-governed relations encapsulated in the terms 'cause' and 'effect'? Is it the Aristotelian-Scholastic idea of efficient cause as the agent of change, the Wolffian subordination of causality to the principle of sufficient reason, the type of mathematical determinism which soon reached its most explicit expression in Laplacian determinism, or simply the general idea that any change in the state of a system must be explained in terms of rule-governed relations to earlier states and other systems?

In his *Inaugural Dissertation* Kant had treated the problem of causality as it applies to intelligible (or, to use his later terminology, noumenal) beings:

> Again, the possibility of all changes and successions—the principle of which, so far as it is sensitively known, resides in the concept of time—presupposes the persistence of a subject whose opposite states succeed one another.[67]

When Kant returned to this problem in the "Second Analogy," he was precluded by his own methodology from presupposing as knowable the causes he had treated in the dissertation, God, noumenal beings, or physical influx. The succession of arguments he gives hinge on the distinction between a subjective and an objective ordering of experience. A purely subjective ordering, seeing one room after another while inspecting a house, is arbitrary and changeable. An objective ordering, watching a ship move down a harbor, involves an ordering of experience which is only intelligible in terms of lawlike relations between successive states of objects.

Physics, as a science, is concerned with phenomenal objects, but not with an inquiry into what it means for anything to be an object. Accordingly, within the discipline of physics, one treats causality as a purely objective relationship, i.e., a real relation obtaining between objects having an existence, independent of the observer, in the spatiotemporal world. The precise relationship involved in particular cases can only be determined by an empirical investigation. What Kant's transcendental argument is intended to supply are a priori constraints guiding any such empirical analysis. An alteration can be understood as such only through lawlike relations between the present and preceding states.

In this interpretation, Kant's use of 'cause' like his use of 'substance' must be given a minimal and purely formal sense. In both cases the content is to be filled out by empirical inquiry. The formal sense imposes a priori constraints on the to-be-discovered empirical content. First, what one is looking for in terms of the questions imposed upon nature is a lawlike relation between the states. Whether these changes of state are to be explained by agent causes, such as the physical forces Kant discussed earlier, cannot be determined a priori. Second, since both the "First" and "Second Analogies" are determined by analyzing the a priori significance of schemata imposing temporal determinations on the manifold of experience, the two analogies must be coherently interrelated. Kant clearly seems to have thought that physics treats alterations as changes of state in something (substance in the vague sense) perduring.

The "Third Analogy" is the principle of coexistence in accordance with the law of reciprocity or community: "All substances, insofar as they can be perceived to coexist in space, are in thoroughgoing reciprocity."[68] This is sometimes interpreted as an attempt to give an a priori justification for Newton's third law of motion. Though, as the *Metaphysical Foundations* show, Kant did intend to justify the necessity and universality attributed to these laws, he had something a bit more general in mind in the present context.

As we have seen, Kant's growing awareness of the crucial significance of the distinction between merely logical inference, or conceptual entailment, and real inference, or physical entailment, played a pivotal role in the development leading to the first *Critique*. These real inferences are not to be justified by different laws of logic—for Kant, there was only one logic. They are, rather, justified by general, often implicit, presuppositions which ground the framework of physicalistic reasoning.

The method of synthesis as Kant came to interpret it involves inferences from parts to a whole and ultimately to an ordered universe. From reflections on the general practice of science and, more particularly, on the scientific work he himself had done, Kant had become clear on *how* such synthetic inferences proceed. An obscure recognition of the world as a whole, an ordered totality of interrelated objects, supplies a background horizon against which individual objects are discerned. Existence is not a property of a thing, as shape or size are. To attribute existence to a physical object is to claim for it a spatiotemporal location in this world of interrelated objects. Both analysis and synthesis operate with this as an implicit background presupposition. One task of analysis is to begin with an inadequately understood whole and resolve it into parts. The method of synthesis goes from parts to whole. The ultimate goal is a more developed account of the ordered world presupposed by both methods. Even in the context of introducing logic, Kant insisted on the subordination of school philosophy, the development of a consistent system, to world philosophy, a coherent account of a world in which man is part, knower, and moral agent. The real philosopher is the wise man, not the system builder.[69]

Kant felt reasonably clear on how synthesis worked. He had not been very clear on how this practice is to be justified. What little justification he had offered was essentially a mixture of psychology—the unit of discrimination is an object against a horizon—and theology—any world fashioned in accord with an infinitely wise plan must be unified. The "Third Analogy" supplies an a priori justification to this presupposition of the method of synthesis. Things coexist when in empirical perception the perception of one can follow on the other, or vice versa. Such coexistence is not given by sensation, for time is not given by sensation. Nor is the simple concept of coexistence sufficient. Logic does not validate inferences from the concept of one existent to a separate existent. The concept involved here is that of a reciprocal sequence which, as schematized, gives a temporal determination of perception. The result is the perception of a world of coexistent objects reciprocally influencing each other.

The postulates of empirical thought in general are of interest here primarily to illustrate Kant's brand of empiricism.[70] In the Wolffian tradition the possible is prior to and conceptually grounds the actual. Though the system ultimately rests on metaphysical assumptions, its functioning revolves about the priority accorded to logical analysis. Any concept having noncontradictory notes is possible. The principle of noncontradiction serves as the ultimate ground for deducing the possibles. The actual is the special case of the possible that God chooses to create.

Kant distinguishes between such mere logical possibiilty and real possibility. The actual for us is what is bound up with the material conditions of experience. It need not be directly experienced. Yet, nothing can be accepted as actual unless it is part of the physical world in the sense of 'physical world' clarified by the three analogies of experience. Unobservable entities must either be inferred, in accord with the type of inference sanctioned by the "Second" and "Third Analogies," or at least have observable consequences.

The possible for Kant is parasitic upon the actual. Mere noncontradiction of notes never gives the real existence of any object. A physically possible object is one that is, in principle, empirically detectable, or, in Kant's terms, is in accord with the formal conditions of experience. Necessity is also defined by relation to the conditions of experience. Neither 'possible' nor 'necessary' as employed in Kant's treatment of empirical thought has the meaning given it in contemporary modal logic.

Can the atomic composition of matter be inferred as actual? Kant seems to treat this in the "Antinomies," one of the longest and most disputed parts of the *Critique*. Fortunately, we can skip most of the disputes. The only point I wish to establish is that Kant's treatment of the "Second Antinomy" leaves open the possibility of accepting some form of atomism as a *physical hypothesis*.[71] The "Second Antinomy" is:[72]

Thesis: Every composite substance in the world is made up of simple parts, and nothing anywhere exists save as the simple or what is composed of the simple.

Antithesis: No composite thing in the world is made up of simple parts, and there nowhere exists in the world anything simple.

Kant's resolution of this antinomy like his resolution of the "First Antinomy" is that both the thesis and the antithesis are false.[73] If two statements are *contradictory*, then the falsity of one implies the truth of the other. If, however, two statements are *contrary* (All men are wealthy; no men are wealthy), then the falsity of one does not imply the truth of the other. Both may be false. If the thesis and antithesis of the "Second Antinomy" concerned things in themselves, then they would indeed be the contradictory statements they appear to be. If, however, they both rest on an incorrect presupposition, then they are like contraries in that the falsity of one does not imply the truth of the other. One cannot argue from the falsity of "Square circles have four corners" to the truth of "Square circles do not have four corners." Both are false since both presuppose that there are square circles.

The incorrect presupposition of the "Second Antinomy" is much less obvious. It is so hidden that, Kant argues, only his noumenon-phenomenon distinction can clarify the conceptual confusion involved. Both the proof of the thesis and the proof of the antithesis are in the form or *reductio ad absurdum* arguments, proving a position by refuting its contradictory. Yet, the defense of each rests on the presupposition that the idea of infinite divisibility may validly be applied to objects. It could, if objects were things-in-themselves. By 'object', however, one should mean 'phenomenal object', the result of a synthesis of the given and various sensual, imaginative, and intellectual impositions. The question of divisibility of an object really concerns the divisibility of an object of knowledge. Hence, 'divisibility' has a critical sense only relative to the mental constructions involved. Such divisibility cannot be infinite; it can only be indefinite.

Two conclusions flow from this treatment: an obvious one consequent upon this antinomy; and a not so obvious one, following from the antinomies in general, especially the "Third" and "Fourth Antinomies." The obvious conclusion is that we can never reasonably, i.e., in accord with the *Critique of Pure Reason*, either affirm or deny the existence of absolutely simple substances of which all things are composed. This, of course, does not prohibit one from speaking of atoms as phenomenal objects or, within the discipline of physics, simply as objects.

The not so obvious conclusion flowing from Kant's resolution of the antinomies is the admissibility of complementary conceptual frameworks. Thus, in the "Third Antinomy," the Thesis affirms, while the Antithesis denies, that in addition to causality considered as invariable sequence in accord with natural law, there is also the freedom of spontaneity. In this case Kant's resolution is that both may be true, but that the conflict is only apparent, because they are really about different things. The thesis could be true of noumenal beings. Though this cannot be established, it is reasonable to accept its truth as a postulate of ethics. Similarly, the antithesis could be true of phenomenal being. This hint that contradictory conceptualizations might be reinterpreted as complementary takes a more explicit form and a more foundational role in Kant's later writings.

Metaphysical Foundations of Natural Science

Kant's distinction between inner and outer sense led to a distinction between two complementary systems of metaphysics. This is expressed most clearly in the preface to the *Metaphysical Foundations of Natural Science* (hereafter referred to as *Foundations*).[74] After distinguishing between 'nature' in the formal sense—the nature of some thing—and 'nature' in the material sense—the sum total of all things insofar as they can be objects of our senses—Kant says of the latter: "Nature taken in this signification of the word has two main parts according to the main distinction of our senses: the one contains the objects of the external sense, the other the objects of the internal sense. Therefore a twofold doctrine of nature is possible: a doctrine of body and a doctrine of soul. The first considers extended nature, and the second, thinking nature."[75]

The significance Kant attaches to 'metaphysical' will be considered shortly. Now I wish to explain why I think it reasonable to adapt Bohr's term and call these two systems, stemming from the subject-object duality of our experience 'complementary'. In the *Foundations of the Metaphysics of Morals* (1785), *The Critique of Practical Reason* (1788), and *The Metaphysics of Morals* (1797), man is considered as an agent who must be represented as free and autonomous. Following Rousseau, Kant does not attempt to demonstrate man's freedom. Unless man has free choice, there can be no ethics. The notion of free choice is effectively contained in the concept of man as moral agent. In the *Foundations* and in the *Opus Postumum* the motions of all observable bodies, not excluding man, are explained through their determination by forces. These doctrines would be contradictory if both were interpreted as explaining things as they exist in themselves. They are complementary when interpreted as clarifications of the properties which different *representational systems* must have.[76]

A second type of complementarity is also implicit in the *Foundations*, a complementary relation between functioning science and science as reconstructed in accord with philosophical ideals. A science in which the basic laws are simply generalizations derived from experience lacks the apodeictic certainty which, Kant insists, principles of science should have. For this reason he thought it more appropriate to call chemistry a systematic art than a science.[77] As he explained in the third *Critique*, written some three years later, the laws scientists discover are contingent with respect to our way of knowing them. Though scientists may treat such discovered laws as laws of nature, they should be considered approximations to laws which are determined be nature.[78]

A functioning empirical science generally falls short of the ideals of necessity, certainty, and intelligibility, or overall coherence, which science should have. The way Kant attempted to fill this gap between the ideal and functioning science was through *rational reconstruction* of established science. Today 'rational reconstruction' is almost synonomous with 'axiomatic reconstruction'. Kant had something different in mind, a reconstruction of the *concept* of matter which plays a foundational role in the science of mechanics. A simple example might illustrate the type of reconstruction intended. After its initial discovery oxygen was effectively defined in terms of its properties: colorless, odorless, tasteless, constituent of air, supportive of breathing and combustion, etc. The much later discovery that normal oxygen has eight protons and eight electrons had the status of a discovered, and therefore epistemologically contingent, law. When, however, this atomic composition became accepted as the basis for defining 'oxygen' as the term is used in chemistry,[79] then all oxygen atoms necessarily have eight protons. In Kant's terminology, the analytic concept of 'oxygen' has been replaced by a synthetic concept. That any instance of oxygen have the atomic composition that supplies the basis for the definition is now a matter of a priori necessity.

Kant's basic purpose in the *Foundations* was to replace the analytic concept of matter by a synthetic concept and thus confer an a priori necessity on the principles of mechanics. However, it is much more difficult to replace a foundational concept, like 'matter', than it is to replace a specific concept, like 'oxygen'. The replacement is subject to three different constraints. First, the synthetic concept must be able to replace the analytic concept in its basic uses *salve veritate*. Second, the synthetic concept must supply an intelligible basis for mechanics as a mathematical science of nature. Third, (the most comlex requirement of all), the synthetic concept must fit in with a coherent account of the universe as man knows it, i.e., with the rest of Kant's philosophy.

Kant's lectures on logic explained what is involved in meeting the first requirement. In any discussion of things, whether things in general or particular kinds of things, one can distinguish between a real and a nominal essence. The real essence, what a thing is in itself, is unknown to us. The nominal essence is clarified by a process of analyzing the predicates attrib-

uted to the thing in judgments about it. One begins with a diverse collection
of representative judgments and works toward a definition by attempting to
determine which characteristics are essential and which are derivative. In
one of his early works Kant explained the relation between use analysis and
meaning clarification:

> Linguistic usage [Der Redegebrauch] and the association of a
> given expression with various contexts, where we consistently
> encountered the same characteristics, will attach to the ex-
> pression a definite meaning. Consequently we are able to dis-
> cover this meaning only by a comparative listing of different
> examples in practical usage in order to ascertain which of them
> correspond and which do not, and thus we may succeed in
> clarifying an obscure concept.[80]

No *given* concept ever admits of a strict definition. However, the process of analytic clarification leads to an approximative definition. The clarification of the concept of matter depends, not just on the characteristics
attributed to matter, but also on the conceptual ordering involved. Kant's
derivations are generally unpackings of what is implicit in some concept.
The concept of matter admits of different characteristics and different
conceptual orderings. As noted in the last chapter, the Aristotelian-
Scholastic concept of matter had the conceptual ordering substance—
quantity—quality—relation—and so on. Though this might be interpreted
as an explication of ordinary language usage, it was not the concept of
matter that played a foundational role in mechanics. Kant's own views on
this issue were reinforced by the writings of Euler, which Kant seems to have
accepted as authoritative in physics, though not in philosophy.[81] Implicit in
the notion of body are the further notions of extension, impenetrability, and
figure.[82] Thus, the judgment that a body is extended is an analytic judgment.
In terms of the role that the concept of matter plays in physics, the notion of
impenetrability must be considered as basic: "Matter when placed into a
given space, resists the intrusion of other matter, and is thus called impenetrable. That this is so we know by experience, and by abstracting from
all experiences, we obtain the general concept of matter."[83]

Further characteristics such as 'weight' are synthetic in that they depend
on experience as well as on the concept 'body'. While such a process of
analysis may clarify the use of 'matter' proper to mechanics, it does not
suffice to make the concept of matter and its role in mechanical explanations
truly intelligible. The protracted debates over forces in general, the force of
inertia, action at a distance, the transmission of motion, and the living versus
dead force controversy brought that out. The "Second Antinomy" pinpointed the concept of matter as the center or conceptual confusion. A

synthetic concept of matter should reproduce the foundational role that the characteristic 'impenetrable' plays in mechanics while also supplying an intelligible basis for the other properties attributed to matter in general.

Philosophy is concerned with knowledge through concepts, mathematics with knowledge through the *construction* of concepts. If the to-be-constructed concept of matter is based on a priori intuition, it will automatically supply a ground for mathematics. The type of intuition involved requires some comment. The schematization of the categories in imagination and the principles consequent upon this schematization were determinations of *inner* sense. They concerned the temporal structuring of the manifold of experience and gave the possibility of an object in general.

Any determinate physical object must have extension and shape. Such spatial determinations are a matter of outer, rather than inner, sense. Though particular determinations are not known a priori, the possibility of such determinations must be determined a priori. Constructed intuitions supply such an a priori basis. Thus, if motion is represented by a line, the result of two simultaneous motions can be represented by the vector addition of two lines. This geometrical intuition supplies an a priori basis for a mathematical treatment of motion in space.

Besides replacing the analytic concept and supplying a conceptual ground for mathematical laws, the synthetic concept of matter must accord with the general doctrine of the a priori aspects of experiential knowledge developed in the first *Critique*. The dominant idea in Kant's whole development is knowledge as synthesis, as a constructed unification of a manifold of elements. This leads to a doctrine of hierarchically ordered synthesizing activities and concomitant principles. At the apex is the transcendental unity of apperception. Subordinate to this in descending order are the logical functions of judgment, the pure categories of understanding, the schematization of categories in imagination, and the transcendental principles consequent upon this schematization. This supplies the transcendental a priori basis for the objectivity of knowledge.

To know any determinate object rather than the mere possibility of an object, these a priori principles must be applied to something given. If the given is given through internal senses, one has a metaphysics of soul. This, for Kant, is chiefly a question of the grounds of morality. If the given is given through external sense, then one is concerned with corporeal nature.

The metaphysics Kant repudiated in the first *Critique* was a metaphysics concerned with objects that transcend possible experience: God, the world as a whole, the soul as immortal. Kant's concern now is with things that can be objects of experience through outer sense and inferential reasoning. Here, he sees something of a division of labor between physics and metaphysics. Metaphysics is concerned not with the actual intuition of things in space, but with the a priori possibility of such intuitions. Everyone makes such metaphysical presuppositions. No one had yet attempted to isolate these a priori aspects and give them a consistent systematic development. This Kant intends to do by focusing on the a priori steps involved in the construction of a concept of matter which corresponds to possible intuitions and supplies an intelligible basis for a science of matter. Not surprisingly, the construction of the concept of matter has four steps corresponding to the fourfold division of the categories into quantity, quality, relation, and modality.

Phoronomy gives the first step in the construction of the concept of matter: "Matter is the moveable in space."[84] Space is the form of outer intuition. The simplest determination through which space can affect the senses is through the motion of one space relative to some other, e.g., an enveloping space. Since internal structure is not considered, the moving space can be treated as a point when one is considering problems of motion. Since space is extensional, one space can be added to another. This type of geometric addition carries over to motion only if one makes the additional assumption that two equal velocities, or the components in the same direction, admit of being combined in the same way as two equal spaces. Unlike Einstein, Kant did not examine the consequences that would flow from a denial of this assumption. As a consequence of this assumption, the motions of bodies admit of quantitative determination.

Dynamics, the next stage, adds one further characteristic: "Matter is the moveable insofar as it fills a space."[85] To fill a space means to resist everything moveable, i.e., any other matter that tries to fill the same space. This requires a special moving force. The forces Kant postulates are essentially the same as those he postulated in his *Monadologia Physica* thirty years earlier. All matter has an inverse square attractive force. There must also be repulsive forces. Again, Kant assumes that these are inverse cube forces, but does not assign any foundational role to such speculations. The balance of attractive and repulsive forces determines the natural boundaries of a body.

The advance over the earlier treatise is not so much in the concept of matter itself, but in the way in which the concept of matter now fits into Kant's overall system. Kant's basic problem is to develop a concept of matter with the greatest *relative* intelligibility. By 'relative intelligibility' I mean that for Kant the fundamental factor in the constitution of matter is simply assumed. If this fundamental assumption cannot be derived from anything more basic, then it can only be tested by the consequences flowing from it.

There are only two alternatives that Kant takes seriously, atomism and dynamism. Atomism, i.e., the traditional doctrine of hard, impenetrable atoms moving in a void, has some explanatory advantages, particularly in explaining varying densities. Yet, it renders unintelligible the relation of force to matter as well as the concept of action at a distance. Kant pointed out that, in spite of Newton's disclaimers, action at a distance did play an indispensable role in Newton's system.[86] With the assumption of atomism, furthermore, the "Second Antinomy" remains unresolved, leaving reason in conflict with itself.

If forces, rather than particles, are assumed to be basic, then the other properties of matter can be coherently reconstructed without any intelligibility gap. Kant's concept of matter is, in modern terms, a field theory of matter, but one based on a severely limited concept of fields. The only forces having the intuitive intelligibility that Kant's theory of construction seemed to require were central field forces between points, forces which attracted and repelled each other only along the lines joining them.

The monads Kant had postulated thirty years earlier were individual point-centers, surrounded by short-range repulsive and long-range attractive forces. Also, this was a speculative model about the real nature of ultimate matter. Kant's new doctrine is not based on physical monads. The points discussed are not individual centers for individual atoms or monads. They are idealizations, helpful in treating fields. This is essentially the treatment given in Euler's *Mechanics*. Euler begins with space and then treats bodies as collections of spatial points,[87] and only later speaks of these as point-masses.[88] Furthermore, Kant's new doctine is not a speculation about what ultimate matter really is. It is, rather, an attempt to construct a concept of matter which can make the science of mechanics intelligible and render its laws truly necessary and universal.

The concept of matter constructed on this basis can be treated in three stages: supplying a ground for mathematics, synthetic reconstruction of the properties basic to the analytic concept of matter, and preparing a domain for empirical investigation and determination. The first point is easily treated. Kant's synthetic concept of matter is constructed out of space through attractive and repulsive forces. This immediately resolves the "Second Antinomy." Since space admits of infinite division, matter constructed out of space also admits of infinite division. This does not mean that matter as it is in itself is infinitely divisible. Rather, the representation of matter as a movable region of space has the potential for infinite divisibility built into it. This representation also grounds mathematics. The boundaries of a particular piece of matter are determined by the balance of attractive forces, which are universal, and repulsive forces, which are particular. These repulsive forces admit of degrees. As in the *Anticipations of Perception*, this explanation of quantities of qualities, or degrees, grounds the quantitative mathematical determination of qualities.

In explaining the properties contained in the constructive concept of matter Kant distinguishes between the properties that can be determined a priori and those requiring a posteriori determination. The only pure a priori properties are those that follow from the force assumptions without any further assumptions. Gravitation, the universal attractive force, proportional to the quantity of matter, leads directly to the notion of weight.[89] The addition of a repulsive force leads to the characteristic of elasticity.

Though other properties are dependent upon experience, reason can supply a priori guidelines for the systematization of experience. This is done in two ways. The first is preparing a domain for investigation through the guidance of the a priori notions of *genus* and *species*. Thus, one investigates a new domain by looking for universal common features and characterizing differences.

The second way is through *definition*. Though the analysis of concepts may lead to an approximate definition, no given concept can be strictly defined, for further characteristics may always be discovered. Since this is not true of synthetic concepts, one can use such concepts to give strict definitions. Kant's examples bring out the significance of such definitions. "A *body*, in the physical signification, is a matter between determinate boundaries (and such matter therefore has a figure)."[90] The synthetic notion of 'body' is constructed from the elements proper to Kant's theory of matter: 'content of space' and 'density which admits of degrees.' The adequacy of such construction is evaluated both by its ability to replace the analytic concept in its scientific usages and by its usefulness in giving theoretical explanations. This second aspect is clearly shown in Kant's definition of cohesion: "Attraction, insofar as it is thought as active merely in contact, is called *cohesion*. (Indeed one proves by very good experiments that the same force which in contact is called cohesion is found to be active also at a very small distance.[91]

What Kant is doing here can best be understood by situating it between the older methods of physics that he was familiar with and the newer methods that to some degree Kant anticipated. Both Newton and Euler began their treatises on mechanics in Euclidean fashion with a series of definitions. Neither, however, had a theory of the nature and function of definitions. When Newton defined 'quantity of matter' and 'quantity of motion,' he did not discuss the relation between stipulating meanings and using terms to explain reality. After theories had developed sufficiently so that there was a clear distinction between a conceptual framework used to report phenomena and the theory used to explain the same phenomena, physicists introduced rules coordinating theoretical terms with established observational terms. Thus, in kinetic theory one coordinates the average kinetic energy of molecules with the temperature of a gas and momentum transfer with pressure. In Kant's case, the problem is not the correlation of relatively theoretical (kinetic energy) and observational (temperature) concepts. It is, rather, the correlation of analytic and synthetic versions of the same concept, e.g., 'body,' 'cohesion,' 'solid,' 'fluid,' etc. In neither case is the meaning of a concept directly determined by what things are in themselves. The meaning of the analytic concepts is determined by their use in ordinary and especially in scientific discourse and through an analytic clarification of that use. The synthetic concepts are defined in terms of the terms taken as basic in the process of reconstruction, e.g., 'force,' 'occupy space,' etc. A necessary, though not sufficient, condition for the replacement of analytic by synthetic concepts is that the synthetic concept have the characteristics which were seen as basic in the analysis of the given concept.

What this method does not explain is what Kant takes to be the foremost problem of natural science, the specific variety of types and properties of matter. The attempt Kant made to get at the a priori aspects of this in the *Opus Postumum* will be considered briefly later. What should be noted now is that, in modern terminology, this gives the sufficient condition for the synthetic concept of matter. Here again, the parallel with Bohr is revealing. From his 1913 trilogy to his 1923 paper on the periodic table he thought that one of the two basic challenges his atomic theory must meet was to explain the chemical properties of the elements as revealed in the periodic table. The other challenge was to explain observed spectral lines. This in Kant's terminology is an explanation of specific properties and activities.

The "Metaphysical Foundations of Mechanics" adds the next characteristic in the concept of matter: "Matter is the moveable insofar as it is something having a moving force."[92] Mechanics in Kant's ordering differs from dynamics inasmuch as it is concerned with the force one body exerts *on other bodies* rather than with the internal forces that give a body its definite extension. Kant's principles of mechanics are essentially Euler's principles with a different conceptual ordering. For Euler 'mass-point' effectively functioned as a primitive concept. For Kant 'force' is the primitive concept. 'Matter' is explained in terms of 'force' and 'occupy space,' and then 'mass' is defined in terms of 'quantity of matter': "That the quantity of matter can only be thought of as the number of its moveable parts (external to one another), as the definition [his Explication 2] expresses it, is a remarkable and fundamental statement of universal mechanics."[93]

Considering the vagueness of the term 'moveable parts external to one another,' one might find this statement neither fundamental nor remarkable. When Kant has a self-congratulatory aside, as here or after the deduction of the categories, it is usually because he has found some part or method that seems to hold his system together. The significance of this statement comes from the overall coherence it gives Kant's interpretation of science. Within mechanics in the special sense in which Kant uses 'mechanics' in contrast to 'dynamics,' one can treat bodies as extended bits of matter. Kant's new foundation yields the operative presupposition of Newtonian mechanics. One can accept Kant's reconstruction without changing the way physics is actually done. From a larger perspective, however, two things are required to make this practice *intelligible*. First, one must have some basis for understanding how one body can move another body. If bodies are basic, as in traditional atomism, then this remains unintelligible. If force is basic, as in Kant's dynamics, so that a body's boundaries are simply equilibrium points for opposing forces, then such a moving force is quite intelligible.

The second requirement is one that recurs at each stage of the development. One must justify the applicability of mathematics to physical concepts. Kant now argues that any doctrine of physical monads would require intensive quantities. Though he does not cite it in this context, Kant was quite familiar with Euler's *Letters to a German Princess* and the attack on monads contained in this, as well as in other, writings of Euler. The point that Kant stresses is the reliance in mechanics on extensive rather than intensive magnitudes. Now, Kant's definition of mass in terms of quantitative extension supplies an intelligible, i.e., internally consistent, basis for the practice of integration.

The further significance of Kant's distinction between dynamics, concerned with intelligible foundations, and mechanics, concerned with the practice of mathematical physics, is most evident in his new treatment of the *vis viva* controversy. Within mechanics itself, the terms 'living' and 'dead' are no longer appropriate. The difference is just a question of whether one uses *mv*, which is directly correlated with motion, or mv^2, which may be determined either by measuring a body's motion or indirectly, e.g., from the height which a body has fallen.[94] In either case all the forces treated within mechanics are dead forces in Kant's former sense: bodies act only to the degree they are acted upon.

In Kant's familiar fourfold distinction, mechanics comes under the general heading of "Relation." In the first *Critique* the temporal schematization of this category led to the three analogies of experience. Now Kant is concerned with the significance of schematizing this category for *spatial* intuitions. This yields three laws of mechanics which both serve as quantitative determinations of the three analogies for material bodies and supply an intelligible ground for Newton's laws of motion.

The "First Law" for Kant's mechanics is, "With regard to all changes of corporeal nature, the quantity of matter taken as a whole remains the same, unincreased and undiminished."[95] When 'substance' in the vague sense proper to the "First Analogy" as presented in the B edition is rendered extensive through application to spatial intuition, the result is a law of mass conservation. Though this had been vaguely anticipated by others, Kant was the first to give a clear explicit statement of mass conservation as a general law. Subsequent scientists, however, were much more strongly influenced by Lavoisier's measuring pans than by Kant's transcendental deduction in coming to accept mass conservation as a law.

Kant's "Second Law of Mechanics" clearly has an intermediate role between the "Second Analogy" and both Newton's "First" and "Second Laws of Motion": "Every change of matter has an external cause. (Every body remains in its state of rest or motion in the same direction and with the same velocity unless it is compelled by an external cause to forsake this state.)"[96] Similarly, Kant's "Third Law of Mechanics" relates the "Third Analogy" to Newton's "Third Law of Motion": "In all communication of motion, action and reaction are always equal to each other."[97]

Once again, Kant's emphasis is on making concepts fit into a coherent system. Hume's famous billiard ball example had shown that physics did not really explain how one moving billiard ball causes another billiard ball to move. Nor, in Kant's judgment, did Newton's own doctrine of a *vis inertiae* succeed in making the communication of motion intelligible. When, however, one presupposes that matter, whether at rest or in motion, already presupposes moving forces, then reciprocal communication of motion is intelligible a priori.

The final section, the "Metaphysical Foundations of Phenomenology," relates to the category of modality and adds the final characteristic to the concept of matter: "Matter is the moveable insofar as it can be an object of experience."[98] Something becomes an object of experience only through the unification of a given, forms of sensibility, schematization in imagination, and concepts of understanding. Kant has constructed a concept of matter as the movable in space and explained how this a priori construction can structure intuition. To determine an object of experiential knowledge, the concept must be predicated of a subject which involves the representation of a given.

The crucial issue here is the question of how motion inheres in a subject. As indicated in the last chapter, there was prior to Kant's work a long tradition of metaphysical disputation concerning the way in which motion understood as an accident inheres in a substance. This dispute involved such notions as antiperistasis, impetus, inertia, force of inertia, living force, dead force, transmission of accidents, and divine concurrence with secondary causes. Kant's systematic construction of a concept of matter from movable space effectively handled all the metaphysical issues involving composition. The only conceptual issue left is how movable space, which now supplies a conceptual foundation for the experiential knowledge of a moving object, is to be understood relative to other spaces.

Kant's answer to this, while interesting, is deliberately incomplete. The rectilinear motion of a body can only be understood relative to some other body. A consideration of the motion of a body relative to a container which is moving relative to a larger container leads in the familiar way to the problem of absolute space. Kant accepts absolute space as an *idea*, in his technical sense of 'idea.' It plays a regulative role, but does not determine an object.

Circular motion, however, seems to be a proper rather than a relative predicate. Newton's rotating bucket and other similar experiments lead to the conclusion that circular motion is intelligible only with respect to absolute space. This conclusion, if accepted, would seem to make absolute space into a real object rather than a regulative idea. Elsewhere, especially in the *Opus Postumum*, Kant developed a doctrine of ether both to banish void space and also to explain the transmission of forces, particularly those involved in explaining coherence. In the present context, he merely notes that, though physical grounds might be found for banishing empty space, there is no merely logical ground for rejecting the possibility of empty space.

Concluding Observations

Kant's growing awareness of the significance for theories of scientific explanation of the difference between purely logical inference based on deductive rules and real, or existential, inference which is content-dependent has played a pivotal role in the present survey. In view of this stress it is interesting to consider Kant's final public statement of philosophy, his open letter to Fichte:

> I hereby declare that I regard Fichte's *Theory of Science* [*Wissenschaftslehre*] as a totally indefensible system. For the pure theory of science is nothing more or less than mere logic, and the principles of logic cannot lead to any material knowledge. Since logic, that is to say, *pure logic*, abstracts from the content of knowledge, the attempt to cull a real object out of logic is a vain effort and therefore a thing that no one has ever done.[99]

Kant, as seen through such public pronouncements, might seem to be suffering from the ailment that besets many an older philosopher, hardening of the categories. Yet, privately Kant continued to work on the completion of his system and may even have attempted, at the age of seventy-six, a reconstruction of the whole critical philosophy. The *Foundations* provided, in Kant's opinion, a complete system of metaphysical principles for the a priori aspects of the cognition of material bodies. Yet, this still left a gap in Kant's overall system, the *Transition from the Metaphysical Principles of*

Nature to the Physical Principles of Nature. This topic concerned Kant from at least 1790 and seems to have received his undivided attention from 1797 to his death in 1803. The scattered and repetitious notes gathered together as the *Opus Postumum* are open to a range of interpretations, from a proof of senility to a bold attempt to revise the whole system. Since I have not read this rather massive collection of notes, I am relying on some published summaries.[100] All that I wish to do here is to indicate the bearing of Kant's final efforts on the problematic we have been considering.

On a plausible reconstruction, Kant's late work can be divided into two periods. From 1797 to 1800 he worked on the transition from the metaphysical principles of nature to the physical principles. He hoped through the construction of tables of forces and properties to supply an a priori framework which could serve in characterizing the specific types of properties and activities proper to different types of matter. This effort carried Kant past the bounds he had assigned a priori reasoning in the first *Critique*. Such reasoning was now seen as contributing to the *content* of empirical knowledge, rather than the pure form. Partially for this reason and partially because all the bright young men defending the critical philosophy were tending toward idealism as the only consistent basis for defending Kant's system, Kant himself moved in the direction of idealism. This move, however, concerned Kant's interpretation of the noumenal self and the way in which this self posited the forms of experience, rather than his interpretation of science. To the end Kant insisted on empirical realism *within* science. Even Kemp Smith, who repeatedly criticized Kant for excessive subjectivism and who judged the *Opus Postumum* to be a badly misguided effort, concludes:

> Thus, in regard to Kant's final positions, as revealed in the *Opus Postumum*, whatever else be doubtful, two points at least are abundantly clear: first, that he definitely commits himself to a realist's view of the physical system in space and time, and of the manner in which we acquire knowledge of it; and secondly, that he is willing to go almost any lengths in the way of speculative hypothesis regarding the noumenal conditions of our sense-experience, if only thereby the difficulties which stand in the way of this empirical realism can be successfully dealt with.[101]

The subsequent development of physics did not build on the foundations Kant supplied or follow the guidelines he suggested. Even those scientists who were influenced by Kant's writings of the foundations of physics, such as the early nineteenth-century *Naturphilosophen*[102] and, somewhat later, J. R. Mayer and H. Helmholtz in developing the law of energy conservation, used only fragments of Kant's thought: the primacy of the force concept, the use of teleological explanations in biology, openness to speculation instead of pure Baconian inductionism. The edifice from which these were stripped was too massive and cumbersome to supply a framework for functioning physics.

Yet, Kant's contribution to the interpretative problem we are here examining cannot be judged simply in terms of the degree to which he influenced the subsequent development of physics. Rather, he should be judged primarily for his contribution to an interpretative problem inevitably generated by a successful conceptual revolution in the explanation of matter and its properties. A relatively phenomenological level that is accepted as at least descriptively adequate is explained through a deeper level, which involves the postulation of unobserved entities, forces, and processes. Thus, Kepler's laws are explained through Newton's mechanics and theory of gravity, the observed properties of matter through atomic structure and intermolecular forces, the properties of baryons through quarks and gluons, the Einsteinian account of gravity through a quantum theory of gravity and supergravity. The development of such explanations is the task of the physicist. The philosopher has an epistemological cleaning-up operation: how to explain the grounds of meaning of the concepts basic to the theory and the ontological significance of the referential use of such theoretical concepts.

Kant knew a physics which he accepted as functionally adequate and essentially complete. For this reason he was able to focus on the epistemological problem in a unique way. Others, both before and after Kant, have treated the epistemological problems in piecemeal fashion, attempting to explain particular concepts that emerge as problematic. Only Kant has attempted to analyze the problematic concepts generated by the advance of physical science in the light of a comprehensive theory of what concepts are and how they relate to other aspects of knowledge and experience in constructing a consistent account of a universe in which man is part, knower, and moral agent.

The positive significance I attach to Kant's methods and doctrines will be manifested by the use I make of them. Accordingly, it seems more appropriate to conclude by appraising the significance of Kant's major limitations. If we ignore his historical limitations, the frequent but relatively minor inconsistencies so dear to the defenders of the patchwork thesis, and the aspects of his thought not related to the present problematic, there still remain at least three major limitations which deserve consideration. They concern his psychologism, his architectonic structure, and his a priorism. The first two are so interrelated that they are best treated together.

I have presented Kant's doctrine in a way that minimizes his dependence on faculty psychology. His *practice* of conceptual analysis is, I believe, essentially independent of such psychology. His *theory* of concepts, however, is open to criticism as representing some form of private-language theory, at least implicitly. This is not simply a post-Wittgenstein reflection. Kant's own student, J. G. Hamann, brought out the same point very strongly, accusing Kant of turning words into ghosts. Modern critics insist that the meaning of a concept is essentially public and only derivatively private. In his characteristically baroque language, Hamann in his *Metacritique of the Purity of Reason* also insisted on the primacy of public language over private thought in questions concerning meaning:

> Accordingly, one principal question yet remains: how is the ability to think possible?—the ability to think through on the right and left, before and without experience, with and beyond experience? Really no deduction is necessary to demonstrate the genealogical priority of language for the seven sacred functions of logical premises and conclusions and their heraldry. Not only is the entire possibility of thinking founded in language, according to the unappreciated sayings and lasting achievements of the distinguished Samuel Heinke; but language is also at the center of reason's misunderstanding of itself.[103]

By Kant's architectonic I refer especially to the structuring role which dividing the mind into sensibility, memory and imagination, understanding, reason, and will and which the table of twelve categories plays in Kant's overall system. As many critics have pointed out, Kant's reliance on faculty psychology is not really consonant with the foundational role he assigns the critical method, while even Kant's staunchest supporters find it difficult to defend the deduction of the categories and the necessary role this particular division is accorded in all acts of understanding.

The architectonic is easily criticized, but not so easily dismissed or replaced. It has a subjective necessity. Without this architectonic Kant could not develop his philosophy into a unified system. Without a sympathetic understanding of the architectonic's role, the interpreter of Kant's system cannot grasp its unity. Can it be redeveloped? Wilfred Sellars has made a truly heroic effort to do so, beginning with language as the locus of meaning and explaining how words which are used to refer to mental acts, states, or faculties could come to play the role they do.[104] Yet, he has never succeeded in developing a system with anything like Kant's overall coherence. For our more limited purposes there is fortunately no need to attempt to redevelop the Kantian system on the basis of a semiotic transformation of the categories. What is needed is to understand the unity of Kant's system well enough to understand how particular categories and concepts function as parts of a whole and then be able to adapt his methods well enough so that particular concepts, e.g., the concept of matter, can be handled in a critically acceptable way.

That knowledge in general, and scientific knowledge in particular, has a priori aspects seems altogether beyond doubt. That these a priori aspects can be developed as an independent system which supplies a foundation for empirical science is not at all certain. Kant's justification of the independence of his systematization of a priori knowledge depended on his architectonic and his psychology. It seems more reasonable to take a sort of scissors approach to this issue, cutting down from above with general considerations such as Kant's and from below with detailed examinations of the a priori aspects of different scientific theories. In such a perspective, Kant's failures may prove almost as valuable as his successes. He thought he was supplying a permanent a priori foundation for mathematical physics *überhaupt*. Any aspects of his foundation that have been undermined by subsequent conceptual revolutions should plausibly be reinterpreted as a priori aspects of Newtonian mechanics rather than of physics as such. Because of the complementary relation obtaining between quantum and classical mechanics, this demarcation criterion should prove valuable in assessing what is distinctively new about the conceptual foundations of quantum mechanics.

These are the major limitations in Kant's philosophy that I think relevant to the present study. Others will undoubtedly have different appraisals. Such limitations notwithstanding, Kant was and remains the philosopher of science par excellence.

3. Atoms and Conceptual Change

Newton and Kant had each attempted to develop a concept of atoms which could play a foundational role in a science of matter. In Newton's later writings and in Kant's whole career, the emphasis was on such second-order conceptual questions as, What are the grounds of meaning of the terms ascribing properties and activities to atoms? Such questions come to the forefront only after the first-order questions concerning the existence of atoms and the science of matter seem to have been sufficiently well answered. Yet, as indicated in chapter 1, a doctrine of atomism ceased to play a foundational role in the mechanics of Euler and the French Newtonians. The speculative efforts of Kant and of Roger Boscovich to develop a dynamic concept of atoms which could make Newtonian forces intelligible never won wide acceptance as supplying a necessary foundation for a science of mechanics. The first-order questions concerning the existence, types, and properties of atoms remained in the foreground.

Speculation about atoms and their role in scientific explanation did not cease. It simply changed focus: from the foundations of mechanics to newer emerging sciences, from second-order questions about the nature of knowledge and the grounds of meaning of crucial concepts back to first-order concepts and questions. Are there atoms? Are they all the same or of different types? What forces and activities do they have? Which of many competing accounts should be accepted as correct? In the course of attempting to answer such questions the term 'atom' was retained while the concept of what an atom is underwent a profound change and the requirements that any theory of atoms must meet became much more sharply specified.

What I wish to focus on in the present chapter is this conceptual change and the factors that shaped it. The treatment of history will be pointillistic, selecting and arranging particular points which may blend together to give an overview of this development. The survey is far from complete. In the interests of brevity some topics such as the study of crystals, organic chemistry, and the attempts to explain the diffusion of solutions through permeable membranes will be omitted, though they do have a bearing on the development of atomism. Since the aim of this chapter is to present an overview of a long, complex development, or series of developments, technical details will for the most part be omitted. Fortunately, for most of the points selected, more detailed studies are available presenting a more complete historical account.

Before considering particular points, there are two background features which should be considered. First, the older idea of a philosophy of nature still perdured. It not only provided the matrix from which new disciplines gradually emerged, but also presented something of a general world view which supplied the initial qualitative basis for interrelating these new disciplines. In retrospect, the most significant role this philosophy of nature played was that of providing a framework for the development of a minimally adequate conceptually consistent descriptive account of some of the basic ingredients that went into the new scientific disciplines.

The second background feature is the general pattern of development of the new disciplines we will be considering. The initial stage, when the nascent discipline is still subordinate to a philosophy of nature, is generally characterized by an ontological luxuriance. Different speculative hypotheses about particles, powers, pervasive media, and various occult qualities are introduced and given something of an initial qualitative test. Basically, this test comes from the ongoing dialogue among the community of inquirers attempting to appraise what any new assumption really explains, how consistent it and its implications are with accepted doctrines, and whether the assumption is superior to the competition. After a new discipline emerges with apparently adequate principles and methods, there is generally a shift from qualitative physicalistic reasoning to formal deductive reasoning. Thus, if Maxwell's theory of electromagnetism is interpreted as Maxwell's equations rather than as a model of electromagnetic force fields existing in the ether, then developing the theory consists primarily in deducing the consequences of the mathematical equations. This, in turn, leads to a pruning of the ontological luxuriance characterizing the formative stage, such as the speculations about the stretching and rotating of magnetic tubes of force. What such disciplines as chemical atomism and kinetic theory really have to say about the existence of atoms, their properties, and their forces is ultimately determined, not by the initial speculations that launched the enterprise, but by the assumptions that prove indispensable in keeping it afloat.[1]

Natural Philosophy as a Matrix

Natural philosophy is a rather amorphous subject with a rather checkered history. As part of the Aristotelian-Scholastic division of the speculative sciences into natural philosophy, mathematics, and metaphysics, natural philosophy employed, but left it to metaphysics to justify, presuppositions about objects and their properties, causes and their effects, and general laws of nature. The rise of mechanics modified the operative presuppositions, while the philosophical revolution initiated by Descartes and culminating in Kant shifted the burden of justification from a metaphysics of ultimate reality to a critique of knowledge. Yet, as long as the tradition flourished, it rested on presuppositions about objects, properties, causes, and regularities in nature. Though in the eighteenth and early nineteenth centuries natural philosophy was not the dominant influence on the Continent that it remained in British and Scotch universities, it was still regularly taught as an introduction to philosophy and science. As such it supplied something of a common background and shared language for scientific investigators.

Before focusing on British natural philosophy, some mention should be made of a continental contribution that became quite influential, especially in nineteenth-century British science. Roger Boscovich, a Jesuit from Croatia, wrote *A Theory of Natural Philosophy* in 1758.[2] In this work he developed a theory of the atom as a point-center surrounded by alternating short-range repulsive and attractive force fields which blend into a long-range inverse square attractive force. For most practical purposes this theory was functionally equivalent to Kant's point-center atoms surrounded by inverse cube repulsive and inverse square attractive forces. Both schemes attempt to confer a Leibnizian intelligibility on Newtonian physics by a conception of the atom in which force is basic and mass derivative. Yet, it was Boscovich's conception rather than Kant's that supplied the historically influential alternative to Newton's conception of hard, massy, inpenetrable atoms. It seems that then, as now, Kant was more respected than read, especially by scientists.

Schofield's detailed study of British natural philosophy brings out a somewhat surprising shift of emphasis.[3] In the generation after Newton, British natural philosophy was generally *mechanistic*, attempting to explain observed phenomena in terms of ultimate atoms whose only explanatory properties are size, shape, combination, and forces. After about 1745, speculation tended toward a neo-Aristotelian particularism, attributing essentially different types of properties to different substances or states. My superficial examination of a few of the more successful textbooks from the end of this era indicates something of a common pattern in the way the properties of matter were treated and the sort of presuppositions this involved. To cite the introduction of a four-volume text published in 1803:

> Agreeably to this, the reader will find in the course of this work, an account of the principal properties of natural bodies, arranged under different heads, with an explanation of their effects, and of the causes on which they depend, as far as has been ascertained by means of reasoning and experience; he

will be informed of the principal hypotheses that have been
offered for the explanation of facts, whose causes have not yet
been demonstratively proved; he will find a statement of the
laws of nature, or of such rules as have been deduced from the
concurrence of similar facts; and lastly, he will be instructed in
the management of philosophical instruments, and in the mode
of performing experiments that may be thought necessary
either for illustration of what has already been ascertained, or
for the further investigation of the properties of natural
bodies.[4]

These are the types of presuppositions about objects and properties,
causes and effects, that Kant had sought to justify. Within the natural
philosophy tradition, there was very little concern with justifying such
presuppositions. The emphasis was on determining and explaining the
properties of matter. The properties common to all matter, extension,
divisibility, impenetrability, mobility, inertia, and gravitation, were
assumed to have been explained or, in the case of gravity, at least to have
been systematized by Newton, Active investigation was more concerned
with particular properties of matter, those which depend on the material or
state of a body. I. B. Cohen lists the particular types of properties studied
during this post-Newtonian period under six general headings:[5]

1. *electrical properties*, whether a conductor or an insulator, whether
exhibiting resinous or vitreous electrification when rubbed with silk or fur;
2. *thermal properties*, whether and how good a conductor of heat,
melting and boiling points, etc.;
3. *thermochemical properties*, whether or not combustible, combustion
temperature, etc.;
4. *general chemical properties and behavior*, kind of substance, types of
reaction with other substances;
5. *optical properties*, color, transparency, dispersion, refraction, bi-
refringence;
6. *mechanical properties*, hardness, elasticity, density.

From a contemporary perspective, most of the work done looks like
very low-level science, uncritical, minimally theoretical, making only incidental use of mathematical reasoning. Yet, this work supplied the background for nineteenth-century theorizing and discipline development by
attempting to fashion as detailed and precise a descriptive account of matter, its states, and its properties as was then possible. The attempts to
explain these properties generally followed the pattern set by Newton of
roundly condemning (Cartesian-type) hypotheses and extolling induction as
the method of science, while freely indulging in (Newtonian-type) hypoth-

eses. These hypotheses involved the postulation of such theoretical entities as corpuscular atoms, caloric atoms, frigorific atoms, pholgiston, one or more electric fluids, magnetic effluvia, and an all-pervasive ether.

Such a summary makes the effort seem totally lacking in discipline and restraint. Yet, constraints gradually emerged. Eventually, as Cohen has shown, scientists developed explicit criteria which any proposed hypotheses should meet.[6] In addition to the explicit criteria posed within particular traditions, there were also criteria recognized, at least implicitly, in the debates between competing traditions. The two most important were coherence and adequacy.

By 'coherence' I do not now refer to the logical or conceptual consistency of a particular system. In a conflict between competing systems what counts as a sign of success is the consistency found in a descriptive account of natural phenomena and their causes. Though more an ideal than a realized goal, this criterion nevertheless exerted an abiding pressure to produce a coherent world view. The second criterion, empirical adequacy, took on increasing importance as more extensive and precise data became available and required explaining. Two particular conflicts, in which overall descriptive coherence and empirical adequacy served as decisive criteria, played a role in the changing concept of atoms. One concerned the theory of hard body collisions, the other the thermal properties of gases.

Wilson Scott has traced the development of the first problem from Descartes's statement of a conservation law for total motion in 1644 through the general acceptance of a resolution of the conflict between atomism and conservation theory in 1860.[7] Some of the factors he treats, particularly Dalton's atomism and the rise of kinetic theory, will be treated in a different context. All that I wish to do now is to bring out the problems involved in the historical attempts to fashion a descriptive account of a collision between two hard bodies, an account that is both conceptually consistent and empirically adequate.

Though Descartes supplies Wilson's starting point, the core of the conflict is most easily understood in terms of a clash between Newtonian and Leibnizian mechanics. In Newton's account, atoms are hard, impenetrable, and unbreakable. A perfectly hard body does not bounce on impact. Two such bodies colliding with each other would, it was argued, each come to an instant stop. Their motion must be described in a way that is totally discontinuous, by, in modern parlance, step functions. This position, if accepted, implies discontinuities in nature and no conservation laws. Its strongest support came from the fact that motion (or energy, in later terms) is demonstrably lost in any process involving dissipation.

The position stemming from Leibniz stressed the conservation of *vis viva* and rejected the conception of atoms as hard particles. Apparent losses of *vis viva* were explained by a transfer of *vis viva* from the visible to the invisible realm. While this interpretation preserved the conservation laws, it had difficulty in explaining the stability and invariance of properties of matter.

This debate could not be settled by any observational test, partially because it concerned unobservable atoms and partially because each side had recourse to position-saving assumptions that were in principle unobservable. Newton's assumption of divine intervention to restore lost motion was balanced by Leibniz's assumption of the transfer of *vis viva* from the visible to the invisible realm. Like many similar developments, this ongoing debate was involved in conceptual confusion and in many side issues which, in retrospect, seem irrelevant. There was, however, one criterion that advocates for each side implicitly accepted. The correct position should give an account of the process of the collision of two atoms that could be accepted as a description of what happens and that would be consistent with established physical principles.

The attempts to resolve this conflict varied from conceiving of hard bodies as the limit approached when inelastic bodies became less pliable to that of elastic bodies which have extremely large internal forces. Various modifications were worked out, treating atoms as something between perfectly rigid and perfectly elastic bodies, treating impact as involving compression and restitution in insensible degrees. One could preserve Newtonian orthodoxy while obviating its difficulties, as Dalton did, by postulating perfectly rigid matter atoms surrounded by perfectly elastic caloric atoms or, as Maxwell did, by arguing that a gas is made up of elastic molecules whose ultimate constituents may or may not be elastic.

Such descriptive accounts of collisions between atoms seemed, one and all, to be theory-saving devices. Eventually the conflict was resolved through the gradual adaption of such new concepts as action, energy, and work, and corresponding principles. If a collision conserves energy, then one need specify only the initial and final states. No descriptive account of the collision itself is needed. Instead of describing the deformation and restitution of a hard body's shape, one could treat the product of pressure and volume change as a form of work. A broadening of the conservation of *vis viva* to the more general idea of energy conservation led to the idea that lost energy is transformed into the work ($P\Delta V$) required for compression. If compression and restoration are considered part of a cycle, like a Carnot cycle, then detailed balancing of energy losses and gains is not required.

The eventual result of this protracted debate was an approach to collision processes that allowed the preservation of both a doctrine of atomism and the laws of conservation of energy and momentum. What was not so obvious was the fact that this resolution imposed severe restrictions on any new doctrine of atomism. No descriptive account of the atom had yet been found which accorded with these conservative principles and also gave a convincing, or even plausible, account of the properties atoms must have. A reconciliation of the opposing views was obtained only by obviating the need for such a description.

Kinetic Atoms

The development of the kinetic theory of gases involved a collection of initially disjoint parts which were eventually assembled to form a coherent whole. The present sketchy account only includes the aspects helpful in understanding the contribution of this development to the changing concept of an atom.

Francis Bacon demonstrated his inductive method by proving that heat is a form of motion.[8] Both Boyle and Newton treated a gas or elastic fluid as if it were a collection of essentially static particles maintained in an equilibrium state by mutual repulsion. Yet, neither presented this as anything more than a speculative hypothesis. As Newton put it:

> But whether elastic fluids do really consist of particles so repelling each other, is a physical question. We have demonstrated mathematically the properties of fluids consisting of particles of this kind, that hence philosophers may take occasion to discuss that question.[9]

Philosophers did take occasion to discuss that question, particularly after the development of caloric theory in the 1770s.[10] The repulsive forces thought to exist between stationary atoms were attributed to the presence either around or between the gas particles of atoms of a weightless fluid, caloric, which was postulated to explain heat. The manner in which the caloric theory was introduced, developed, and eventually abandoned is a good illustration of the way in which ontological hypotheses supply a scaffolding for theory construction.

It is somewhat simplistic to speak of *the* caloric theory. The general assumption that there are atoms of heat was developed in different ways to explain radiation, temperature, specific heat, expansion, change of state, and the repulsive force between gas particles. The view that came to be dominant stemmed from the work of Lavoisier, Laplace, Poisson, Bertholet, Gay-Lussac, and other French scientists. The core of the doctrine is seen in Lavoisier's original position where caloric was thought of as a fluid composed of particles that repel each other (hence heat's diffusion), but are attracted to particles of ordinary matter with the degree of attraction depending on the nature and state of the matter. With the latter assumption it was possible to incorporate Black's ideas on latent and specific heat into caloric theory. Caloric is neither created nor destroyed. This principle of conservation of caloric grounded the application of mathematics to heat transfer. Each particle of ordinary matter was thought to be surrounded by an atmosphere of caloric whose density increases with temperature, causing particles of matter to repel each other at small distances. This, as Lavoisier saw it, explained the fact that volume increases linearly with temperature. It can also explain the transition from solid to liquid state which occurs when caloric molecules so surround matter molecules that each is separated from the sphere of influence of surrounding molecules.[11]

These assumptions concerning caloric not only supported a coherent account of the phenomena associated with heat, but also led to distinct quantitative advances. By assuming a force law proper to caloric repulsion, Laplace was able to derive the ideal gas law and Gay-Lussac's law. He also used this assumption to give a theoretical calculation of the velocity of sound which accorded well with experimental values, a problem that had been outstanding since physicists recognized the arbitrariness of the correction terms Newton had introduced to make theory agree with practice.[12] Sadi Carnot used caloric theory in developing his fundamental ideas on heat cycles and ideal engines. Caloric theory also supplied the conceptual under-pinning for Avagadro's systematic research on the relation between chemical affinities and electrical properties; for Regnault's meticulous investigations of specific heats, which showed that all the ideal gas laws were merely first approximations requiring correction terms; and for much of the research on chemical atomism and thermal properties of bodies. It was a very successful theory.

Yet, few scientists, especially after the 1820s reaction against Laplacean orthodoxy, believed that the truth of the caloric theory had been conclusively established. Count Rumford and Humphry Davy argued convincingly that the caloric theory was unable to give an adequate account of frictional heating. Neither really held a kinetic theory, for neither held the key postulate of kinetic theory, the free motion of gas molecules. Also, it is somewhat misleading to speak of Rumford's views in Baconian terms as a *motion* theory of heat. As Rumford realized, no motion theory could explain the heat the sun radiated through empty space. For this reason Rumford postulated ether waves.[13] Yet, neither Humphry nor Davy developed a heat theory that offered caloric theory any serious competition.[14]

Besides internal difficulties of explaining away frictional heating, there were also external difficulties in the attempt to fit caloric theory into a coherent overall account of natural phenomena. Since the sun radiated both heat and light through empty space and, following Newton, light was considered corpuscular, it seemed reasonable to assume that the heat radiated was also corpuscular. The gradual acceptance of the wave theory of light, following the researches of Young and especially of Fresnel, undercut this indirect support of the caloric theory. Cardwell cites an 1818 paper by H. Meikle which expressed the situation through a singularly infelicitous choice of metaphors: "The nature of heat, like that of light, is still in darkness."[15]

The textbooks of the era present the practical solution that was generally followed.[16] The real nature of heat is not known. It may be caloric or a form of motion or something else. Yet, the caloric theory, whether true or not, is the best theory to teach. It supplies a physical basis for a mathematical formalism which proved successful in treating most of the known thermal phenomena apart from frictional heating and, later, Joule heating. To be acceptable as a replacement, any competing theory of heat would have to give a plausible account of the thermal phenomena explained by the caloric theory and supply a foundation for the successful quantitative treatment of heat flow and related phenomena.

Kinetic theory attempted to do this, but with little immediate success or acceptance. Long before the rise of caloric theory and its application to the Boyle-Newton static theory of gases, Daniel Bernoulli had given a mathematical formulation of the kinetic theory of gases. On the assumption that a gas is composed of an enormous number of particles in very rapid chaotic motion, he argued that gas pressure could be explained as the result of particle impacts on the walls of the container and that an increase in heat would increase the velocity of the gas particles.[17] In spite of the reputation of the Bernoulli name, this theory won virtually no immediate acceptance.

The kinetic theory of gases was revived by John Herapath in 1821 and by John Waterston from 1845 on.[18] Waterston's theory corrected some errors in Herapath's account and extended it by giving the first statement of the law of the equipartition of energy among different types of gas molecules. He also derived the correct relation between the pressure exerted by a gas and the kinetic account of this in terms of the number of molecules per unit volume, their masses, and the mean squares of the velocities. Neither theory won any acceptance. Waterston's paper was effectively buried in the archives of the Royal Society until Lord Rayleigh exhumed it in 1891. Even if his paper had been given a fair hearing, it is hardly likely that it would have convinced Waterston's contemporaries any more than it convinced the two referees for the Royal Society. The new theory rested on such implausible assumptions as extremely high molecular velocities and perfectly elastic collisions between molecules. As noted in an earlier section, at this time (1845) there was no nonarbitrary way of reconciling the assumption of elastic collisions between molecules and the application of Newtonian mechanics to molecules. Finally, the new theory could not supply a quantitative account of all the thermal phenomena successfully handled by caloric theory.

The path that led to the development and acceptance of the kinetic theory of gases can be summarized in two stages. The first was the emergence of thermodynamics as a distinct discipline from around 1850 on. The most important aspect of this development in the present context is the formulation of the ideas of the mechanical equivalent of heat and the more general law of the conservation of energy.[19] The doctrine that caloric is neither created nor destroyed had grounded the equations balancing heat loss against heat gain when an isolated system attains equilibrium. If energy is conserved and if heat is a form of energy, then the new principle of energy conservation also implies that heat is conserved in an environment in which heat is neither generated nor dissipated.

The second stage depended on a more detailed physical account of the behavior of gas molecules. The simple assumption of translation motion did not account for the thermal properties of differing gases and differing states of the same gas. In 1857 R. Clausius argued that in addition to translation energy (or *vis viva*), there must also be energy associated with rotational and vibrational states of the molecules.[20] Yet, as C. H. D. Buijs-Ballot pointed out, Clausius's model of a gas still did not seem to fit all the phenomena. If, as Clausius held, gases were made up of molecules which move very rapidly through largely empty space, then two gases should diffuse through each other in the very short time it takes molecules to travel the length of the gas container. To meet this objection Clausius developed another fundamental concept of kinetic theory, mean free path. Later, this served to explain heat conduction and viscosity as well as diffusion.[21]

Like the other kinetic theorists, Clausius assumed that all gas molecules of the same type move with the same velocity. Though J. C. Maxwell did not at this time believe in the kinetic theory of gases, he found it interesting "as an exercise in mechanics" to work out the statistical distribution of velocities resulting from random collisions.[22] This led to the surprising and, as Maxwell then thought, unreasonable conclusion that the viscosity of a gas is independent of its density. The static molecules of the caloric theory led to the conclusion that viscosity should increase with increasing gas density. Maxwell set up an elaborate experiment in the attic of his home and proved that the viscosity of air at fixed temperature remains constant when the pressure varies between one-half an inch and thirty inches of mercury.[23] This and similar experiments by O. E. Meyer converted most of the remaining scientists who had not yet accepted the kinetic theory.

Kinetic theory is a theory about the motion of molecules. Accepting kinetic theory as explanatory entails accepting molecules as real. This argument, so familiar to philosophers, seems simple and straightforward. But it does so chiefly because we can separate out the points we now consider basic and, with the clarity of hindsight, feel assured that the complications we omit will not jeopardize the conclusion. Yet, the question of the reality of atoms and molecules and the properties to be attributed to them was neither simple nor straightforward for the men who developed kinetic theory. Before considering the interpretative difficulties they encountered, we will briefly consider two other scientific disciplines in which assumptions about atoms or molecules seem to have played a foundational role.

Chemical Atomism

Antoine Lavoisier consciously initiated a revolution, fashioning a new science of chemistry by introducing fundamental changes in both the concepts and methods of chemistry. In place of the speculative elements assumed by the Aristotelians, the Iatrochemists, and others, Lavoisier insisted on defining 'element' in terms of the methods available to chemists:

> I will, accordingly, content myself with saying that, if by the name of elements we are understood to designate the simple indivisible molecules which make up bodies, then it is likely that we do not know them; if, on the contrary, we attach to the name of elements, or to the principles of bodies, the final term at which our analysis arrives, then all the substances which we have not yet succeeded in breaking down by any means are elements for us . . . [24]

The new method he most insisted on, precise measurements of weight while performing new or repeating old experiments, led to his law of mass conservation. Its general acceptance facilitated the emergence of quantitative chemical laws in the decade after his death.

John Dalton did not begin his career as a follower of Lavoisier or even accept Lavoisier's chemical definition of an element. This was partially due to British opposition to Lavoisier's chemistry. Humphry Davy's discovery of the elements potassium and sodium and the realization that they combine with oxygen to form a base rather than an acid undercut Lavoisier's oxygen theory of acids. Lavoisier's caloric theory, which had played a significant role in his chemistry, was rejected by Davy and also by Count Rumford, who married Lavoisier's widow. An even more decisive factor in Dalton's case seems to have been the fact that what little formal education he had oriented him more toward Newtonian physics than chemistry.[25] Dalton began his research on gases, then called 'elastic fluids', with Newton's idea that a gas is composed of stationary atoms that repel each other. In Dalton's quaint terms, in the elastic state ultimate particles support their dignity by keeping encroachers at a respectful distance.[26]

What was initially unique about Dalton's approach was that he thought of an atmosphere containing water vapor as a mixture rather than a chemical compound. To explain the separate behavior of the different gases mixed together in the atmosphere, Dalton assumed that atoms of different chemical substances form different species which repel like atoms, but may attract others. This conception of a gas gradually emerged through the development leading to his law of partial pressures (1802) and his law of multiple proportions (1804). Unlike Lavoisier, whose ideas he used somewhat uncritically, Dalton believed the elements of his chemical philosophy to be ultimate particles, Newton's hard massy atoms. He did, however, strongly endorse the new emphasis on precise weighing of the ingredients that enter into a chemical combination. He thought that the fixed ratios he obtained supported his contention that ultimate elements enter into binary, ternary, quaternary, etc., combinations. To other chemists, however, it was not at all clear that these fixed ratios referred to anything more than relative combining weights.[27]

At about the same time, Gay-Lussac was demonstrating that gases combine chemically according to the ratio of small whole numbers of volumes. Dalton considered the attempted explanation, in which volumes rather than weights were basic, incompatible with his own atomic theory. In 1811, Avagadro proposed a solution to this conflict based on the assumption that equal volumes of different gases at a given temperature contain the same number of molecules. Earlier, Dalton had held a similar view and had drawn the conclusion from it that relative atomic weights could be determined simply by measuring relative vapor densities. He then rejected it on the grounds that the vapor density of a product is often more than that of the constituents. The hypothesis as developed by Avagadro seemed to be arbitrary and to lead to conclusions inconsistent with observations. A simple example illustrates the nature of the difficulty. If a given volume of nitrogen contains n nitrogen atoms and the same volume of oxygen contains n oxygen atoms, then a mixture of the two gives two volumes of gas containing $2n$ atoms. When the oxygen and nitrogen combine chemically, there is little volume change. If the simplest formula is assigned for nitrous gas, one oxygen atom combining with one nitrogen (or azote) atom, then the $2n$ atoms originally present should form n molecules of nitrous gas and occupy only half the 'mixture' volume.

Avagadro's solution to this and similar objections was to assume that oxygen and nitrogen units are split in two before combining or, in Avagadro's terms, that an integral molecule contains two constituent molecules. This assumption was one that leading chemists rejected. Not only would it imply that chemical atoms are not ultimate indivisible units, but also it would contradict the basic assumption stemming from Newton, that like atoms repel each other. Avagadro's hypothesis did not win general acceptance until after 1860, when S. Cannizzaro circulated a tract at the 1860

Karlsruhe chemical coference promulgating the polyatomic hypothesis. An acceptance of this position also implied that many of the entities that chemists had been calling 'atoms' were molecules. There was no guarantee that any of the as yet unresolved entities were ultimate units in any stronger sense than Lavoisier's final term at which chemical analysis has so far arrived.

As these problems concerning the number of atoms in different chemical combinations began to be cleared up, new problems emerged. Humphry Davy and especially Michael Faraday made it clear (e.g., in Faraday's electrolysis experiments) that electrical attraction somehow plays a basic role in binding atoms into molecules. How this happens was not at all clear. The leading view was Berzelius's dualistic theory attributing molecular combinations to a union of oppositely charged atoms. While this assumption fitted the experiments decomposing compounds by electrical means, it seemed totally incompatible with the Avagadro-Cannizzaro assumption that basic molecules like oxygen or nitrogen are formed from the union of like atoms.

A final point is in retrospect one of the crowning achievements of nineteenth-century atomism, the system of the periodic elements generally attributed to Mendeleev.[28] As the charts came to be filled in, the number of distinct chemical elements grew rapidly. To many who pondered the problem of the composition of matter it seemed clear that the multitude of chemical elements could not be the ultimate building blocks of matter. Thus, J. C. Maxwell writing on 'Atom' for the *Encyclopaedia Britannica* carefully distinguished the molecules of the chemists (which they might call 'atoms') from the molecule of the physicists: "The definition of the word molecule, however, as employed in the statement of Gay-Lussac's law, is by no means identical with the definition of the same word as in the kinetic theory of gases."[29] The difference ultimately stemmed from the chemists' reliance on laws of combination versus the physicists' reliance on dynamical laws of motion as the basis for the use of 'molecule.'

Spectral Lines and Atomic Theory
In the early nineteenth century instrumental optics developed rather independently of the new wave theory of light. In 1800, Sir William Herschel established the existence of infrared heat radiation. In the following year Johann Ritter showed that invisible rays also exist beyond the violet end of the spectrum, rays capable of such chemical action as blackening silver chloride. In 1802, William Wollaston discovered seven dark lines in the sun's spectrum while trying to determine the number of primary colors in sunlight. This passed without much notice until it was rediscovered in an independent and more ample fashion by Joseph von Fraunhofer. In 1814–

15, Fraunhofer presented to the Munich Academy a map of the solar spectrum showing a multitude of dark lines, the chief of which he distinguished by letters of the alphabet. Fraunhofer invented both the spectroscope and the diffraction grating. With their help he eventually detected 574 dark lines in every type of sunlight and measured the wavelength of some of the more prominent lines. These measurements led to the realization that the D lines in the yellow part of the sun's spectrum coincide with two bright lines in the spectrum of the sodium lamp.[30]

From these early studies emerged the new discipline of spectroscopy. In 1826, W. H. Fox Talbot began the application of spectroscopy to chemical analysis. Absorption experiments by William Miller in 1833 and by the astronomer Sir John Herschel led to the idea that Fraunhofer lines are due to absorption in the outer layers of the sun's atmosphere. Léon Foucault later made a more precise determination of how the sodium D line is absorbed and emitted. Subsequently, J. Angström established the identity of the emission and absorption lines. These developments culminated in the very successful cooperation of the physicist Gustav Kirchhoff and the chemist Robert Bunsen. A joint paper they published in 1859 gave the fundamental principles for spectral analysis and its application to astrophysics.[31] Relying on the mechanical theory of heat, Kirchhoff analyzed the conditions for an equilibrium between absorption and emission of radiation. He established the general law that, for rays of the same wavelength and sources at the same temperature, the ratio of the emissive to the absorptive power is the same for all bodies. Kirchhoff and Bunsen also established the fact that chemical combinations have no effect on the lines characterizing different metals.

If each element has its own characteristic spectral series, then it should be possible to explain spectral lines in terms of the elements producing them. This obvious suggestion was not so easily implemented. As a matter of convenience we will consider first the qualitative attempts to explain the origin of spectral lines and then the quantitative attempts to develop formulas for spectral series. In 1868, G. J. Stoney worked out a molecular theory of spectral lines based on the kinetic theory of gases and on Maxwell's explanation of molecular motions. In a gas at normal temperature and pressure, molecules collide every 1.4×10^{-11} seconds. Brief as this interval is, it is from 50,000 to 100,000 times as long as the duration of one vibration of light in the visible part of the spectrum. The combination of these two facts suggested a plausible mechanism for spectral emission. A collision shakes up a molecule and gives it excess energy. After the collision, this absorbed energy is radiated away through the vibration of the parts composing the molecule. These characteristic vibrations can be decomposed into a series of simple vibrations through Fourier analysis. Stoney argued that this decomposition has a physical analogue, that the multiplicity of observed spectral lines is produced by a natural decomposition of the complex vibrations characterizing each type of molecule. Though the amplitude of these vibrations depends on the strength of the collision, the frequency should depend exclusively on the characteristic structure of each type of molecule.

No amount of revision and refinement seemed to enable this theory to cope with an underlying difficulty. Neither a Lucretian hard atom nor a Boscovichean point-center admits of vibrations. The vibrations characterizing different chemical elements must be due to combinations of atoms or larger units. J. Lockyer proposed a chemical evolution of increasingly complex molecules. A leading spectroscopist, H. Kayser, was willing to assume that iron molecules are composed of 5,000 atoms as a means of explaining iron's highly complex spectrum. Others postulated smaller molecules with allotropic forms.[32] The failure of all such attempts made the basic choice clear. Either one must admit highly complex molecules as the source of each element's characteristic spectra or one must introduce new models which allow the atom internal degrees of freedom. Before treating the second alternative, we will consider the attempts to formulate quantitative laws for spectral lines.

Granted the plausibility of the basic idea that spectral lines are due to some sort of internal vibrations, it seemed reasonable to look for quantitative relations between the spectral lines in a series in terms of the frequencies and overtones produced by vibrations. Stoney, in fact, saw this as the strongest supporting evidence for his model. Drawing an analogy between the harmonic overtones of a fundamental frequency in sound and a series of lines in hydrogen, he argued that the first, second, and fourth lines in the hydrogen spectrum were the twentieth, twenty-seventh, and thirty-second harmonics of a fundamental vibration whose wavelength is 131,000 Å (angstroms). Ten years later, A. Schuster discredited this argument by showing that this coincidence is no better than might be expected on the basis of pure chance.

In 1855, Johann Balmer was the first to succeed in discovering a mathematical formula that fit an observed series of spectral lines. He presented his law with no explanation of how he achieved his result. To the best of my knowledge, only Stanley Jaki has succeeded in uncovering the peculiar path Balmer followed in reaching this result.[33] Balmer was not a physicist, but a geometrician with somewhat mystical feelings for natural harmonies. His doctorate (1849, Basel) was for a dissertation on cycloids. He obtained a position as a math teacher in a girls' secondary school and wrote a *Habilitationschrift* which he presented to the University of Basel in 1865. It was entitled, "The Prophet Ezekiel's Vision of the Temple Broadly Described and Architectonically Explained." Here, as in his later attempts to analyze

Gothic cathedrals and ancient buildings, he exhibited his underlying conviction that any regular structural order must be a manifestation of a mathematical relationship. When his friend E. Hagenbach told him of the regular
lines in the hydrogen spectrum, he began a search for the mathematical
order implicit in this regularity.

On the basis of the values assigned to the four lines in the hydrogen
spectrum, Balmer suggested the formula[34]

$$\lambda = b(m^2/(m^2 - n^2)), \text{ where } b = 3{,}645.6 \text{ mm}/10^7. \tag{3.1}$$

When n is assigned the value 2 and m the values 3, 4, 5, 6 the observed lines
are rather precisely reproduced. The value $m = 7$ led Balmer to the prediction of another visible line which, as he soon learned from Hagenbach, had
already been observed. In a short time nine more lines observed in stellar
spectra were also accommodated by his formula. Balmer noted that his
formula also allowed for other series by letting, e.g., $n = 3, m = 4, 5, 6, \ldots$

Balmer's work provides a classic example of the discovery of a purely
phenomenological law. Neither Balmer nor his immediate successors,
Kayser, Runge, Rydberg, and others, could supply any basis beyond trial
and error adjustment of formulas to data for the formulas they developed.
In addition to the formulas for spectral series, other regularities were
discovered and fitted into similar phenomenological laws. Two years before
Balmer introduced his law, W. N. Hartley discovered that the components
of a doublet or triplet series have the same separation when measured in
terms of frequencies rather than wavelengths. J. R. Rydberg, some ten
years after Balmer, showed that Hartley's law of constant frequency difference was applicable even when doublet or triplet components are widely
separated. His study of such lines led to the classification of spectral series as
sharp, principal, diffuse, and fundamental. In each case a Balmer-type
formula was found proper to each type series and a new law, the Ritz
combination principle, was found interrelating the fixed terms. Following
Rydberg, the general formula is usually written in the form

$$1/\lambda = R_H(1/n_f^2 - 1/n_z^2), \tag{3.2}$$

where R_H, the Rydberg constant, is 109.677 cm^{-1}, n_f has a fixed integral
value for each spectral series, and n_z takes on integral values greater than
n_f.[35]

None of these formulas were supported by a physical explanation of why
the observed regularities obtained. Nor did any of the attempts to develop
an analogy between spectral lines and acoustical overtones prove helpful.
There was no successful theoretical explanation of any spectral regularity
until 1896 when H. A. Lorentz explained the newly discovered Zeeman
effect.

New Theories of the Atom

As usual, the problematic situation is clearer in retrospect than it was in the confusion and complexity of developing science. An acceptable theory of the atom should have internal consistency and empirical adequacy, and should supply a basis for a coherent descriptive account of physical reality. In the latter part of the nineteenth century this meant that any theory of atoms and molecules should supply at least a potential basis for understanding the molecular assumptions of kinetic theory, should supply the chemical properties of the elements and their periodicity, and should shed some light on the spectral lines characterizing each element. These were the basic external constraints for any atomic theory. All of them could not, of course, be met at once. Yet, any theory devised to explain some of the phenomena, e.g., kinetic theory, could not be considered a candidate for the ultimate theory if it positively excluded other pertinent phenomena, e.g., spectral lines. In addition to these general requirements, there was some factual information about atoms that should also be considered. We will indicate its basis and plausibility.

Both chemical atomism and spectroscopy systematized macroscopic observable properties. The only operative assumptions about the properties of atoms that played a role in these sciences concerned their relative rather than their absolute weights. Thus, J. T. Merz, writing at the turn of the century, could say that, for most chemists, atoms had become convenient symbols for the formulation of various combining laws: "Although, therefore, chemical research was governed all through the century by the atomic view of matter, it does not appear that philosophers considered the existence and usefulness of chemical formulae as a proof of the physical existence of atoms, or of smallest indivisible particles of matter, in the oldest sense of the theory."[36]

Of the new developments considered, only kinetic theory seemed to rest directly on assumptions concerning the physical properties of atoms and molecules. On the plausible assumption that acceptance of kinetic theory as explanatory rationally entails acceptance of the assumptions indispensable to its success, it is reasonable to investigate what these assumptions were. Here, fortunately, the founders of kinetic theory supply helpful guidance, for they were well aware of the difference between helpful models and indispensable assumptions.

Clausius was, with Rankine and Kelvin (William Thomson), one of the founders of thermodynamics. Independently of Kelvin, Clausius had introduced internal energy U in the formula linking changes in heat Q with changes in work W: $dQ = dU + dW$. Somewhat earlier, Carnot had introduced and Clapeyron had developed the idea of a state function. By a detailed analysis of an infinitesimal Carnot cycle, Clausius proved that the internal energy of a gas is a state function or is independent of particular molecular configurations. Its successful use accordingly revealed nothing about molecular configurations.

The size attributed to kinetic molecules was more a matter of mathematical convenience than of descriptive precision. It was effectively the distance at which attractive and repulsive forces balance. How this relates to inherent size depends on the model used. Shape attributions seemed to have a better basis. The postulation of rotational and vibrational degrees of freedom involved some sort of dumbbell model of molecules. Since, however, these models led to a discrepancy between theoretical and experimental values for specific heats, there was doubt about their objective validity.

Because the kinetic explanation of the perfect gas law implicitly assumes point particles, this explanation gives no information about the size of molecules. Deviations from the gas law do depend on molecular size. Using this information as well as information from viscosity measurements and the estimates of the number of molecules in a standard volume, it was possible to make reasonable estimates of the mass, diameter, and velocity of molecules. Thus, Maxwell gave values for hydrogen, oxygen, carbonic oxide, and carbonic acid as well as tables of diffusion.[37] Some of these values, given in modern notation, may be compared with modern values (given in parentheses):

	Hydrogen	*Oxygen*
Mass		
Relative	1	16
Absolute	46 (34)	$736 (532) \times 10^{-25}$ g
Velocity		
(mean square at 0° C)	1859 (1830)	465 (461) m s^{-1}
Diameter	5.8	7.6×10^{-10} m

His velocity values are closest to contemporary values, since these were determined from experimental data while the other values were adjusted. All his values are within an order of magnitude of currently accepted values.

This minimal and somewhat insecure factual basis supplemented the theoretical considerations we have been considering in setting constraints on any theory of atoms. Meager as this basis was, it sufficed to eliminate both the Lucretian (hard, impenetrable mass) and the Boscovichean (point-center plus force fields) models of the atom. As B. G. Doran has shown, this rejection was part of a trend away from a concept of particles in a void toward a conceptualization in which fields are basic.[38]

Against this background it is illuminating to sketch the struggles of some leading theorists to frame a theory of atoms. Michael Faraday presented a theory, something like Kant's, in which atoms were pictured as fields of force in space rather than material particles.[39] Yet, he never succeeded in developing this in any detail. The new speculative model that seemed to offer the greatest promise of meeting the constraints listed was the model developed by Lord Kelvin. He summarized the difficulties with the earlier model, particularly the Lucretian-Newton model:

> The idea of an atom has been so continually associated with incredible assumptions of infinite strength, absolute rigidity, mystical action at a distance, and indivisibility, that chemists and many other reasonable naturalists of modern times, losing all patience with it, have dismissed it to the realms of metaphysics, and made it smaller than 'anything we can conceive.'[40]

What was needed, as Kelvin saw it, was a physical model that might supply the basis for a mechanical-mathematical explanation. In 1867, after witnessing P. G. Tait's experiments with smoke rings, he studied and eagerly adapted some mathematical theorems Helmholtz had developed showing that vortex motion in an ideal frictionless fluid has peculiar conservation properties. A vortex ring has an invariable volume, retains a uniform strength in the sense that the product of the velocity of rotation and the area of any section is constant, and is closed to other parts of the fluid. No vortex tube can pass through any other. Two tubes linked together can never be separated. Relying on these mathematically established properties, Kelvin made the bold assumption that atoms can be explained as vortex rings in the ether, conceived of as an incompressible frictionless fluid. This theory would entail one primordial matter, ether, in which move different types of rings and linked loops. These vortex rings would be quantitatively permanent in volume and strength, qualitatively permanent in implication—knottedness within itself, linkedness with other rings, or a combination of both. Yet, it would be flexible enough to allow vibrations. The degree of, and perseverance in, complexity gave a promise of explaining chemically different atoms, while the ability of each different type of vortex ring atom to have its own complex set of characteristic vibrations supplied a potential basis for explaining spectral lines. Finally, it could be demonstrated mathematically that encounters between vortex atoms would have the elastic properties kinetic theory required.[41]

Before considering the eventual denoument of this ingenious theory, we will consider one other theorist. Maxwell, who had initially been skeptical of kinetic theory, had gradually developed a very sophisticated view of the role of models and analogies in scientific explanation.[42] As late as 1873, he was willing to accept the Lucretian model of the atom, presumably on the grounds that it supplied an adequate analogy for kinetic theory. In 1875, he addressed the Chemical Society of London on the dynamical evidence for the molecular constitution of bodies. There, he explained the greatest difficulty which the molecular theory had yet encountered. The ratio of the specific heat of a gas at constant pressure c_p to the specific heat at constant volume c_v is

$$c_p/c_v = (2 + n + e)/(n + e). \tag{3.3}$$

Here n is the number of variables needed to express the internal configuration of a molecule (or the number of internal degrees of freedom), and e is a positive quantity depending on the law of force which binds the constituents together. For point particles, $n = 3$ and $e = 0$, "and the ratio of specific heats is 1.66, which is too great for any real gas."[43] When internal degrees of freedom are allowed, n is at least 6, so that the greatest value of the ratio is 1.33. For many gases, the observed value is 1.408. Since this is not compatible with any model of molecules, it seemed difficult to take any model seriously.

Two years later, in the article "Atom," Maxwell began by distinguishing the atoms of the chemists, which are really molecules, from the atoms of the physicist. To explain the latter, he gave a detailed account of Kelvin's vortex model and concluded that it satisfied more of the conditions such models must meet than any atom hitherto imagined.[44] He expressed his hope that this model might serve to explain both mass and gravitation, though he did not know how this could be done. Why this sudden acceptance after Maxwell's skepticism concerning all molecular models?

I believe that McGucken is correct in his claim that the change was brought about by Kundt and Warburg's discovery in 1876 that the ratio of c_p/c_v for mercury gas is 1.666, the exact value that Maxwell had predicted for a monatomic gas. According to the arguments surveyed in the preceding section, a monatomic gas, at least in the Lucretian or Boscovichean model, should have no spectral lines, for it has no internal degrees of freedom. Mercury, which had just been shown to behave dynamically like a monatomic gas, exhibits a series of spectral lines. The implication seemed inescapable. Even a monatomic gas must have internal degrees of freedom. The vortex model, which pictures atoms that can bend, bulge, and vibrate, allows internal degrees of freedom.

Maxwell's article excited interest, at least in England, in the vortex model of the atom. Maxwell's own enthusiasm, however, quickly faded for reasons that are worth considering. L. Boltzmann, who will be considered subsequently, had introduced a model of a gas molecule as an aggregate of material points held together by forces which are arbitrary functions of the separation of atoms.[45] This seemed to supply a basis for explaining both the

Kundt-Warburg results and the fact that mercury has a spectral series. If the internal forces are sufficiently strong, then the mercury molecule acts like a perfect elastic sphere with three degrees of freedom. The line spectrum is now explained as virbrations produced *during* collisions, rather than through radiation emitted subsequent to collisions.

Maxwell considered this new Boltzmann model in his review of Watson's *Treatise on the Kinetic Theory of Gases*, a work which followed Boltzmann's methods.[46] The net result of the full following of this method, Maxwell concluded, is "perhaps ultimately to drive us out of all the hypotheses in which we have hitherto found refuge into that state of thoroughly conscious ignorance which is a prelude to every real advance in knowledge."[47] The difficulty Maxwell now finds with both Boltzmann's and Kelvin's model of the atom as a rigid elastic body is that such a body is capable of an infinite number of internal vibrations and consequently should have an infinite number of degrees of freedom and an infinite specific heat. The only suggestion he offered as a solution was a model involving an atom of Boscovich surrounded by some sort of force field which repels bodies that come within a very short distance. It is tempting to interpret this suggestion as an anticipation of the nuclear atom. If so, it was an anticipation that Maxwell does not seem to have taken seriously, for, as he noted, this model also fails to explain spectral lines.

After Maxwell's death, even Kelvin's support for his own vortex model began to waver. There were two difficulties which he could not seem to get around. The first was the tendency of vortex rings to dissipate to infinite distances and then move with infinitely small velocities. The second was the inability to explain mass.

In June 1896, the city and the University of Glasgow had a three-day jubilee celebration to commemorate Kelvin's fifty years as professor of natural philosophy. In his own address at the jubilee banquet, Kelvin made a startling admission:

> One word characterizes the most strenuous of the efforts for the advancement of science that I have made perseveringly during fifty-five years; that word is FAILURE. I know no more of electric and magnetic force, or of the relation between ether, electricity, and ponderable matter, or of chemical affinity, than I knew and tried to teach my students of natural philosophy fifty years ago in my first session as Professor. Something of sadness must come of failure.[48]

By this time Kelvin was concluding that both his theory of the ether and his theory of the atom must be abandoned. In 1898, in a letter to S. W. Holman, Kelvin said:

> I am afraid that it is not possible to explain all the prop-
> erties of matter by the vortex-atom theory alone, that is to say,
> merely by motion of an incompressible fluid; and I have not
> found it helpful in respect to crystalline configurations, or elec-
> trical, chemical, or gravitational forces. . . . We may expect
> the time will come when we shall understand the nature of an
> atom. With great regret I abandon the idea that a mere con-
> figuration of motion suffices.[49]

Even this setback did not quench Kelvin's sanguine enthusiasm for progress. After retiring as a professor, he had himself enrolled as a research student and continued to ponder the problem of the atom. He accepted the idea developed by J. Larmor and H. A. Lorentz that the atom consists of positive and negative parts bound together by electrical forces. In 1901, in a paper entitled "Aepinus Atomized," he developed a model of the atom as positive and negative components held together by electrical forces.[50] Electrions (Kelvin's suggested term for electrons) are considered to be small spherical particles of resinous (negative) electricity which can move freely through the ether or can penetrate atoms, considered as much larger spheres of positive electricity. The neutralizing quantum of electrions for any atom, or the quantum number, is the number of electrions required to balance the oppositely charged atom.[51] Two different types of configurations are then considered. The first is the neutral atom in the proximity of a charged particle. A one-electrion atom would be in stable equilibrium if the electrion were in the center. When a positively charged particle or atom approaches, the electrion would be attracted in the direction of the charged atom, making the atom into something of a dipole. The other type of configuration considered is for multielectrion atoms. Kelvin calculated the conditions for stable equilibrium for different types of configurations. Two electrions would be at the ends of a diameter, three at the vertices of an equilateral triangle, four at the corners of a tetrahedron. These were stable configurations. Though Kelvin was unable to prove stability of more complicated cases, he suggested various possible configurations. This is the general model that was adapted by J. J. Thomson. As new discoveries were made— radium, polonium, three different types of radioactive decay—Kelvin kept introducing new models of atoms to try to accommodate them.[52] What is more significant than these particular and, it must be admitted, rather different models is the fact that Kelvin lent his unique prestige to the idea that purely mechanical models of the atom must be dismissed as failures. Something radically new was needed. It is one of the ironies of history that Kelvin, who even in his seventies was remarkably open to new developments, would later be used in popular histories of quantum theory as the symbol of opposition to progress and change.

Boltzmann and the Atomic Debates

At the close of the nineteenth century Ludwig Boltzmann became the pivotal figure in the debates about the reality of atoms. To appreciate the significance of the conflicts and the ambiguities in Boltzmann's position we will review Boltzmann's scientific work and outline the opposing positions before considering Boltzmann's contribution to these debates. Ludwig Boltzmann (1844–1906) did outstanding work in physics, particularly in kinetic theory, until about 1896. The last years of his troubled life were more devoted to the defense of atomism and the clarification of related philosophical problems than to technical science.[53]

According to Elkana, the physics taught at the University of Vienna, where Boltzmann received his doctorate and taught during three separate periods, was characterized by a strong belief in the reality of atoms and by the attempt to explain observable phenomena in terms of the mechanical properties attributed to atoms. Boltzmann's initial acceptance of this orientation formed the background to his most significant contributions to kinetic theory, the explanation of entropy in terms of the tendency of systems of molecules to approach the most probable distribution. This explanation encountered serious conceptual difficulties with the reversibility argument J. Loschmidt advanced in 1876 and the recurrence paradox E. Zermelo developed in 1896. In both cases, Boltzmann's defense hinged on the idea that his theory relied on probability considerations as well as mechanics. Both were needed to give a molecular account of irreversible processes. Mechanics alone gave only reversible processes.

Boltzmann's kinetic theory depended on assuming the existence of molecules, but not on a completely realistic description of their properties. Thus, following Maxwell, Boltzmann assumed an inverse fifth power repulsive force between molecules, not on the grounds that there was any evidence for such a force law, but because it simplified some integrations. By the time Boltzmann wrote volume 2 of his lectures on gas theory (1898) he was well aware that the assumptions he had introduced about gas molecules were neither physically realistic nor empirically adequate nor totally self-consistent:

> In order to take into account this indubitable composite structure of the gas molecule necessitated by spectral and chemical considerations, one will have to consider it an aggregate of a definite number of material points, held together by central forces. One does not obtain very good agreement with experiment in this way; on the contrary, for many gases the thermal phenomena, at least, are better interpreted by assuming that the molecules are rigid nonspherical bodies. Thus it appears that the connection of the parts of these molecules is

so intimate that they behave like rigid solids with respect to
thermal phenomena, even though in other cases the constituents appear to vibrate against each other.[54]

Faced with such difficulties, Boltzmann attempted to keep assumptions
about molecules as minimal and general as possible. Kinetic theory should
not depend upon particular models of molecules, but on the general assumption that molecules have parts whose configurations are determined by
general mechanical laws.

The scientific developments which we have been considering rested on
some programmatic assumptions which were widely shared. Chief among
them were the assumptions that the measurable inorganic properties of
observable bodies could be explained in terms of their ultimate constituents
and that the forces, motions, and structures of molecular parts are governed
by mechanical laws. By the end of the nineteenth century, many were
drawing the conclusion that this program must be considered a failure. A
new program with new foundations was needed. Since these countercur-
rents do not figure directly into the present account, we will simply sketch
three schools of thought that Boltzmann recognized as competitors to the
atomistic-mechanistic view.

First were the *phenomenologists*. This was a general term Boltzmann
used to cover those, particularly G. Kirchhoff and E. Mach, who thought
that the proper method for physics was to pick a strictly circumscribed area
and then to systematize the observable phenomena through differential
equations without introducing hypotheses.[55] In countering this position,
Boltzmann seems to have felt that he was pushing back an incoming tide.
Positivistic interpretations of science were clearly manifested in the writings
of such contemporaries as J. B. Stallo, G. Kirchhoff, E. Mach, A. Roy,
P. Duhem, K. Pearson, and others. Of particular interest here is Mach,
Boltzmann's predecessor in his final position at Vienna, who strongly influenced Boltzmann's epistemology. Mach stressed the primacy of sense
experience and, through his principle of economy, explained scientific laws
and theories as useful summaries of actual and possible experiences: "It is
the object of science to replace or *save* experiences, by the reproduction and
anticipation of facts in thought."[56]

The second group were dubbed *energeticists*. Wilhelm Ostwald and
Georg Helm led this short-lived movement.[57] If the laws of thermodynamics
are interpreted as laws which not only are universally valid but do not
depend on the particular material any system is made of, then it seems
plausible to consider thermodynamics more basic than mechanics. Ostwald
and Helm crusaded for a conceptual revolution in which 'energy' would
replace 'substance' as the basic concept and the first law of thermodynamics
would be the most basic law in physics.

In this conflict E. Mach and Max Planck split their loyalties in interesting ways. Though Mach opposed Boltzmann's atomism on epistemological grounds, he rejected the attempt to reduce physics to a science of energy. Max Planck was at this time a strong believer in the science of thermodynamics as a science based on the most general laws yet discovered. He rejected Boltzmann's statistical explanation of entropy as an attempt to reduce a general law to a particular instance. Yet, he also rejected the attempt of the energeticists to give the concept of entropy a derivative status and to reduce mechanics to thermodynamics. Planck, the conceptual conservative, wished to keep mechanics and thermodynamics essentially unchanged.[58]

The final group could, for want of a better term, be called *electrodynamicists*. Mechanics in the classical tradition was generally presented as resting on a few indispensible concepts: point-masses, forces, absolute (or at least objective) space and time, and action at a distance. In the nineteenth century the development of electromagnetism by Faraday and Maxwell led to the notion of electrical and magnetic forces being transmitted through extended fields rather than through action at a distance.[59] Yet, Maxwell still thought it desirable to attempt, chiefly through theories of the ether, to reduce electromagnetism to mechanics. His treatment of mechanics puts it on essentially the same footing that Euler gave it.[60] In explaining the various orders of explanation, he has no doubt which order is the most basic:

> On the other hand, when a physical phenomenon can be completely described as a change in the configuration and motion of a material system, the dynamical explanation of that phenomenon is said to be complete. We cannot conceive any further explanation to be either necessary, desirable, or possible, for as soon as we know what is meant by the words configuration, motion, mass, and force, we see that the ideas which they represent are so elementary that they cannot be explained by means of anything else.[61]

Maxwell's electromagnetic theory was not generally accepted or even widely studied until after H. Hertz's 1888 paper proving that discharge of an induction coil produces detectable electromagnetic radiation.[62] Hertz, a theoretician as well as an experimentalist, gradually swung over to a field concept in the course of the development leading from a Weber-Helmholtz view of electricity in terms of charges and currents to Maxwell's field concept.[63] Maxwell had originally developed his position in terms of a mechanical model of tubes of force rotating and stretching in the ether. Though this prop was gradually discarded, Maxwell never abandoned the idea that the principles of mechanics should supply a foundation for electromagnetic field theory. Hertz took this step, identifying Maxwell's electromagnetic theory with Maxwell's electromagnetic equations.[64]

A concomitant development that also contributed to the electrodynamicist position was the growing realization that the forces binding atoms into molecules, and also the force binding the parts of atoms together, is electrical in nature.[65] Faraday's researches on electrolysis suggested that electrical forces play some part in binding molecules together, but was open to differing interpretations. Further developments eventually led to the contention of S. Arrhenius in 1887 that in a very dilute solution the electrolyte is completely dissolved into charged ions.[66] This strongly supported the position that the forces binding atoms into molecules is electrical in nature.

That the forces binding the parts of an atom together are also electrical was a much more speculative thesis with rather ambiguous support. After the development of the Geissler vacuum tubes in 1855, research advanced on cathode rays. The work of J. Plücker, W. Hittorf, E. Goldstein, C. Varley, W. Crookes, and others on the straight-line trajectories of cathode rays and their deflection in a magnetic field led to the 'molecular-torrent' theory of cathode rays. In this theory cathode rays were explained as negatively charged particles, though it was not yet clear whether these particles were ionized molecules or something different.

The leading opponent of this theory, H. Hertz, demonstrated that cathode rays were not deflected by an electric field, as charged particles should be. He suggested that cathode rays are a new form of light. This conflict was finally resolved by J. J. Thomson, who explained away Hertz's negative results as due to the influence of induced space-charges. His decisive experiment (1897) of using crossed electrical and magnetic fields to determine the ratio of mass to charge for cathode rays definitely established electrons as charged subatomic particles.[67]

These experimental discoveries were complemented by some theoretical speculations whose significance has been, at least until recently, widely misinterpreted. A theory of electricity stemming from W. Weber had given mathematical expression to the old idea that electrical currents are to be explained in terms of a flow of vitreous (positive) and resinous (negative) electrical particles. This tradition continued to flourish even after the publication of Maxwell's equations. B. Riemann and R. Clausius accepted the idea of attributing the phenomena of electrodynamics to the agency of moving electrical charges. Each assumed that the forces on these charges depend both on their positions and on their relative velocities.

The man who merged this with the competing tradition stemming from Maxwell was the Dutch physicist H. A. Lorentz.[68] In his thesis he tried to generalize Maxwell's theory by relating it to molecular theory. This union should pave the way to a new synthesis in physics.

If it is true that light and radiant heat are constituted by electrical vibrations, it is natural to admit that the molecules of bodies, which give birth to such vibrations in the surrounding milieu, are also the seat of electrical oscillations, whose intensity rises with the temperature. This conception . . . seems to me very fertile.[69]

This projected synthesis soon came to pivot on two speculative theses. The first was Lorentz's doctrine of a static ether which retains the same properties in intermolecular space that it has in free space. The second is the thesis that ponderable matter contains charged particles whose harmonic oscillations produce radiation.

Instead of the action-at-a-distance interaction proper to the Weber tradition, Lorentz assumed that charged particles interact through the mediation of the stationary ether and that the behavior of this ether is governed by Maxwell's equations. This meant, in effect, expanding Maxwell's theory to include the forces acting on charged particles. This theory was soon extended from ether to dielectrics. On the assumption that the molecules in a dielectric are composed of vitreous and resinous particles in a neutral balance, Lorentz argued that an electric field transforms a molecule into an electrical doublet whose moment is measured by the product of charge and displacement. This explained the magnetic field produced when an uncharged dielectric is in motion at right angles to a constant electrostatic field. When P. Zeeman discovered that a magnetic field splits some spectral lines into triplets, Lorentz used his model of a vibrating dipole to explain it.

Lorentz's theory of metals and the problems associated with the Zeeman effect will be given more detailed discussion in later chapters. What we wish to consider now is Lorentz's pivotal role in a projected conceptual revolution.[70] Lorentz's early work accorded something of an equal status to mechanical and electromagnetic concepts. Oscillating charged particles obey mechanical laws, while the electromagnetic field is governed by Maxwell's laws. However, Lorentz hoped to explain charged particles through some relation to singularities in the ether. Electromagnetic concepts, accordingly, gradually assumed a more basic role. At this time, however, the focus of Lorentz's concern was not so much the basic concepts as it was the set of transformation equations he had introduced to explain away the failure of the Michelson-Morley experiment.

These Lorentz equations related in a still controverted way to the origins of the special theory of relativity. Of more immediate interest is the bearing of Lorentz's work on the then current speculations concerning the nature and mass of charged particles. In 1881, J. J. Thomson had argued from an application of Maxwell's equations to the conclusion that the effective mass of a charged body varies with velocity. Others had improved and extended this calculation. When Lorentz related this to his transformation equations, he found that consistency required extending this effect from electrical forces to whatever forces bind molecules together. This, in turn, seemed to imply that all mass, not merely the mass of charged particles, should exhibit a velocity dependence.

Earlier, when Lorentz and his contemporaries spoke of charged particles, they referred to whatever particles might be constituents of molecules, without assuming a uniform charge or mass. Thomson's discovery of an electron which seemed to have the same charge regardless of what molecule it was associated with brought these speculations into sharp focus. Some younger physicists, Theodor Des Coudres, Wilhelm Wien, Arnold Sommerfeld, Alfred Bucherer, and especially Max Abraham, developed different, but related, electromagnetic concepts of mass. If the mass of a body, or at least one component of the electron's mass, could be explained through some sort of self-induction produced in the ether, then electrodynamics might supplant mechanics in supplying a foundation for physics. Walter Kaufmann and others began extensive tests to determine whether the electron's mass exhibited a velocity dependence and if so whether this dependence fitted any one of the proposed formulas.

Lorentz was intrigued by this concept of an electromagnetic explanation of mass and sought to include it in a more comprehensive theory. The role he aspired to was that of a master architect fashioning a comprehensive theory which not only unified electromagnetism and mechanics, but could also supply a coherent basis for explaining the new discoveries about the behavior of electrons, the contraction of rapidly moving bodies, and the puzzle of blackbody radiation. The explanations of these phenomena that ultimately proved successful did not fit the game plan Lorentz had devised. Yet, in the period from the mid-nineties until about 1911, Lorentz was at the center of what initially seemed to be the most exciting and revolutionary movement in physics, the attempt to restructure the whole science in such a way that electromagnetic theory and action through fields would be more basic than mechanics and action at a distance.

Boltzmann is often remembered as the intransigent defender of atomism trying to stanch the tide of a rising positivism. His actual position was, I believe, more complex and ambiguous. He was not an outright opponent of conceptual revolution in the foundations of physics. He admitted that eventually this must happen. But, he insisted, it could not come about as easily or as simply as the proponents of conceptual revolutions proposed. The task of articulating criteria for judging revolutionary success devolved by default on Boltzmann's troubled shoulders. Though Boltzmann was a superbly competent theoretical physicist who combined breadth of knowledge with a passion for precision, he was but an amateur philosopher, sketchy, inconsistent, deeply suspicious of established systems, yet incapable of working out a systematic formulation of his own. We will consider his general position on replacing the foundations of physics before considering his reaction to the three systems considered.

Boltzmann's basic position was, I believe, determined more by his understanding of how physics worked than by his properly philosophical considerations. On the one hand he did not believe that either mechanics in general or the mechanical conception of the atom in particular would provide a permanent foundation for physics: "If one wants to bother at all about future centuries or even millennia I readily admit that it would be presumptuous to hope that our present day mechanical picture of the world will be preserved for all eternity even only in its most essential features."[71] He admitted the possibility that purely mathematical, nonmechanical models of the atom might one day prove basic:

> Nor should we lose from sight the possibility that it (present day atomism) might one day be displaced by quite different pictures, let us say, to avoid appearing small-hearted, ones taken from manifolds that lack even the properties of our three-dimensional space, so that for example simple geometrical constructions of atomism would have to be replaced by manipulations with numbers forming a complicated manifold.[72]

Replacing foundations is, however, no easy task. There was a loose, yet mutually reinforcing, alliance between the doctrines of atomism and the foundational role attributed to mechanics. Mechanics did not rest on an atomistic foundation. Yet, its foundational concepts of point-masses and forces acting at a distance could easily be interpreted as an idealized first approximation to an atomic account. No mechanical model of the atom had been found adequate or even consistent in a large perspective. Yet, any proposed substitute would inevitably involve matter, force, and motion. Mechanics, as generalized by Lagrange and Hamilton, could treat these abstractly with a minimal reliance on models. The established foundations of physics accordingly had a consistency and adequacy acquired through some two hundred years of sustained effort. Boltzmann insisted that these could not simply be discarded without an adequate replacement:

> I merely wish to work against the thoughtless attitude that declares the old world picture of mechanics an outworn point of view, before another such picture is available from its first foundations up to the applications to the most important phenomena which the old picture has for so long now represented so exhaustively, especially where the innovators have not the least understanding of how difficult it is to construct such a picture.[73]

With this general background we may consider Boltzmann's reaction to the three types of proposed revolutions previously considered. Phenomenology clearly presented the most serious problem. The underlying difficulty here was that Boltzmann had come to accept an epistemological position essentially the same as Mach's. All human knowledge is based on sensations, connections between sensations, and relations between sensations received and consequent actions.[74] A doctrine of atomism is a hypothesis that goes beyond the evidence. This Boltzmann admitted. But, he countered, so does phenomenology. The differential equations basic to phenomenology are extrapolations from the arbitrarily small units used to set up differential equations. Thus, a particulate view is conceptually prior to a continuous view of matter. Since both approaches introduce hypotheses which go beyond the evidence of the senses, the only real basis for judging between them is their simplicity and fruitfulness. By this standard, atomism is superior.

Boltzmann was trying to hoist Mach by his own petard, the principle of economy. Rather than explode, the charge sputtered out. A doctrine of atomism is at a much further remove from sensory input than any acceptance of perceived quantities as continuous. Boltzmann attempted to supplement Mach's doctrine of sensationalism by a picture theory of knowledge derived, it seems, chiefly from Hertz's *Mechanics*. Hertz claimed:

> We form for ourselves images or symbols of external objects; and the form which we give them is such that the necessary consequents of the images in thought are always the images of the necessary consequents of the things pictured. In order that this requirement may be satisfied, there must be a certain conformity between nature and our thought.[75]

Boltzmann rejected Hertz's formulation of mechanics, especially his idea of hidden masses. Yet, Boltzmann's later writings manifest an increasing reliance on this picturing theory. This provided, at best, a shaky support for Boltzmann's doctrine of atomism. Boltzmann had, as he clearly realized, no adequate or even coherent picture of the atom: "What the atom of each element is, whether it is a movement, or a thing, or a vortex, or a point having inertia, all these questions are surrounded by profound darkness."[76] A picture theory of knowledge without coherent pictures is obviously in deep trouble.

Hertz's position had some plausibility because, following Kant, he held for a conformity between the mind and reality based on the role of synthetic a priori judgments. Boltzmann did not accept this doctrine. The only innate conformity between reality and our thought that Boltzmann would admit was that resulting from the evolutionary adaption of man to his environment. In one essay, Boltzmann suggested a way around these difficulties, to take as a basis for objectivity the world view given in language, rather than subjective experience:

> If therefore I am to make myself understood, I must adopt a language in which all exist on the same footing ('objectively'). This adherence to the language of others which is given to me in experience (because learnt) I call the objective point of view, in contrast to the subjective one so far described.[77]

Boltzmann was unfortunately unable to develop the idea of a language-centered objectivity into a coherent alternative to Mach's sensationalism.

The net result of this debate was something of a draw. Boltzmann's arguments to the effect that physics required a doctrine of atomism were convincing. Their strength came from Boltzmann's profound understanding of how physics actually functioned. When Boltzmann treated second-order questions concerning the nature of the concepts physics used and the relation between concepts, percepts, and physical reality, he was in unfamiliar and, for him, uncongenial territory. His attempts to give an epistemological justification to his doctrine of atomism must be judged a total failure.

The energeticists provided Boltzmann with more of a personal problem, through public clashes with the ebullient Ostwald, but much less of an intellectual problem. The rebuilding of the edifice of physics on the foundation of the energy concept never really got beyond the stage of slogans, proposals, and public agitation. The program in Boltzmann's view was deceived by superficial and purely formal analogies; its law lacked the clear and unambiguous formulation customary to classical physics; its inferences were fuzzy; its stress on energy was really a disguised metaphysics; and its repudiation of point-masses involved an essential circularity.[78] Here Boltzmann was on surer grounds, discussing the way science is actually structured rather than trying to justify this structuring on epistemological grounds.

The electrodynamicist program was much more to Boltzmann's liking. It involved, not a rejection of atomism, but a promising new approach in the attempt to explain what atoms are. In discussing the problems involved in developing a consistent reformulation of the foundations of mechanics, Boltzmann had set the requirement, "Above all, if one wants to avoid the picture of material points, one should not later introduce them into mechanics at all, but one should start from individuals or elements of different constitution, with properties that can be described as clearly as those of material points."[79] In revising this introduction to mechanics seven years later (in 1904) after the discovery of the electron, Boltzmann added, "However, the advantage of being able to derive all mechanics from other ideas that are in any case necessary for explaining electro-magnetism would be just as great as if conversely electro-magnetic phenomena could be explained mechanically. May the former succeed and my requirement of seven years ago be fulfilled!"[80]

Conclusion

The developments and debates we have been considering reveal something of a growing consensus on the requirements an acceptable atomic theory must fulfill. If macroscopic bodies are composed of atoms, then an account of what atoms are should explain the measurable properties of macroscopic bodies. To John Dalton it seemed possible to fulfill this requirement in a fairly simple and straightforward way. One simply postulated different atoms with the properties requisite to explain each different type of element. In the course of the century the properties atomism should explain were developed with an ever-increasing precision and growing wealth of detail, chiefly through advances in chemistry and spectroscopy. Concomitantly, there was a growing clarification of the requirements which any atomic account must meet to be both self-consistent and empirically adequate. By the end of the century the leading physicists had reached something of a consensus that the original program of providing an atomistic account of physical reality could not be completed along the lines originally envisaged. As a final summary, it might be helpful to present the reasons for this conclusion as simply and starkly as possible.

Both chemical atomism and spectroscopy began with observable macroscopic properties and postulated that atoms or molecules or whatever is ultimate must supply a basis for explaining these properties. Kinetic theory began with specific assumptions about molecules, their sizes and speeds, their shapes and interactions. These assumptions not only supplied a systematic basis for explaining gas laws and thermodynamic properties, but also led to new predictions, some of which were verified. For these reasons, kinetic theory was reasonably judged to be the most basic source of information about the nature of atoms and molecules.

In appraising the information kinetic theory supplied, it is important to distinguish between the models of molecules employed and the mechanical assumptions on which the science rested. The models of molecules as, for example, dumbbell-shaped objects which could move in a trajectory, rotate, and vibrate were known to be approximations rather than realistic pictures. The basic assumption on which the science rested was that the motions of molecules and their constituent parts are governed by the known laws of mechanics. This was not intended as an approximation. It was a foundational assumption.

The basic requirements that must be met if kinetic atoms are to play a foundational role in explaining the properties and activities of bodies are internal consistency and empirical adequacy. The latter condition was given its most succinct expression by J. W. Gibbs in 1901 in explaining why he tried to prescind, as much as possible, from hypotheses about molecules in setting up statistical mechanics:

> In the present state of science, it seems hardly possible to frame a dynamic theory of molecular action which shall embrace the phenomena of thermodynamics, or radiation, and of the electrical manifestations which accompany the union of atoms. Yet any theory is obviously inadequate which does not take account of all of these phenomena.[81]

These were the minimal requirements for adequacy. Gibbs seems to have shared the then common assumption that an atomic explanation of the periodic properties of the chemical elements was a task relegated to the remote, rather than the near, future.

By the end of the century there seemed to be little hope that the dynamics of molecular action, presupposed in kinetic theory, could meet even these minimal requirements for empirical adequacy in a self-consistent way. It was also becoming increasingly clear that the underlying difficulties somehow hinged on the assumption that the dynamics of molecules, atoms, and their constituent parts are governed by the known laws of mechanics.

The only means mechanics had provided to account for radiation was some sort of internal vibration. The difficulty here was not just the inadequacy of the particular models proposed to explain radiation. It was the very assumption of mechanical vibrations as a cause of radiation. Ramsey, Rayleigh, and others had extended the earlier Kundt-Warburg results and had shown that the newly discovered noble gases also had a value of 5/3 for the ratio of specific heat at constant pressure to the specific heat at constant volume. This indicated that they were also monatomic gases. Yet, each had a complex series of spectral lines. If internal vibrations were postulated to explain these lines, then, in accord with the principle of equipartition of energy, the internal vibrations should have a share of energy proportional to the number of degrees of freedom postulated. This would lead to quite a different value for the ratio of specific heats. Though this was the most extreme case, the difficulty was general, covering all molecules. No satisfactory way was found to reconcile the inconsistencies between the observed values for the ratios of specific heats and those predicted on the basis of the mechanical assumptions of internal vibrations adequate to explain spectral lines and the general principle of the equipartition of energy.

Some fundamental assumption had to be abandoned. The most plausible candidate for abandonment was the assumption that internal molecular activity could be explained on the basis of the known mechanical laws. In addition to the internal difficulties of inconsistency and empirical inadequacy, there were other reasons favoring such an abandonment. The two strongest stemmed from electromagnetic considerations. Action at a distance, one of the traditional foundational concepts of mechanics, had been suspect from the beginning. It was generally accepted, not without reluctance, on the grounds that there was no reasonable alternative and that mechanics based on this assumption worked. The development of electromagnetic theory provided an alternative conception with a much greater degree of intuitive plausibility, action transmitted through extended, physically real fields. Though this too presented difficulties related to the ether assumption, it worked well enough to suggest that this assumption, proper to electromagnetism, might be more basic than the assumption of action at a distance, proper to mechanics.

The second reason stemmed from the realization that intermolecular and interatomic binding forces are electrical in nature and from the discovery of the electron as a constituent of all atoms. A theory of electronic behavior in atoms led to the first successful explanation of spectral regularities through atomic assumptions, Lorentz's theory of the normal Zeeman effect.

We concluded our survey of the century with an account of the conflict between Boltzmann and his opponents. This was chosen, not merely for its relation to atomism, but even more because these conflicts reflect some implicit standards concerning what an adequate scientific explanation should be and do. Here, through a certain amount of oversimplification, I will attempt to make these implicit standards explicit.

Contrary to the picture presented in many popular histories of quantum physics, turn of the century physicists were not for the most part conservative Newtonians. Most judged that the earlier program of developing a theory of atoms on the basis of classical mechanics had proved to be bankrupt. Something radically new seemed needed, new foundational concepts, a new methodology, or a drastically revised account of what scientific explanations actually accomplish. Mach's phenomenology and Ostwald's energetics soon proved too shallow and sterile to spark the revolution. The electromagnetic explanation of mass seemed to supply a much more promising conceptual pivot.

Boltzmann was the sharpest critic of these projected conceptual revolutions. Yet, even Boltzmann conceded that the physics of the future would undoubtedly require radically new foundations. However, he continued to insist, the task of replacing the present foundations of physics with something radically new could not be accomplished as quickly and easily as the sanguine would-be reformers hoped. The existing physics, for all its deficiencies, had already achieved considerable success in mechanics, in electromagnetism, in kinetic theory, and in other branches of physics. His own efforts to bring out the underlying coherence of this physics through an epistemological justification proved to be more an embarrassment than a support. Yet, the very attempt to meet this need highlighted a requirement which any new concepts or methods must meet if they are to be accorded a foundational role. They must supply a basis for giving a coherent account of physical reality. Boltzmann clearly saw the problem. Yet, till his suicide in 1906, he manifested no suspicion that the needed reconstruction had already begun in the way his conservative young supporter Max Planck was using the new concept of the quantum to treat the problem of blackbody radiation.

4. The Atom and the Quantum

Until their fusion in Bohr's 1913 atomic theory, atomism and quantum theory were only loosely interrelated. They will accordingly be considered separately. We will first consider atomism and then the origins of quantum theory, focusing on the interpretative problems it generated, particularly on the ambiguous relation between classical physics and the quantum hypothesis and on wave-particle duality. We will not consider other developments which, though they were related to atomism and quantum theory, did not have an immediate bearing on the interpretative problems we are considering. This includes such topics as X-rays, radioactivity, and the special and general theories of relativity.

Quantum theory at almost every stage of its development had one peculiar feature. Formulas were introduced and found to fit the experimental data, though, more often than not, the justification originally given was eventually found to be incorrect or misleading. Heisenberg summed up the situation with the expression, "The equation knows best." The reasoning that led scientists to accept a formula while modifying or rejecting the arguments originally given for the formula's introduction reflect in a distinctive way changing norms of scientific explanation. For this reason it will be necessary in this and the next three chapters to consider the reasoning that led to the introduction of many key formulas, even though the arguments involved are not now considered adequate justifications. The treatment given here will have one other distinctive feature, extreme selectivity. The developments of atomic physics and quantum theory that came to fruition in the mid-twenties involved the contributions of many people. Only a few, however, shaped the interpretation that emerged. Our account focuses on those few. Adequate general histories are available to complement this specialized presentation.[1]

Atomic Physics before 1913

The turn of the century physics-cum-philosophy debates discussed in the last chapter had called into question the real existence of atoms and molecules. From a scientific, rather than an epistemological, point of view, two props seemed to quiver under the weight the Boltzmann program imposed on them. The first was the principle of the equipartition of energy as applied to a molecule's different degrees of freedom. Maxwell and Boltzmann had both defended this on theoretical grounds. Kelvin led the opposition, developing special models which seemed to supply counterinstances. Lord Rayleigh (John W. Strutt) tried to develop a middle-of-the-road position. Though he could counter Kelvin's examples, he felt obliged to admit that the principle did not seem to apply to any model of the atom with an immense number of degrees of freedom.[2] He assumed the validity of the equipartition theorem in deriving his radiation formula.[3] Yet, in 1905, when forced to acknowledge that Planck's radiation law fitted the data of blackbody radiation better than his own law with Jeans's numerical correction, he concluded that the law of equipartition fails in some cases.[4] A contemporary survey of this difficulty by a leading physicist is given in Joseph Larmor's 1909 Bakerian lecture, "The Statistical and Thermodynamical Relations of Radiant Energy."[5] He too concluded that, though the abstract arguments for this principle seemed valid, its applicability to molecules seemed to depend upon the model of molecules used.

The second wavering prop involved a difficulty of a different sort. Boltzmann had introduced his *H*-theorem to explain the second law of thermodynamics in terms of the statistical probabilities involved in attaining molecular equilibrium. In Germany and Austria this was opposed by the energeticists considered earlier. In France the opposition came from a somewhat different source. Duhem's famous description of the differences between the English scientific mentality, with its predilection for mechanical models, and the French spirit, favoring abstract logical deduction, brought out a characteristic of French science that was widely recognized.[6] Innate Gallic Cartesianism seems a much less likely explanation of this trait than the strongly centralized control of French higher education, which accorded decisive influence to a few well-established leaders, usually of the older generation. The ideal of science which the French educational system then inculcated accorded general laws a more foundational role than models, mechanical or otherwise. The tradition of giving even a technical report a stylistically correct and logically ordered presentation reinforced the primacy accorded deduction from general laws. In such an intellectual milieu Boltzmann's explanation of a general law, the second law of thermodynamics, through a mechanical model seemed like a perversion of the proper order. Even Poincaré, who was quite familiar with Boltzmann's work, hesitated to accord atoms any foundational role until 1911 and the decisive Solvay Conference. These were the intellectual currents Jean Perrin had to swim against in defending atomism.

In terms of the now accepted distinction between classical and quantum mechanics, the consideration of kinetic atoms was in the domain of classical physics. What complicates the issue and blurs the distinction is the fact that the pivotal theoretical analysis was given by the man who did the most to overthrow classical physics, Albert Einstein. Einstein's contributions to quantum physics and his distinctive style of doing physics will be considered later. What we wish to treat now is his contribution to the resolution of the problem of the reality of the molecules assumed by kinetic theory.

Einstein's earliest papers manifest a belief in the atomic hypothesis coupled to a willingness to experiment with different methods of implementing this belief. His first published paper, "Inferences from the Phenomenon of Capillarity,"[7] begins with thermodynamic considerations about fluid flow in capillary action to get a general formula characterizing the potential energy. The question then treated is, Can this potential energy be explained in terms of the attractive fields of molecules? Einstein's analysis supports the conclusion that there are such attractive molecular fields. However, he was unable to determine the precise form of the field, $\phi(r)$. He next attempted a similar type of inference from observable phenomena to unobservable forces using the phenomena of the electrical potential of dissolved electrical salts.[8]

Apparently, he found this approach unsatisfactory. He abandoned the method of attempting to infer molecular forces from observable phenomena and focused on a redevelopment of the foundations of kinetic theory. Since the main thrust of his work essentially duplicated results already obtained by Gibbs, this early work of Einstein's is not well known. Yet, it clearly manifests that the youthful Einstein was already a master of statistical reasoning. The brief summary presented here is intended both to supply a background for Einstein's treatment of Brownian motion and to indicate some distinctive aspects of Einstein's early scientific style.

In his general treatment of the foundations of thermodynamics[9] Einstein has the same basic idea of the interrelation between observable phenomena and the unobservable forces responsible for it as in his earlier papers. The state of a physical system observed by us is characterized by n scalar magnitudes, $p_1, p_2, \ldots, p_n$, effectively a state-space representation. The variation of this system in time, dt, is determined through the variations $dp_1, dp_2, \ldots$ If the system is isolated, then its state at one time determines its state at future times. Experience shows that such isolated physical systems eventually reach a state in which the macroscopic variables characterizing the system no longer change. This Einstein calls a stationary state. Earlier he had attempted to infer molecular forces from such observable phenomena. Now, instead of such inferences, he is relying on definite, though very general, assumptions concerning the behavior of the variables characterizing the state. In particular he assumes that the state-space point representing a system in a stationary state returns to the neighborhood at a given point with a definite frequency, effectively a quasi-ergodic hypothesis for stationary states.

To make this more precise he considered an isolated system whose energy at some time is E. If one considers a particular region Γ of this state-space or a time interval T, then one could in principle observe the portion of that interval τ during which the state-space point lies in Γ. If, for each region Γ, τ/T approaches a limit with increasing T, then the system possesses fixed observable properties. After thus defining probabilities in terms of time averages for a particular system, Einstein switched to averages over an ensemble of N identical systems. If m is the number of systems from the ensemble in the region Γ, then $m/N = \tau/T$. If all the systems in the ensemble have energies within a narrow region, then, Einstein argued, the number of systems dN in an infinitesimal region g of the state-space characterized by the n variables p_1 is given by

$$dN = \text{const} \int_g dp_1 dp_2 \ldots dp_n . \tag{4.1}$$

What equation (4.1) means is that the probability of finding the type of system being considered in an infinitesimal volume of state-space is directly proportional to the volume. This simplifying assumption allowed Einstein to develop statistical definitions of temperature, entropy, and fluctuations. Temperature is defined by dividing each of the n systems into a large system Σ, the subsystem whose temperature is being measured, and a small subsystem σ, the thermometer. For this case Einstein developed a new constant, $\chi = R/2N$, where R is the gas constant. It is equal to $k/2$, the Stefan-Boltzmann constant that Planck had introduced in a way that will be considered shortly. Einstein showed that the inverse of χ behaves like temperature. With energy specified and temperature defined, it was then possible to define entropy and to show that the entropy so defined fulfills the basic condition of increasing as the system moves from a less to a more probable state.

Our immediate concern is with Einstein's treatment of fluctuations. He considered a relatively small system in contact with a much larger heat bath and concluded that the energy fluctuations for the small system are given by

$$\overline{\epsilon^2} = \overline{(\bar{E} - E)^2} = 2\chi T^2 dE/dT. \tag{4.2}$$

The size of the fluctuations characterizes the stability of the system. Thus, Einstein's new parameter 2χ (or k) effectively defines the scale of fluctuation phenomena.

After Einstein developed this general formulation of statistical mechanics and fluctuation phenomena he tried to devise a thought experiment in which his treatment of fluctuations could be correlated with observable phenomena. This approach led to the peculiar situation in which Einstein developed the first successful theory of Brownian motion without realizing that the phenomenon had long been known and was then being extensively studied.[10]

Einstein's professed purpose in the series of papers he wrote on Brownian motion was to determine the existence of molecules of definite finite size. By viewing an idealized physical system from two different perspectives, he showed that the assumption of finite molecular size does make a difference in the theoretical determination of osmotic pressure.[11] He considered a nonelectrolyte dissolved in a volume V^* forming a part of a quantity of liquid of total volume V and assumed that V^* is separated from V by a partition permeable for the solvent but impermeable for the dissolved solute. Does the presence of such a solute cause osmotic pressure on the membrane?

If the solvent and solute are both treated as continuous fluids, then the answer is affirmative. The pressure p exerted on the membrane by z gram molecules of solute would be

$$p = RTz/V^*, \qquad (4.3)$$

where R is the universal gas constant and T is the temperature. If, however, one treats the solvent as a continuous fluid and the solute as a collection of particles, then the answer is negative, since the free energy, the crucial parameter, is independent of the position of the partition and of the suspended particles. As Einstein saw it, the resolution of this contradiction really hinges on the issue of whether the solvent, e.g., water, should be treated as a continuous fluid or as a collection of molecules. Since the pioneering work of Euler, discussed in chapter 1, fluids had been treated by continuous mechanics. A treatment of a fluid as a collection of molecules required an extension of kinetic theory from gases to liquids. This Einstein did.

If the solvent is viewed as a collection of molecules and the solute as a collection of particles much larger than the molecules, then, according to the molecular kinetic theory, the only significant physical difference is size. Suspended particles should produce the same osmotic pressure as the same number of molecules. These general considerations supplied a framework for the determination of the conditions of dynamic equilibrium. Suppose that a constant force K acts on each particle in a direction x, e.g., the force of gravity. Then one may first determine equilibrium conditions by treating this as a diffusion problem. This leads to a coefficient of diffusion D which, for equilibrium conditions, depends only on the particle size (assumed to be spheres of radius P) and on the viscosity q of the solvent:

$$D = RT/(6\pi qPN), \qquad (4.4)$$

where N is Avogadro's (or Loschmidt's) number, the number of molecules in a gram-molecular weight of a substance.

One may also get at diffusion by considering the random motions of a solute particle. This leads to a diffusion equation and also to λ_x, the mean value which a particle diffuses in time t along the direction x in which the force K is acting:

$$\lambda_x = 2Dt. \qquad (4.5)$$

This, Einstein thought, is a measurable magnitude. For parameters proper to sugar, $\lambda_x = 0.8$ microns. He concluded his paper, "It is to be hoped that some enquirer may succeed shortly in solving the problem suggested here, which is so important in connection with the theory of heat."[12] As he soon learned, such experimental inquiries were already being actively pursued, inquiries that stemmed from a historical background different from Einstein's abstract theorizing.

In 1827 Robert Brown observed through a microscope that cytoplasmic granules extracted from pollens and suspended in water manifested an unending chaotic motion. This activity was at first attributed to vital forces. When comparably sized particles of inorganic matter were shown to have the same motion, other explanations were attempted involving electricity, evaporation, minute temperature differences, caloric atoms, and magnetism. Though others suggested kinetic accounts, the decisive figure in relating Brownian motion to kinetic theory was Léon Gouy of Lyons. Unless a kinetic interpretation of entropy were accepted, he argued, one could ideally use Brownian motion to construct a perpetual motion machine.

In 1903 A. F. Siedentopf and R. A. Zsigmondy developed the ultramicroscope. By illuminating suspended particles with an intense pencil of light at right angles to the optical axis of the microscope, one could produce diffraction patterns centered around particles. This resulted in an effective magnification of up to 150,000 diameters, revealing minute Brownian motion of very small particles. Of the various experimental studies performed by T. Svedberg in Uppsala, M. Seddig in Marburg, V. Henri at the Collège de France, and Jean Perrin at the Sorbonne, the most decisive were those of Perrin.[13]

If the molecular kinetic account of Brownian motion is correct, then Brownian particles of uniform size in dynamic equilibrium in a solution should have a density that decreases exponentially with height. This follows from Einstein's theory, with the force of gravity as the force K. Perrin began examining this phenomenon before he was familiar with the implications of Einstein's theory, or the slightly later theory of Maryan Smoluchowski. Perrin emulsified particles of dried vegetable latex gamboge, used a centrifuge to precipitate particles of similar size in definite regions where they could be collected separately, prepared emulsions with uniform size particles, and then began the meticulous task of counting numbers of particles as a function of height. In some experiments he and his graduate assistants counted as many as 17,000 particles. This definitely established an exponential height relation. In later experiments he confirmed Einstein's theories on the magnitude of Brownian motion and also on the rotation of Brownian particles. The latter result was taken as confirming at least one application of the equipartition principle to molecules.

For most scientists, including Ostwald, these experimental results supplied the final convincing proof of the reality of molecules and the truth of the molecular-kinetic theory. Perrin's widely read work *Les atoms* not only gave a readable summary of his own work, but also marshaled and simplified the theoretical and experimental arguments for the reality of atoms. Mach and Duhem remained nonbelievers, but more for epistemological than scientific reasons. Such skepticism notwithstanding, Perrin's experiments crowning Einstein's theorizing effectively marked the end of one phase in the problematic we are considering. There no longer seemed to be any reasonable ground for doubting the essential validity of the molecular-kinetic-theory or its explanation of heat through the motion of molecules.

One aspect of this development should be noted. The reasoning involved was all within the framework of classical physics. When Perrin or Einstein introduced quantum theory in this context, it was only to indicate that Planck's work led to an independent estimate of Avogadro's number. The convergence of the values assigned to this number by essentially different methods was one of the strongest arguments Perrin advanced for the reality of molecules:

> Our wonder is aroused at the very remarkable agreement found between values derived from the consideration of such widely different phenomena. Seeing that not only is the same magnitude obtained by each method when the conditions under which it is applied are varied as much as possible, but that the numbers thus established also agree among themselves, without discrepancy, for all the methods employed, the real existence of the molecule is given a probability bordering on certainty.[14]

The Origins of Quantum Theory

It is now eighty years since Max Planck introduced the radiation formula that inaugurated the quantum era. Yet, the way in which this discovery should be understood remains a topic of considerable debate. Many popular histories of quantum theory and introductory texts still present the idea that Planck discovered that radiant energy comes in little bundles of value $\epsilon = h\nu$. In his original papers Planck introduced the idea of discrete energy values, not for radiant energy itself, but for fictitious oscillators in equilibrium with this radiation. This seemed puzzling until historians of science, especially Martin Klein, gave a plausible reconstruction of Planck's work showing why he focused on fictitious oscillators.[15] Their more or less standard reconstruction has recently been challenged by Thomas Kuhn's reconstruction of Planck's work.[16] Before discussing Planck's development, it will be helpful to isolate the crucial issues in the current controversy.

An ideal blackbody is one which absorbs all the radiant energy incident upon it and then reradiates energy as it heats up. The energy given off by radiation depends in the ideal case only on the temperature of the blackbody, not on the material of which it is composed. The law which Stefan suggested on an experimental basis and which Boltzmann developed theoretically linked the energy radiated E to the absolute temperature T by

$$E = \sigma T^4, \tag{4.6}$$

where σ is the Stefan-Boltzmann constant. From a consideration of the Doppler shift in the energy radiation, Wien established the functional relation known as Wien's displacement law:

$$E_\lambda = \lambda^{-5}\phi(\lambda T). \tag{4.7}$$

These two laws were established on the basis of very general considerations of energy balance and did not involve any presuppositions concerning the mechanism of radiation. On the assumption that the wavelength of the radiation emitted by a moving molecule is a function of the molecule's velocity and that the molecule's velocity distribution is given by the Maxwell-Boltzmann curve, Wien derived a radiation law which attempted to specify the form of ϕ in equation (4.7). For convenience of reference we will express Wien's radiation law both as a function of wavelength λ and as a function of frequency ν:

$$E_\lambda = c_1 \lambda^{-5} \exp(-c_2/\lambda T), \tag{4.8a}$$

$$u_\nu = \alpha\nu \exp(-\beta\nu/T), \tag{4.8b}$$

where c_1, c_2, α, and β are constants.

Since black radiation, as it was then called, is not a function of the type of substance involved, one could substitute for the material of the actual emitter some other substance which has a mathematically more tractable form. Planck postulated simple harmonic oscillators in equilibrium with blackbody radiation. According to the standard interpretation of Planck's development, the decisive breakthrough occurred in two stages. First, in October 1900, Planck presented a new radiation law as a modification of Wien's law. When this was found to fit all the data remarkably well, Planck spent three months of intensive work and then presented a theoretical justification of the new law. The crucial point in this justification was the idea that these fictitious oscillators absorb and emit radiant energy only in discrete bundles of value $\epsilon = h\nu$.

According to Kuhn's reinterpretation Planck did not quantize the energy absorbed or emitted by individual oscillators. The discreteness was simply a calculational device applied to collections of oscillators. For individual oscillators Planck still believed in continuous energy distribution. This initially implausible thesis draws support from a rather surprising source, Max Planck himself. Whatever his limitations as a conceptual revolutionary, Planck was certainly a man of meticulous honesty. In his autobiographical accounts he does not claim credit for introducing the idea that energy comes in discrete units.[17] This he attributes to Einstein. What he does claim credit for is the discovery of the law that bears his name and for the two constants h and k. Furthermore, Kuhn's account makes Planck's second and third quantum theories intelligible as an advance (from Planck's perspective) rather than a regression.

Kuhn's book was published after what I then took to be the last draft of this chapter was completed. This section has been rewritten, not without reluctance, because much of Kuhn's reconstruction of Planck's development seemed convincing. Accordingly, what follows is essentially a summary of Kuhn's reconstruction. The interpretation accorded Planck's breakthrough, however, differs from that given by Kuhn. In this sense it is something of a via media between Klein and Kuhn.

As noted in the last chapter, Planck joined Boltzmann in combating Ostwald's energetics. Planck, however, was in the fray because he thought the energeticists distorted thermodynamics, especially entropy. He was definitely not defending atomism:

> After all that I have related, in this duel of minds I could play only the part of a second to Boltzmann—a second whose services were evidently not appreciated, not even noticed, by him. For Boltzmann knew very well that my viewpoint was basically different from his. He was especially annoyed by the fact that I was not only indifferent but to a certain extent even hostile to the atomic theory which was the foundation of his entire theory.[18]

Planck saw science as a search for absolutes. The goal of science, as the youthful Planck saw it, is to present a coherent world view based on general laws which have an objective validity. His ideas on the type of pictures science supplies changed somewhat after his adaption of Boltzmann's statistical methods and his clash with Mach. This change will be treated briefly in chapter 9. Here we need only note one norm, one with a vaguely theological background, which perdured through all of Planck's work. The order of the universe should ultimately be explained through laws whose validity is objective. It should not be explained as a merely phenomenal order arising out of some deeper chaos of random events. The first and second laws of thermodynamics stood as prototypes of the laws to be sought. They are not simply conventions. Their objective validity is manifested through the impossibility of constructing perpetual motion machines of the first or second kind.

Planck had proposed a new definition of the entropy principle: "The process of heat conduction cannot be completely reversed by any means." Though this formulation won virtually no acceptance, Planck retained it and focused on a peculiar problem it entailed. The general laws of physics, such as Newton's laws and Maxwell's laws, are time reversible. Entropy in Planck's definition is characterized by irreversibility. In this perspective the problem that emerged as "the fundamental task of theoretical physics" was to reduce unidirectional changes to conservative laws.[19]

Planck felt that a statistical explanation of basic laws could not fulfill this task. Zermelo was Planck's assistant when Zermelo presented the recurrence paradox mentioned in the last chapter. This reinforced Planck's conviction that a statistical explanation based on random ordering of individual events could not explain unidirectionality. Apart from entropy, where Planck's position was already formed and hardened, black radiation presented the most interesting challenge for anyone attempting to explain an irreversible approach to equilibrium through reversible laws. Maxwell's equations, describing the electromagnetic field, are invariant under time reversal. If, however, one were to assume a frictionless vibrator absorbing energy from the electromagnetic field and reradiating it, then one would have an effect that is not invariant under time reversal. The resonator absorbs randomly directed radiation and emits spherical waves.

Planck's original way of treating this was to consider a spherical surface surrounding the hypothetical resonator. For equilibrium the energy coming through the sphere to the resonator must equal the energy emitted from the resonator through the sphere. If the resonator is absorbing energy, then the energy balance involves a time-dependent term. Though the general scheme seemed straightforward, the details of its development involved obscurities. Planck treated the radiation by means of a Fourier expansion. He assumed that the resonators were damped. They could only absorb radiation in the neighborhood of their natural frequencies. Yet, to make the energy balance, Planck had to make special assumptions concerning radiation components that allowed the variable terms in his expansion to balance. This, in turn, required apparently arbitrary assumptions concerning radiation states. Thus, he could not allow an initial situation of incoming spherical waves converging on the resonator.

Kuhn claims that it was at this point, in 1898, rather than in 1900 that Planck began a systematic adaption of Boltzmann's methods. Planck was, in effect, trying to develop an electromagnetic equivalent of Boltzmann's *H*-theorem. At one stage in his development Boltzmann had introduced the concept of molecular disorder, a concept that was used to prohibit the type of initial states that led to paradoxical consequences. Boltzmann used this concept to equate the actual rate of molecular collisions to the average rate given by the theory. Planck introduced an electromagnetic analogue with the concept of natural radiation.[20] This concept allowed him to derive the energy balance he needed and led to a formula[21] linking the radiation field energy of frequency v (uv) to the energy of a resonator (Uv):

$$u_v = (8\pi v^2/c^3)U_v. \tag{4.9}$$

Planck now had an energy balance. He did not yet have the approach to irreversibility he sought. To get this he needed a function which is determined by the instantaneous state of the system and which changes in only one direction. He developed such a function and in spite of some ambiguities identified it with the entropy of the spherical shell containing a resonator interacting with the radiation field through absorbed and emitted radiation. When he later generalized this for arbitrary shapes and many resonators, he had a general expression for the entropy S_t involving a sum over resonator entropies S_i and an integral over the radiation entropy density s:

$$S_t = \Sigma^{S_i} + \int s d\tau. \tag{4.10}$$

Using e, the base of the natural logarithms, and two as yet undetermined natural constants a and b, Planck defined the entropy of an individual resonator as

$$S = -(u/av)\log(U/ebv). \tag{4.11}$$

At equilibrium the total entropy must be constant. By considering a virtual transference of energy between resonators of different frequencies, Planck derived a formula for the energy of a resonator

$$U = bv \exp(-av/\theta). \tag{4.12}$$

By using equation (4.9) linking the energy of a resonator to radiant energy, Planck derived the formula

$$u = (8\pi b v^3/c^3)\exp(-av/\theta). \qquad (4.13)$$

With the appropriate identifications of constants, this is equivalent to Wien's law in the form (4.8b), a law that then became known as the Wien-Planck law.

The problematic identification in equating (4.8b) to formula (4.13) is the identification of Planck's θ with the absolute temperature T. Neither Maxwell's laws governing the electromagnetic field nor Newton's laws governing the behavior of resonators involved the concept of temperature. To identify θ with T, Planck again adapted some methods Boltzmann had used in developing his H-theorem and relating this theorem to entropy. From Planck's definition of the entropy of a resonator it followed that $\partial S/\partial U = 1/\theta$. In thermodynamics $\partial S/\partial U = 1/T$. Planck concluded on this ground that his θ represented the only possible electromagnetic definition of temperature.

This argument, summarizing Planck's position as of mid-1898, involves at least two obscurities. First, there is an ambiguity in the use of the term 'entropy.' Traditionally, thermodynamic properties are divided into *intensive* properties, such as pressure and temperature, which are not dependent on the amount of material involved, and *extensive* properties, such as volume, which do depend on the amount of material. Entropy is an extensive property. One might accordingly define the entropy of a collection of oscillators, divide the value of this entropy by the number of oscillators, and call the result the entropy of an individual oscillator. This, Planck's procedure, involves an ambiguity when large-scale concepts are applied to small-scale phenomena. If, for example, one were to divide the total volume of a gas by the number of molecules, the result would be a volume which one might treat as the volume of an individual molecule in some calculations. This cannot, however, be called the 'volume of a molecule' *simpliciter*.

The way in which one should speak about the entropy of an individual oscillator depends on the way in which the concept of entropy is understood. In the later papers in which he developed his own radiation formula Planck explicitly identified entropy with disorder. When entropy is so understood, 'entropy of an individual oscillator' can only be interpreted as a convenient way of speaking about a collection, much as one might speak about the average American or the typical middle-class family. However, this identification of entropy with disorder hinged on an acceptance of Boltzmann's statistical interpretation of entropy, something Planck had not yet adopted. For this reason his earlier use of 'entropy of an oscillator' is best interpreted as ambiguous rather than incorrect.

The second obscurity stemmed from the fact that Planck's definitions of oscillator energy and entropy were not the only possible ones. In early 1900,

Max Thiesen showed that a family of solutions is compatible with Wien's displacement law. Some of the members of this family were not easily accommodated to the definition of temperature Planck had used in identifying his θ with absolute temperature $(1/T = \partial S/\partial U)$. Planck treated this problem from a more general perspective by asking what characteristics $S(U)$ must have if it is to possess local maxima. If an individual oscillator is displaced from equilibrium and then returns, its energy and entropy must, Planck concluded, be related by

$$3/5(\partial^2 S/\partial U^2) = -f(U), \tag{4.14}$$

where $f(U)$ is any positive function. If n resonators considered as independent units are similarly displaced, then one must have, by virtue of the independence assumption,

$$f(nU) = (1/n)f(U). \tag{4.15}$$

This is satisfied only when $f(U)$ is proportional to $1/U$, or when

$$\partial^2 S/\partial U^2 = -\alpha/U. \tag{4.16}$$

This formula, based on the concept of the entropy of an oscillator, yields Wien's radiation law. Planck published it in February 1900. At this time the status of the black radiation problem was beginning to change both experimentally and theoretically. Rayleigh suggested a simple modification of Planck's law obtained by applying the law of equipartition to radiation (see note 3). Jeans introduced the slight correction of dividing by eight. Later, he gave the formula, which became known as the Rayleigh-Jeans formula, a more theoretical justification. This formula, adapted to Planck's notation, is

$$u_\nu = 8\pi\nu^2 kT/c^3. \tag{4.17}$$

This was soon recognized to be invalid for high values of k/T. Also, the total energy, obtained by integrating over all frequencies, is a divergent quantity.

These difficulties with the Rayleigh-Jeans formula were not recognized until later. However, Rayleigh's original publication had some influence on experimentalists. As Kangro has shown,[22] until early 1900 none of the leading experimentalists were certain enough of their blackbody radiation measurements to interpret discrepancies between theoretical predictions and experimental results as indications that the Wein-radiation law might be erroneous. Rubens and Kurlbaum were stimulated by Rayleigh's publication of an alternative radiation formula to argue that their data indicated that Wien's law might be erroneous, a conclusion they communicated orally to Planck. At about the same time, Lummer and Pringsheim were demonstrating very serious discrepancies between their data and Wien's law, discrepancies too serious and too systematic to be passed off as experimental error.

Planck's derivation of equation (4.16) had assumed independent oscillators, so that the total entropy is simply a sum of the individual oscillator entropies. Planck suspected that a modification of this simple assumption might be the best way to develop a variation of the radiation law. As he later explained it, "I have finally started to construct completely arbitrary expressions for the entropy which although they are more complicated than Wien's expression still seem to satisfy just as completely all the requirements of the thermodynamics and electromagnetic theory."[23] Among the modifications of equation (4.16) the one that appeared the most promising was

$$1/R = \partial^2 S/\partial U^2 = -\alpha/U(\beta + U). \qquad (4.18)$$

In his *Scientific Autobiography* Planck explained the reasons for this choice. For small energies, where Wien's law worked, R, the reciprocal, is proportional to U. The new experimental results seemed to indicate that, for large values of the energy and wavelengths, R is proportional to the square of the energy:

> Therefore, the most obvious step for the general case was to make the value of R equal to the sum of a term proportional to the first power of the energy and another term proportional to the second power of the energy, so that the first term becomes decisive for small values of the energy and the second term for large values. In this way a new radiation formula was obtained, and I submitted it for examination to the Berlin Physical Society, at the meeting on October 19, 1900.[24]

This, of course, was the most obvious step only if one sought to express the radiation law as a second derivative of entropy with respect to energy, something Planck alone was doing. Two integrations of equation (4.18), plus the use of Wien's displacement law, led to a new expression for radiant energy:[25]

$$E = c\lambda^{-5}/[\exp(c/\lambda T) - 1]. \qquad (4.19)$$

This was the formula Planck proposed to the *Berliner Physikalische Gesellschaft* on 19 October 1900. He proposed it for consideration on the grounds that it was simple and fitted the observational data as satisfactorily as any of the other suggested modifications of Wien's law. The experimentalist H. Rubens spent the evening checking Planck's new formula against his measured results and found agreement on every point. So too, after correcting an error, did Lummer and Pringsheim.

Planck's formula, (4.18), was based on an arbitrary modification of formula (4.16). This, in turn, had originally been justified by the fact that it led to Wien's radiation law, a law known to be inaccurate. Yet, the new formula worked remarkably well. Planck began what he called an intensive investigation of its true physical meaning.[26] On 14 December 1900, he gave another paper to the same group, one in which he presented a theoretical justification for his formula. However, as he explicitly admitted, this presentation omitted a full deduction and instead simply tried to present the real core of the theory.[27] A month later Planck sent an article to *Annalen der Physik* which seemed to contain the requisite derivation. However, this paper involved a method of procedure somewhat different from that of the December address. It also involved, as will be shown, some oversimplifications.

Thus, in the space of a few months Planck gave three different presentations of the same law. None of them were complete. The three taken together lacked an overall coherence in methods and presuppositions. Subsequent historians, recognizing Planck's work as a decisive breakthrough, have attempted to fill the gaps and smooth over the inconsistencies—and in doing so have generated new controversies. If we ignore older treatments, which projected later and clearer derivations back onto Planck's original treatment, then the significant dispute is between those (Klein, Kangro, Hermann, Jammer, Whittaker) who interpret the Planck of 1900 as accepting energy quantization for the fictitious oscillators (though not for the radiant energy itself) and Kuhn, who argues that the idea of energy discontinuity was introduced later by Einstein and Ehrenfest.

Here it seems helpful to make a distinction between the historical development of science, considered as something essentially public, and Planck's own understanding of that development, which involves questions of psychology as well as history. I will simply present the basic facts on the public historical development and then, without attempting to settle the controversy, indicate my position on the second question.

Planck's 1901 paper, "On the Law of Energy Distribution in the Normal Spectrum," published in the leading physics journal of the day, *Annalen der Physik*, presented the derivation of the new radiation law that was the historically significant source for further investigators.[28] This paper is rather easily summarized. Experiments show Wien's law to be invalid. To supply the needed improvement Planck will employ an electromagnetic theory based on his concept of natural radiation. From his previous investigations Planck assumed, as an already established relation, that the law of energy distribution is determined when the entropy of a monochromatically vibrating resonator is known as a function of energy. Then the formula dS/dU

$= 1/T$ allows one to determine the resonator energy—and from that the radiant energy distribution—as a function of temperature.

Earlier, Planck had defined the entropy of an individual resonator by dividing the total entropy by the number of resonators. Since that approach led to Wien's law, some modification is necessary if one is to go beyond Wien's law. Accordingly, what he now intends to do is to determine the relation between the energy and the entropy of an individual oscillator. Here, as in his earlier December address, he employed two concepts whose interrelation is not altogether clear. The first is the idea of *monochromatic resonators*. In both papers Planck explains that since the vibrations of a resonator change their amplitude and phase, one must deal with average energy. The average could be either a time average for an individual resonator or an instantaneous average for a large number of identical resonators. It should be noted that the spread in question is in phases and amplitudes. For a classical resonator the energy is proportional to the square of the amplitude. The introduction of a damping factor changes the natural frequency to a new value, but still leaves a fixed frequency. Hence, if one is speaking of classical resonators, there is no contradiction involved in speaking of them as monochromatic, or having fixed frequencies, and also as having varying energies.

This concept of resonators serves as the basis for defining entropy. If S is the average entropy of an individual resonator, then a collection of N resonators has entropy $S_N = NS$. It is at this point that Planck introduces the formula generally attributed to Boltzmann, but which might more properly be called the Boltzmann-Planck formula, linking the entropy S_N to the probability W:[29]

$$S_N = k \log W + \text{const.} \qquad (4.20)$$

The second idea, presented as essential in both papers, is that energy is not treated as a continuously divisible quantity, but as composed of a well-defined number of equal parts.[30] The way both papers are developed, this discreteness applies primarily to groups of oscillators. The idea that individual oscillators have only discrete energy levels is not brought up.

With these assumptions, Planck can then adapt Boltzmann's methods. The number of ways in which P units of energy can be distributed among N oscillators is

$$R = (N + P - 1)!/(N - 1)! \, P! \qquad (4.21a)$$

To handle this, Planck uses what he rather misleadingly takes to be a first approximation to Stirling's formula,[31] $N! = N^N$. This yields for the total number of complexions

$$R = (N + P)^{(N + P)}/N^N P^P. \qquad (4.21b)$$

On the assumption that the number of complexions determines the probability, Planck substitutes this value of R for W, the probability in equation (4.19), to get

$$S_N = kN\{(1 + U/\epsilon)\log(1 + U/\epsilon) - U/\epsilon \log U/\epsilon\}. \qquad (4.22)$$

Once again the entropy of an individual oscillator is determined by dividing the total entropy by the number of oscillators to get

$$S = k\{(1 + U/\epsilon)\log(1 + U/\epsilon) - U/\epsilon \log U/\epsilon\}. \qquad (4.23)$$

To interpret the significance of equation (4.23) Planck first develops what he takes to be the simplest form of Wien's displacement law. It is developed by using equation (4.9) to go from radiant to oscillator energy,

$$S = f(U/v). \qquad (4.24)$$

This general form fits equation (4.23) only if $\epsilon = hv$. This assumption leads to the energy and frequency of an oscillator. If one again uses equation (4.9) to relate oscillator to radiant energy, one has

$$u = (8\pi hv^3/c^3)(1/[\exp(hv/kT) - 1]). \qquad (4.25)$$

This is the standard form of Planck's radiation law.

In the earlier December address Planck had indicated, without developing it, a more complicated proof of equation (4.25). This is based on considering N resonators at frequency v, N' and frequency v', etc., and computing the energy of this distribution. Then the distribution is maximized subject to the constraint

$$E_0 = \Sigma N_v U_v, \qquad (4.26)$$

where E_0 is the total energy of all the resonators. As Kuhn indicates, this could be done by finding the maximum for $\delta(S_{E_0} - \mu E_0)$, where μ, an undetermined multiplier, is then determined to be $\mu = 1/kT$. Though Planck indicates this, it is not certain that he actually worked it out. Even when this distribution is considered, the individual resonators are still treated as monochromatic resonators.

Planck discussed real molecules rather than fictitious oscillators when he attempted to determine the values of the numerical constants he had used. As various commentators have pointed out, Planck thought this to be both the most convincing support for his new formula and a significant contribution in its own right. Planck not only determined the values of the two new constants he had introduced, h and k, but also used these values to obtain more accurate values of the mass of the hydrogen atom and the charge of the electron. Our concern here is not so much with these numerical calculations as with the presuppositions supporting them, especially the relationship between fictitious oscillators and real molecules.

Planck calculated h and k by relating his radiation formula to experimental results.[32] In his December address and in a separate article appended to his derivation of the radiation formula, Planck related this to real molecules by the following argument. The entropy of a total system containing both resonators and radiation must—again following Boltzmann—be proportional to the number of complexions of the combined system. If the resonators are treated as real molecules, then the number of complexions corresponds to the most probable velocity distribution. Since these molecular velocities are, according to electromagnetic theory, completely independent of the radiation distribution, the number of complexions of the overall system is simply the product of the number of complexions of each system. Entropy is a function of the log of the probability, which in turn is proportional to the number of complexions. Hence, Planck can use

$$f \log (P_0 R_0) = f \log P_0 + f \log R_0, \qquad (4.27)$$

where P_0 is the number of complexions of the gas molecules, R_0 the number of complexions of the radiation energy, and f a to-be-determined function.

The form of f has already been determined for radiation in equation (4.20). Neglecting an additive constant, $S = k \log W$, so that $f = k$. The form of f has been independently determined by Boltzmann in the case of a monatomic gas, $f = \omega R$, where R is the gas constant and ω the ratio of the mass of a real molecule to the mass of a mole. Since, by the argument leading to equation (4.27), f must have the same form in both cases, Planck can write

$$\omega = k/R = 1.62 \times 10^{-24}. \qquad (4.28)$$

This value of ω, combined with experimental data, serves to determine the mass of the hydrogen atom and the charge of the electron.

This calculation makes explicit what was implicit from the beginning. Planck's fictitious oscillators are idealized molecules. They are, in effect, streamlined functional models of molecules, effectively identified with their function of absorbing and emitting radiation. What they do, real molecules also do. The results are the same; the internal mechanisms presumably differ.

The historical question of what Planck discovered should be seen in the context of what historians generally count as discoveries. When new scientific concepts are introduced or old terms are given new meanings, the new concepts are often obscure concepts with fuzzy boundaries. There are still debates concerning Galileo's understanding of 'inertia', Newton on 'force', Leibniz's 'vis viva', Mayer's 'energy', and many other novel concepts. This is for simple, fairly straightforward concepts. When the question at issue is one of determining the physical significance of a new equation, the conceptual problems become much more complex. As will be seen, de Broglie and Schrödinger and, to a lesser degree, Dirac and Heisenberg initially gave physical interpretations to the equations that bear their names that were significantly different from the interpretations that subsequently became accepted as normative. Such obscurities notwithstanding, historians rightly attribute to these men the new concepts, whether explicit, implicit, or more or less confused, contained in their work.

So should it be with Planck. He it was who introduced the concept of energy quantization. This is explicit in his text, where groups of monochromatic resonators are assigned discrete energy values proportional to their frequencies. This is implicit in his mathematics. Allowing resonators to have energy spreads destroys the derivation. Ultimately, it can only be the individual resonator that changes by units of $h\nu$. The concept of energy quantization is also implicit, though in a more nebulous form, in the physical interpretation accorded the derivation. Groups of resonators have quantized energy levels. These resonators are, in fact, idealized models of molecules. Planck did not know, and did not publicly speculate on, the mechanism by which molecules absorb and emit radiation. Yet, implicit in his work is the constraint that real molecules must produce the effects that idealized molecules can produce only when they have quantized energy levels.

How well did Planck understand the concept he had introduced? Any reasonable attempt to answer this question inevitably involves rethinking Planck's perspective. For him science centered on the search for laws of objective validity. He had discovered and eventually justified just such a law. This law had no ambiguity in terms of its relation to experimental data. Planck had gone beyond this to discover two natural constants. For him the justified law and the natural constants constituted the fundamental breakthrough. A further clarification of the presuppositions implicit in the derivation he had used had the nature of a mopping up operation. Planck's writings immediately after the papers we have considered were much more concerned with drawing attention to the new breakthrough than with the mopping up operation. This is the perspective within which Planck's understanding of the new concept should be appraised.

As for the concept itself, Planck definitely did not hold that radiant energy comes in discrete units of value $\epsilon = h\nu$. There is no disagreement about that. He did hold that groups of oscillators are quantized. He never explicitly discussed individual oscillators. If the presuppositions he reflects were made explicit, they would manifest an underlying incompatibility. When he speaks of the distribution of the amplitudes and energy of the resonators in a group, he is implicitly presupposing that energy is a classical function of amplitude. This allows a continuous distribution of energies. Yet, when he associates with them the energy $\epsilon = h\nu$, he has an assumption which cannot lead to the same energy distribution.

Planck probably paid little, if any, attention to this issue before Einstein and others made it crucial. In his perspective, probing the internal states of fictitious resonators is a bit like inquiring how Lear functioned as king before he decided to divide his kingdom. The outstanding problem remaining in Planck's mind was to clarify the significance of h, a constant with the units of action, not energy, and the law in which it functioned. Since h was connected with electronic behavior, electrons provided the only plausible candidate for the requisite oscillations. A better understanding of the significance of the new constant should be expected from advances in atomic physics.

The new law presented a different problem. It had joined the growing community of objectively valid laws. Within this community it had the peculiar status of bridging thermodynamics and electromagnetic theory. To understand the significance of the new law, accordingly, one should relate it to the basic laws of thermodynamics and electromagnetism. As Kuhn has indicated, Planck's 1906 lectures on heat radiation did not manifest any understanding of the problem different from that of his earlier papers. However, he did introduce the formula for the energy of an individual resonator,

$$U = h\nu/[\exp(h/kT) - 1]. \qquad (4.29)$$

By the time Planck delivered his 1909 Columbia lectures, the status of the new law and the concept it implied had become much more problematic. Planck's way of attempting to clarify this is to begin with the two laws of thermodynamics and Maxwell's equations and then attempt to see how the new radiation law fits into the interpretative framework they supply. In restrospect it is clear that there is no way in which the new law can fit into the old framework without rupturing the seams. Yet this was not evident a priori. Planck cannot envision resolving the conflict by altering the established foundations of physics. Rather, he hopes that the resolution of the apparent conflict might be obtained "by seeking the significance of the energy quanta $h\nu$ solely in the mutual actions with which the resonators influence one another."[33] Here he clearly accepts the energy quantization implicit in his formula, but only for resonators. By 1911 he is ready to extend this, though only in a limited way, to radiation.

Einstein's Quantum Theorizing

Einstein brought a rather different interpretative basis to the same general problematic. Like Planck, he held the first two laws of thermodynamics plus Boltzmann's interpretation of entropy as securely established. But Einstein definitely did not agree with Planck's evaluation of the Maxwell-Lorentz theory of electromagnetism as being beyond revision, even in its foundations. Einstein's own development of the special theory of relativity had originally been stimulated by his dissatisfaction with Maxwell's treatment of the propagation of light in a vacuum. There was one other difference between the two men which has some significance for the way in which they interpreted the quantum assumption. Though Planck had specialized in thermodynamics, he had long opposed Boltzmann's statistical interpretation of entropy. When, in an act of desperation, he finally accepted this, he got the statistics he needed from Boltzmann's previously discussed 1877 paper. Einstein, on the other hand, had become the acknowledged master of statistical reasoning.[34] Through his own independent development of molecular-kinetic theory, he was acutely aware of the foundations of statistical reasoning in physics. He could clearly recognize foundational inconsistencies where Planck simply saw a need for technical readjustments that did not involve foundational changes.

Einstein's use of the quantum theory in explaining the photoelectric effect and the specific heat of solids has been adequately explained in the general sources cited. Here, accordingly, we will concentrate on some interpretative problems he encountered, particularly those concerning wave-particle duality and those concerning the way in which classical and quantum physics should be interrelated. We will only supply the background physics indispensable to understanding these problems.

Einstein's first paper on quantum theory (1905) began with a novel interpretation of Maxwell's electromagnetic theory.[35] Since optical observations refer to time averages, the continuity proper to Maxwell's equations need not be proper to the behavior of light itself. It seems, he argued, that blackbody radiation, photoluminescence, and the production of cathode rays by ultraviolet radiation can be better understood on the assumption that light energy is distributed in finite energy quanta which can only be emitted or absorbed as wholes.

In terms of the work put in and the results obtained, Einstein's greatest achievement before 1905 was his redevelopment of statistical mechanics according to the method of time averages. He now brought this statistical way of thinking to the problem of radiation. The classical theory of blackbody radiation leads to the divergence which Ehrenfest later dubbed "the ultra-violet catastrophe." Einstein rejected this theory. However, he did not yet commit himself to Planck's formulation. After admitting that it was in agreement with experiment, Einstein simply dropped Planck's formula and thus sidestepped the difficulties it entailed. Wien's formula, equation

(4.8), is valid in the limit of large values of v/T. This, rather than Planck's radiation formula, supplied the basis for Einstein's original treatment of radiation as quantized.

From Wien's law Einstein derived a formula for the volume dependence of the radiation in a cavity:

$$S - S_0 = (E/\beta v)\ln(v/v_0). \tag{4.30}$$

Here S_0 is the entropy proper to radiation of energy E in the range $v-v + dv$, when the volume is v_0; S is the entropy when the volume is v. The constant β is simply presented as having the value 4.866×10^{-11}. It is, in fact, h/k, the ratio of the two constants Planck had determined in his 1900 paper. Einstein's early papers on the quantum hypothesis did not accord Planck's constant any special significance. A perfect gas or a dilute solution has the same formula for the volume dependence of entropy. Therefore, Einstein concluded, monochromatic radiation of low density behaves, at least in the limit in which Wien's radiation formula is valid, as if it consisted of mutually independent energy quanta of energy $R\beta v/N$ (or hv).

Einstein's qualitative explanation of photoluminescence, the photo-electric effect, and the ionization of gases by ultraviolet light is simply based on the model of radiant energy as light quanta rather than on either Planck's or Wien's radiation laws.

When Einstein was eventually and, it would seem, somewhat reluctantly accorded the Nobel Prize, it was for his interpretation of the photo-electric effect. This, plus the significance wave-particle duality soon acquired in de Broglie's work, has had a distorting influence on appraisals of Einstein's early work. The pioneering paper just considered only gave two justifications for the light quantum hypothesis. One was an analogy between an obscure aspect (volume dependence of entropy) of a radiation law (Wien's) which was known to be invalid and an application of Boltzmann's kinetic-molecular theory to fluids, a topic in which only Einstein was then proficient. The second was a qualitative account of some phenomena which experimentalists were just beginning to explore through precise experiments. It is hardly surprising that Einstein's light quantum hypothesis was widely rejected, even by physicists like Planck who were familiar with the problems of blackbody radiation and were willing to accept Einstein's theory of relativity.

Einstein realized that he had to come to grips with Planck's formula. He tried to rederive it in 1907 and again in 1911. In 1925 he personally translated S. N. Bose's new derivation of Planck's law from English into German and arranged for its publication. Einstein accepted Planck's law, but clearly did not accept Planck's account of the law. He seems to have hoped that a more acceptable derivation would clarify the assumptions which were indispensable to this law.

Einstein's 1907 paper shed no new light on the assumptions behind Planck's law. However, it situated the quantum hypothesis in a framework in which statistical arguments were more basic than radiation theory.[36] His paper begins by showing that the kinetic molecular theory applied to a collection of oscillators leads to the average energy per oscillator, $\bar{E} = (R/N_0)/T$ (or $\bar{E} = kT$). If one assumes that this also applies to oscillating ions in equilibrium with radiation and uses Planck's classical oscillator formula, equation (4.9), one obtains the classical radiation law, (4.17).

To avoid the divergence this involves and obtain Planck's radiation formula, something had to be modified. Einstein retained equation (4.9), based on classical radiation theory, and classical thermodynamics, but modified the a priori probabilities assigned phase-space regions. A point in phase space represents a possible state of a system. Instead of Boltzmann's assignment of equal weights to equal regions, Einstein assumed that only those regions of phase space that take on the values infinitesimally near ϵ, 2ϵ, 3ϵ, etc., have nonzero weights. This assumption, Einstein showed, leads to Planck's quantum formula for the average energy of an oscillator and, by virtue of equation (4.9), to Planck's radiation formula.

The novel conclusion Einstein drew was: "If the elementary structures (Elementargebilde) assumed in the theory of energy exchange between radiation and matter cannot be understood in the sense of the present molecular-kinetic theory, must we not also modify the theory for the other periodically oscillating structures which the molecular theory of heat employs?"[37] Planck had introduced idealized oscillators as useful theoretical fictions. Einstein was now insisting that the new quantum assumptions must apply to real oscillators, the vibrating structures postulated by molecular theory.

The familiar image of the latter Einstein as an aloof isolated contemplator of cosmic mysteries, detached from routine ongoing science, tends to distort one's vision of the early Einstein. He kept sufficiently in touch with contemporary developments so that, without getting involved in details, he could finger the pressure points. In this case the pressure point had to be one where the assumption that molecular oscillations obey quantum laws leads to results significantly different from those given by classical molecular theory. Classical statistics had worked well when large numbers of vibrations were involved and quantum differences were negligible. The pressure point, accordingly, should be some low-temperature vibrational phenomena where classical theory does not fit the observed data.

The law Dulong and Petit had announced in 1819, that the atoms of all elementary substances have exactly the same heat capacity, was already known to be incorrect for such elements as carbon, boron, and beryllium. At low temperatures their heat capacity was well below the value 5.94 cal/mole proper to the Dulong-Petit law. Boltzmann's theory had given a simple statistical explanation of this law. If a molecule in a crystal can vibrate in three directions with a total vibrational energy of kT in each direction, then for N ($= R/k$) molecules the specific heat should be

$$c_v = \partial U/\partial T = \partial(3RT)/\partial T = 3R = 5.94 \text{ cal.} \qquad (4.31)$$

What Einstein did was simply to substitute for the U of equation (4.31) the Planck quantum formula for the average energy of an oscillator, leading to the result

$$c_v = 5.94x^2e^x/(e^x - 1)^2, \qquad (4.32)$$

where $x = h\nu/kT$ (though Einstein still did not use Planck's h). Equation (4.32) yields the Dulong-Petit law for large values of x and approaches 0 as t approaches 0.

This was the first successful application of the new quantum concept (not yet a theory) to real atoms. It was in a rather literal sense a superficial application. The quantum had not yet penetrated the interior of the atom. The only assumption so far operative was that the gross vibrations of atoms and molecules were quantized.

After this success with solids, Einstein returned to a quantum theory of radiation. Following his flair for statistical reasoning, he focused on a determination of the fluctuations of radiant energy in an enclosure at temperature T.[38] Again, he coupled his statistical methods to Planck's formula for the average energy of an oscillator in equilibrium with the radiation. This time, however, he introduced a novel thought experiment. He considered two interrelated volumes in the enclosure and a freely moving mirror which only reflects radiation within the narrow frequency range ν–$\nu + d\nu$. The fluctuation Δ in the mirror's velocity in a unit time τ would correspond to the fluctuations in the cavity. The formula he worked out for such fluctuations is

$$\overline{\Delta}^2/\tau = (1/c)[h\rho\nu + (c^3\rho^2)/8\pi\nu^2]d\nu f, \qquad (4.33)$$

where ρ is the density of radiation in the range ν–$\nu + d\nu$ and f is the mirror's surface area.

The first term in the square brackets is the term that would result if Wien's radiation law had been used instead of Planck's formula. It corresponds to the fluctuations one would expect with discrete independent light quanta. The second term is the only term that would be present if the Rayleigh-Jeans law had been used for the radiation. This fluctuation is what one would have from the interference of wave trains. If Planck's law is thought of as applying to the radiation itself rather than merely to the oscillators in equilibrium with this radiation, then radiation must be thought of as having both wave and particle properties.

Independent of Einstein's work, X-ray theoreticians had developed their own wave-particle duality. This, however, was not a unified theory but an apparently temporary conflict. It began when William H. Bragg advanced the idea that X-rays are neutral pairs of α and β particles, while C. G. Barkla argued for a wave theory of X-rays. The debate temporarily subsided after 1912 when Bragg's son, W. L. Bragg, showed that von Laue's diffraction results could be deduced from the wave theory. It was only after the discovery of the Compton effect in 1923 that Einstein's light-quantum hypothesis was related to these conflicts.[39]

It is known that Einstein in 1908–9 attempted to resolve the problem of dualism by developing a new electrodynamics.[40] When this proved unsuccessful, he temporarily abandoned quantum theory and concentrated on research leading to the general theory of relativity. Einstein's retrospective reflections reveal how strongly he felt about his failure to develop a quantum electrodynamics:

> All my attempts, however, to adapt the theoretical foundation of physics to this [new type of] knowledge failed completely. It was as if the ground had been pulled out under one, with no firm foundation to be seen anywhere, upon which one could have built.[41]

The First Solvay Conference

The interpretative problems we have been considering came to the forefront of European physics in the first Solvay Conference. Some other developments, which have been omitted from our selective survey, should be noted to indicate the state of the quantum hypothesis at the time the conference convened. Einstein was in the forefront of those who had come to accept Planck's radiation formula, but were reluctant to accept Planck's derivation of this formula. Equation (4.9), relating the radiation density $u(\nu,T)$ to the average energy of an oscillator U, was based on a classical treatment of both radiation and oscillators. In 1906 Planck derived the formula (4.29) for the average energy of an oscillator, the formula which played a basic role in Einstein's theory of the specific heat of solids. This is a quantum expression incompatible with the classical assumptions which would lead to $U = kT$.

The quantum hypothesis hardly had the status of a theory. Yet, the theories that could complete with it were experiencing greater difficulties. H. A. Lorentz, who had revised Maxwell's theory to include electron interactions and an electrodynamic concept of mass, kept trying to fit the experimental results on blackbody radiation into his theory. In a letter to Wien (6 June 1908) he admitted, "I have struggled almost continuously with this problem in recent years until I finally came to the realization that it would be impossible to reach my goal in this way." By 1909 he reluctantly accepted Planck's formula, but cautioned extreme reserve in interpreting this as contradicting classical physics.[42]

Others besides Einstein had attempted to relate the quantum hypothesis to the atom. Johannes Stark had tried to incorporate the quantum hypothesis into his model of the atom. Arthur Haas had attempted to derive this hypothesis from a theory of atomic structure. Yet, neither man was able to give a convincing or even consistent account of the mechanism of radiation emission. Arnold Sommerfeld interpreted the quantum as a quantum of action rather than energy (h rather than $h\nu$ basic). Yet, he was unable to advance any plausible account of how radiation is produced.[43] The most important theoretical clarification of the quantum hypothesis, Paul Ehrenfest's proof that the quantum assumption was a *necessary* as well as a sufficient basis for Planck's radiation law, received so little notice that it was not even mentioned at the Solvay Conference.[44]

The quantum hypothesis relied on inconsistent assumptions and, especially as interpreted by Einstein, led to paradoxical conclusions. Yet, the experimental support was too strong to ignore. In addition to the extensive and precise measurements confirming Planck's formula for blackbody radiation, a new source of support came from the work of Walther Nernst. Nernst's third law of thermodynamics, that as temperature approaches zero the entropy asymptotically approaches a limiting value, implies that specific heats must also asymptotically approach a limiting value independent of the nature of the material measured. In the period from 1907 to 1910 Nernst and his students performed careful measurements of specific heats at low temperatures. They found, as Einstein had predicted, that the specific heats of different substances approach the limiting value of 0.

Nernst, in conjunction with his student Frederick Lindemann, extended Einstein's approach to the specific heat problem by relating atomic vibrations to interatomic distances at the temperature at which a solid melts. He also extended the idea of quantization from vibrations to rotations of diatomic molecules.[45]

Nernst, a very efficient organizer, persuaded the wealthy Belgian industrialist Ernest Solvay to subsidize a conference where Europe's leading physicists could meet to discuss the significance of the quantum hypothesis. Long position papers on topics assigned by Solvay and Nernst were circulated before the convention, which met in Brussels from 30 October to 3 November 1911. H. A. Lorentz, who presided over all the Solvay Conferences from 1911 through 1927, was, as Einstein put it in a letter written after the first conference, "a miracle of intelligence and subtle tact—a living work of art."[46]

Jeans's paper, "Report on the Kinetic Theory and the Specific Heat according to Maxwell and Boltzmann,"[47] used a hydrodynamic analogy as a model to show how it might be possible to extend classical kinetic theory to include free electrons moving through the interstices of solid matter. The final word on this last-ditch effort to preserve the classical theory by such ad hoc assumptions was pronounced by Henri Poincaré:

> It is clear that by giving suitable dimensions to the communication tunnels connecting the reservoirs and suitable values to the leakage, Jeans would be able to accommodate any experimental situation whatever. But this is not the role of physical theories. They should not introduce as many arbitrary constants as there are phenomena to be explained; they should establish a connection between different experimental facts and, above all, permit prediction.[48]

Poincaré, who came to the conference with little knowledge of the work done on quanta, was quickly becoming a convert.

Emile Warburg summarized the experimental research.[49] The complicating factor was that Planck's formula involved natural constants, such as e and c, which were given somewhat different values by different methods of measurement. Within these limits, however, quite different types of experiments gave values in accord with Planck's law. Rubens's paper reached a similar conclusion.

The survey papers by Lorentz and Planck agreed on one conclusion. Regardless of the model or method used, classical physics inevitably leads to the Rayleigh-Jeans law. Planck simply presented his formula as the one that best fits the experimental data, without giving his much criticized derivation. Then he asked what modifications of classical physics are required by the acceptance of this formula.[50] Planck himself still rejected what he took to be the most extreme modification, Einstein's light-quantum hypothesis, and still favored his own interpretation, linking quantization to Hertzian oscillators. Now, however, Planck admitted that his derivation rested on mutually contradictory hypotheses. His statistics presupposed that the energy levels proper to the oscillators are multiples of $h\nu$, while his law relating average oscillator energy to radiation frequency (4.9) presupposed a continuous variation in energy. This contradiction, Planck noted, also held for the atomic models proposed by Haas and Schidlof. If, following Reinganum's model, one allows the energy of an oscillator to vary in a discontinuous manner, then, Planck argued, it is impossible to understand how radiation is absorbed.

Planck's solution to these problems, his second quantum theory, was to assume that the energy of an oscillator could be expressed in the form

$E = n\epsilon + \rho$. The oscillator possesses n quanta of energy ϵ plus a continuous energy ρ. Absorption of free radiation, governed by ρ, is continuous, while emission, governed by $n\epsilon$, is discontinuous. This new theory led to the zero-point energy $h\nu/2$ which an oscillator should have at absolute zero. Planck argued that his new theory could also accommodate the photoelectric effect without postulating light quanta.

In defending his new theory, Planck claimed that one should simply accept the fact that an elementary domain of probability in phase space simply has an extension h. Rather than pursue a further inquiry into the physical significance of h, one should, at least on this issue, accept the phenomenological point of view. Einstein began the discussion by indicating that he found this treatment of probability a little shocking. If no physical significance is attached to the configurations which enter into the probability calculations, then there is no way to distinguish two different configurations. For Einstein the probabilities of kinetic theory have physical significance only if they can be interpreted in terms of objectively possible states. Planck's answer was that, since it was impossible to retain all the classical laws, one had to make some choice of what was to be modified. He was willing to modify Boltzmann's interpretation of probability.

Einstein faced a similar choice. His papers, summarizing his work on specific heats, was essentially noncontroversial. Einstein began the discussion by indicating the foundational basis he relied on in his interpretation of quantization.[51] The principle of energy conservation seemed beyond dispute. Einstein was still willing to rely on Boltzmann's statistical interpretation of entropy, supplemented by his own interpretation of the probability of a state as the percentage of time a system spends in that state. He now insisted on the provisional character of the light-quantum hypothesis and admitted that it did not fit the experimentally verified wave phenomena proper to light.

The interpretations developed by Planck and Einstein had drawn notably closer. Both accepted Planck's radiation law on the grounds that it fitted the experimentally established data. Both sought a way of showing how this law could be reconciled with the established laws of physics. Both accepted the first and second laws of thermodynamics as established beyond dispute. Though they disagreed somewhat on how probability should be interpreted, each accepted some version of Boltzmann's probabilistic interpretation of entropy.

Planck accepted Maxwell's electromagnetic laws as too well established to admit of serious modification. Einstein did not. Yet, his attempts to develop appropriate modifications had failed so completely that he suspended work on the project. Neither seemed to feel that revised models of the atom supplied a potential basis for resolving the inconsistencies that perplexed them. Einstein asserted in one discussion that the present theory of quanta led to contradictions when applied to any system with more than one degree of freedom.[52] It seemed like a reasonable strategy to try to resolve the inconsistencies generated by Planck's law applied to oscillators, which only required one degree of freedom, before attempting to relate an inadequately developed hypothesis to atomic structures, which were not known even in an inadequate fashion.

The first Solvay Conference marked, to borrow from Churchill, the end of the beginning. It was clear that the quantum hypothesis had entered physics to stay. When Poincaré returned to France, he attempted a mathematical investigation of Planck's law. In December he announced his preliminary results, and in January wrote a long memoir on the quantum hypothesis.[53] Assuming the validity of Planck's law and of the law of equipartition, and also assuming the applicability of probability theory to phase space, Poincaré then inquired into the necessary conditions for Planck's law to obtain. He concluded that radiation in a blackbody cavity could only remain finite if one admitted discontinuities analogous to those postulated by the quantum hypothesis.

His paper, in turn, formed the clinching argument for Jeans. In his report on radiation and the quantum, which effectively introduced the quantum hypothesis to the English-speaking world, Jeans argued that the finiteness of the total radiant energy per unit volume necessitates the abandonment of classical physics and the acceptance of Planck's hypothesis.[54] Jeans's first edition of this influential work stressed the difficulties in quantum theory and insisted that the work so far done must be considered a fragment of a to-be-developed quantum dynamics. It is rather ironic that in the second edition, published just before a quantum dynamics was developed, he dropped his earlier insistence on the need to develop a quantum dynamics.

When Rutherford returned to England, he communicated the consensus of the conference and the enthusiasm of its spirit to Niels Bohr. In France, young Louis de Broglie avidly read the conference proceedings which his elder brother, Maurice, was editing. It seemed evident that further developments were now in the hands of a new generation, of people who accepted the quantum hypothesis as a point of departure for further explorations rather than as a still to be debated hypothesis. The spirit of the times was perhaps most clearly expressed by the man whose radiation law had inspired the original investigations. In his 1911 Nobel Prize address Wien claimed:

> We must admit that the result of radiation theory to date
> is not a very good one for theoretical physics. As we have
> seen, only the general thermodynamic theories have proved

satisfactory as yet. The theory of electrons has come to grief over the radiation problem; the Planck theory has not yet been brought into a definite form. Research is faced with exceptional difficulties and we cannot discern when and how they can be overcome. In science, the redeeming idea often comes from an entirely different direction, investigations in an entirely different field often throw unexpected light on the dark aspects of unresolved problems. We must base our hope in the future in the expectation that the present era which has proved so fruitful for physics may not pass without a complete solution being found for the problem of thermal radiation. Far-reaching and new thoughts will have to be set to work, but the result will be great, because we shall obtain a profound insight into the world of the atom and the elementary processes within it.[55]

5. The Bohr Era

No one has contributed more to the development of atomic physics and the problematic involved in interpreting its significance than Niels Bohr. He was the chief architect of the Copenhagen interpretation of quantum mechanics, which won such widespread acceptance that it is generally referred to as the orthodox interpretation. Yet, the day before his death, he said, "I think it would be reasonable to say that no man who is called a philosopher really understands what one means by the complementary description."[1] A striking statement, yet one that seems to have been justified in 1962 and one that would still have some justification if uttered today. Though the Copenhagen interpretation represents an epistemological residue distilled from Bohr's philosophical position, the two are not at all equivalent. The former does not include Bohr's highly developed, but poorly articulated, ideas on the nature of concepts, the centrality of language to all human thought, subject-object complementarity, and other epistemological and ontological considerations often deemed peripheral to physics.

As Jammer[2] and Holton[3] have shown, the young Bohr was stimulated by a youthful exposure to philosophical ideas stemming from Kierkegaard, Høffding, and indirectly Hegel, and perhaps William James, and later related his own doctrine of complementarity to these sources. Yet, I believe that such sources had almost no influence on his scientific development. The influence they did have will be considered in a later chapter. The present chapter will focus almost exclusively on the development of Bohr's thought. Because of the difficulties involved in understanding, and even more in appraising, Bohr's final position, our special concern will be with the sort of dialectical interaction between Bohr's ideas on the nature of scientific explanation and the advances and setbacks that induced him to modify, and eventually abandon, much of his original interpretative framework.

Thanks to the pioneering research of L. Rosenfeld[4] and the meticulous reconstruction given by J. Heilbron and T. Kuhn,[5] the path that Bohr followed in the initial development of his atomic theory is now fairly well understood. The account that follows will use these sources as well as Bohr's own writings to retrace this history. The emphasis, however, will not be on the historical sequence but on some underlying conceptual problems that gradually emerged. Bohr's original intention clearly seems to have been to make a descriptively accurate account of the atom's real structure the core of his atomic theory. The initial success of this theory convinced him that he attained some parts of the true picture.

The Sommerfeld reformulation of the Bohr theory and Bohr's subsequent attempts to fashion a comprehensive and coherent theory gradually led to deemphasizing descriptive accounts of the orbital motions of electrons and to giving more stress to formal principles. About 1920 Bohr explicitly came to grips with these changing patterns by assigning descriptive accounts and formal principles mutually supporting roles in scientific explanation. Then he went on to use this mode of explanation to treat the basic features of the periodic table, and even to account for the anomalous position and peculiar properties assigned the rare earth elements. This, perhaps the greatest achievement of the old atomic physics, seemed fully to vindicate the explanatory role Bohr had assigned the basic descriptive features of the Bohr-Sommerfeld theory.

Further refinements of Bohr's work, and even more the attempts in the early twenties to tackle the problems the Bohr-Sommerfeld theory left unresolved, soon toppled every descriptive prop Bohr had relied on. Bohr's reaction to these developments was to reinterpret atomic theory as a purely formal account, as a hypothetical-deductive system justified by its accord with experimental data rather than by its accuracy in depicting the actual structure of the atom. This mode of interpretation induced Bohr, in conjunction with Kramers and Slater, to conclude that the formal principles used, including the principles of energy and momentum conservation, did not supply a basis for a descriptive account of individual atomic processes. When subsequent experiments contradicted this conclusion, Bohr began to reexamine the nature of descriptive accounts, but now focusing on the role of language and on the grounds of meaning of the crucial terms employed. In this dialectical fashion he came to assemble the basic ingredients that went into the Copenhagen interpretation even before the development of matrix mechanics and wave mechanics. The struggle to impose a coherence on these diverse ingredients and radical breakthroughs will be postponed till after two further chapters (6 and 7) treating the development of matrix mechanics and wave mechanics.

Bohr's Theory of Metals

The M.A. examination at the University of Copenhagen included a "big problem," a topic assigned by the professor on which the student had to write a report equivalent to a Master's thesis. The topic Professor G. Christiansen set for Bohr was: "Give an account of the application of the electron theory to explain the physical properties of metals." Bohr reviewed the literature and found most of it unsatisfactory, with the exception of Lorentz's theory. On 26 March 1909, he wrote to his brother Harald, who was then studying mathematics at Göttingen: "At the moment I am wildly enthusiastic about Lorentz' (Leiden) electron theory."[6] At that time (1909) there were two conflicting theories of the activities of electrons in metals. Riecke and Drude had assumed that electrons are free to move through metals and that there were both positive and negative electrons.[7] J. J. Thomson was the first to assume that only negative electrons were present. Lorentz treated the electrons as a gas governed by Maxwell-Boltzmann statistics. The greatest achievement of his theory was the derivation of the Wiedemann-Franz law: the ratio of thermal to electrical conductivity is the same for all metals. This strongly indicated that the flow of electrons is responsible for the transport of both heat and electricity and that electrons in all metals are the same.

Bohr's 1909 paper was essentially a survey of the established results. There was one aspect of the accepted theory, however, which Bohr found unsatisfactory. Thomson had assumed that a magnetic field would curve the paths of electrons moving through a metal and that this curvature would produce a magnetic field in the direction opposite to the external field. Thomson and later P. Langevin had used such assumptions to explain diamagnetism. Bohr thought this explanation incorrect and argued that moving electrons should produce no magnetic field.[8] If both an electric field and a magnetic field were present, then the best available theory predicted a potential difference between two separated points which were the same distance along the current flow (the isothermal Hall effect). In this case, as well as for some other thermomagnetic effects, there was a discrepancy between experimental and theoretical results. At the end of his thesis Bohr suggested that these discrepancies might be resolved by considering the effect of bound as well as free electrons.

The unresolved aspects of the electron theory of metals, as well as the discrepancies between theoretical predictions and experimental measurements of some thermoelectric and thermomagnetic effects, seemed to have convinced Bohr that the topic was a suitable one for a dissertation. After two more years of what he later described as "very hard work" he finished his dissertation. By this time he had reached a quite different evaluation of Lorentz's theory:

> Lorentz's theory is based on the following mechanical picture. In the interior of metals both atoms and free electrons are assumed to be present. The dimensions of the atoms and the electrons, i.e., the ranges within which they affect each other appreciably, are assumed to be very small compared to their average mutual distance; thus, they are thought to interact

only in separate collisions, in which they behave as hard elastic
spheres. . . . However, while Lorentz' theory is mathematically
very perfect, they physical assumptions on which it is based
can hardly be expected to be valid, even approximately, for
actual metals. Moreover, on many essential points the agreement between the theory and the experimental results is unsatisfactory. It would therefore be of interest to develop the
electron theory of metals from more general assumptions, and
to investigate which results of the theory are connected with
the special assumptions, and which results remain unchanged
when more general assumptions are adopted.[9]

To see the significance of this evaluation it may help to step back for a
moment and see the problem Bohr is treating in a contemporary perspective. Fortunately, Van Vleck's recent Nobel Prize address has clarified the
peculiar historical status of the special assumption Bohr is criticizing.[10] The
magnetic properties of metals can only be explained on the basis of nonclassical assumptions: quantization of angular momentum, electron spin,
Fermi-Dirac statistics, and the use of an effective electron mass derived
from energy band structure. Yet, some of the nonclassical properties of
metals were apparently explained by introducing special assumptions into a
classical framework. Thus, Langevin in 1905 had explained paramagnetism
by introducing the ad hoc assumption that an atomic or molecular magnet
carried a permanent magnetic moment μ whose spatial distribution is determined by the Boltzmann factor for averaging over angles statistically. To
explain diamagnetism Langevin took into account the Larmor precession of
the electrons about the magnetic field. This led to a formula for the Curie
temperature, the temperature at which ferromagnetism sets in. Lorentz's
treatment of electrons within a metal supplied an apparently general
method capable of incorporating these special results. Yet, the Langevin
assumptions were effectively equivalent to the introduction of the quantization of angular momentum, a nonclassical assumption.
     When Bohr prescinded from special assumptions and treated these
problems on the basis of the general principles of classical physics, he found
that it was impossible to explain either diamagnetism or paramagnetism and
that the established theory gave incorrect results for other thermoelectric
and thermomagnetic effects. I believe that Van Vleck is exaggerating in
calling this dissertation "perhaps the most deflationary publication of all
time in physics."[11] Because Bohr never succeeded in getting this dissertation
published in English translation, few physicists were aware of the conclusions he had derived or of how rigorous and general his derivations were.
Our concern here is primarily with the influence of this work on the development of Bohr's own thought. As the quotation indicated, he was making a

sharp distinction between the physical assumptions and the mathematical formalism, a distinction that came to play an increasingly important role in his subsequent development. As far as the physical assumptions are concerned, Bohr had convinced himself that classical physics, a term he used from at least 1912 on,[12] does not supply an adequate basis for the description of electron motion. In what follows we will give a sketchy indication of how these conclusions were reached.

In place of special ad hoc assumptions Bohr introduced three general physical assumptions:[13]

1. Free electrons are always present in a metal, their number depending on the nature and temperature of the metal.

2. There is a state of mechanical heat equilibrium between the free electrons and the atoms in a homogeneous piece of metal of uniform temperature not subjected to external forces.

3. The properties of individual atoms are on the average isotropic, and this isotropy remains even in the presence of external fields.

As Bohr noted, these assumptions were debatable. Assumption (2) had failed in the treatment of blackbody radiation. Assumption (3) was incompatible with the idea Langevin had introduced that atomic magnets have moments pointing in definite directions. These general assumptions, however, did fit Bohr's stated purpose of first determining which metallic properties could be explained on the basis of standard physics and then using this as a general background for testing the significance of special assumptions.

On the basis of these general assumptions Bohr was able to reproduce the established results on the flow of heat and electricity in metals, including the Wiedemann-Franz law, which Bohr called "one of the most beautiful results of the electron theory."[14] His real concern, however, was with the points and problems where the established theory proved inadequate. One such point, blackbody radiation, was only treated in passing. Though Bohr cited the recognized inadequacy of the Rayleigh-Jeans theory plus later patchwork, he did not yet focus on the quantum assumption as the key to the solution. The underlying difficulty, he thought,

> is presumably due to the circumstance that electromagnetic theory is not in accordance with the real conditions and can only give correct results when applied to a large number of electrons (as are present in ordinary bodies) or to determine the average motion of a single electron over comparatively long intervals of time (such as in the calculation of the motion of cathode rays) but cannot be used to examine the motion of a single electron within short intervals of time.[15]

With this brief comment, blackbody radiation was then dismissed as outside the scope of the dissertation problem.

Bohr's truly novel contribution came in his treatment of the effect of electrical and magnetic fields on the motion of free electrons. Contrary to the positions developed by Langevin and others, Bohr concluded that a piece of metal in electrical and thermal equilibrium should not, according to classical physics, possess any special magnetic properties from the presence of free electrons. His proof came from his generalization of the Lorentz theory. On the assumption that electron-electron collisions are negligible compared to electron-atom collisions, Bohr developed a general expression for the force an atom exerts upon an electron as a function of distance:

$$F(r) = \Sigma_n C_n r^{(n-5)/(n-1)}. \tag{5.1}$$

This general expression effectively reproduced Lorentz's model of hard-sphere collisions for $n = \infty$ and an appropriate choice of two values for C_n. Other values of n corresponded to the assumption that metallic atoms are electrical doublets or elementary magnets. This generalized expression for the force between atoms and free electrons as a function of distance supplied Bohr with a basis for determining which thermoelectric and thermomagnetic properties of metals can really be explained by classical physics and which of the accepted explanations really depend on special assumptions.

The first (and perhaps the most crucial) point was treated quite briefly. The presence of an external magnetic field may cause individual free electrons to undergo a helical motion. This, however, does not change the statistical distribution of the velocity components of the free electrons. Since magnetism is caused by moving charges, the fact that an external magnetic field does not alter the statistical distribution of velocity components means that an external magnetic field does not produce any change in the motion of free electrons that would be responsible for new magnetic properties. Langevin's explanation of diamagnetism was based on the idea that the electrons bound in metallic atoms move in orbits and that the sudden application of an external magnetic field will produce a change in this motion which, in turn, will give rise to an opposing magnetic field. Bohr did not deny that this assumption could explain diamagnetism or that its extension could handle paramagnetism. The point he stressed was the incompatibility between Langevin's special assumption and the general principles of classical physics. If one is assuming that the bound electrons interact with the external field and with each other, then one must also assume that such interactions lead to a state of statistical equilibrium. In this case, the added electronic motion, basic to the Langevin account, would disappear. Langevin had to assume that the extra motion was somehow frozen in, without explaining how this could be reconciled with the principle, fundamental to statistical mechanics, that interacting systems attain a state of statistical equilibrium.[16]

The adiabatic Hall effect turned out to be a particularly revealing test of the general theory. Bohr first showed that the Lorentz assumption of hard-sphere collisions leads to the conclusion that an electrical current in the positive *x*-direction should induce a transverse electrical field along the negative *y*-axis. The same assumption also leads to the conclusion that electrical conductivity decreases under the influence of a magnetic field. Both of these theoretical conclusions were contradicted by experimental results. Of particular significance was the fact, which Bohr demonstrated, that neither the sign of the Hall effect (an electrical field in the negative *y*-direction) nor the predicted decrease in electrical conductivity depended on the particular value assumed for n in equation (5.1), but only on the three physical assumptions listed earlier. Some other discrepancies between the Lorentz theory and experimental results could be removed by special assumptions which were compatible with the three physical assumptions Bohr had listed, e.g., by using $n = 3$ or $n = 5$ in equation (5.1).[17] Yet, no such special assumptions could remove the more fundamental difficulties involved in attempting to use classical physics, together with some general assumptions compatible with classical physics, as a basis for explaining the effect of electrical and magnetic fields on the properties of metals. These highly negative results seem to have led Bohr to two general conclusions. First, in addition to considering the effect of electrical and magnetic fields on free electrons, it is also necessary to consider the effect of such fields on bound electrons.[18] This is impossible without some sort of theory of how electrons behave within atoms.

The second conclusion Bohr drew from his own negative results matched the conclusion he drew from the failure of the Rayleigh-Jeans theory to explain blackbody radiation. His clearest expression of this conclusion came in a summary of his dissertation which he gave to the Cambridge Philosophical Society on 13 November 1911: "The Maxwell equations for the electromagnetic phenomena, are not exactly satisfied with regard to the motions of the single electrons, an assumption which in my opinion is very distinctively shown by the calculations of Lord Rayleigh and Jeans of the law of heat radiation for small times of vibration."[19] Bohr was familiar with, and in fact cited in his dissertation, the 1905 paper in which Einstein had expressed a somewhat similar evaluation of the Maxwell equations. There were, however, some subtle differences. Bohr does not envisage abandoning Maxwell's electromagnetic theory or replacing it by a new electromagnetic theory. He also believed that these equations give some basis for describing the motion of free electrons. He used the equations in this way in his subsequent papers on collision theory. His conclusion is that Maxwell's equations do not supply a basis for a precise descriptive account of the motion of individual free electrons and probably supply less

of a basis for an account of the motion of bound electrons. This appraisal
played a pivotal role in his development of the Bohr model of the atom, the
topic which will be considered next.

The Bohr Atom

The essential ideas involved in the Bohr theory of the atom are now
routinely presented in introductory physics textbooks. Even at this level the
Bohr theory is usually presented as a relatively simple introduction to the
more complex treatments associated with quantum theory. It is accordingly
somewhat difficult for one who learns the theory in this way to realize that
the Bohr model of the atom represents one of the most successful and
stimulating breakthroughs in the history of physics. Yet, this it certainly
was. To appreciate the conceptual clarification Bohr's theory engendered it
is helpful to consider, however briefly, the conceptual confusion that preceded it.

The discovery that electrons are the same, regardless of the type of
atom, clearly indicated that electrons are an essential component in atomic
structure. In adapting his electron theory to explain the Zeeman effect,
Lorentz had assumed that the electron is a charged sphere which can vibrate
about an equilibrium position, but he did not rely on other assumptions
about the structure or ingredients of atoms.[20] It was still not clear what the
bearer of the positive charge was, a positive electrical field, positive electrons, some basic positive unit, or alpha particles. J. J. Thomson adopted
Lord Kelvin's final model of the atom, considered in chapter 4, to develop a
model of the atom which should be called the "Thomson-Thomson" model,
but is more simply referred to as the "Thomson" model.[21] He assumed a
sphere of uniform positive electrification in which electrons are embedded.
He also assumed that the number of electrons in an atom is proportional to
the atomic mass. Initially he determined the allowed configurations by a
combination of mathematical reasoning and mechanical manipulation. The
latter, an adaption of some work by an American named Mayer, involved
inserting uniformly magnetized needles into cork stoppers, letting the corks
float freely on a water surface, and then observing the configurations that
were induced by a magnetic field acting uniformly on the needles. Three-,
four-, and five-needle corks formed the vertices of a triangle, a square, and a
pentagon, respectively. Six formed a pentagon plus an *inner* needle cork.
This procedure was extended as far as one hundred needle corks in seven
concentric rings.

This simplified model of electrons in stationary coplanar rings inside a
sphere of uniform positive electrification could be given a mathematical
formulation. For n electrons in a ring of radius a within a uniform positive
sphere of radius b which has a charge ve, Thomson derived the formula

$$a^3/b^3 = (S_n/4)v, \qquad (5.2)$$

where S_n is a trigonometric function based on the assumption that the n electrons are arranged at equal intervals of $2\pi/n$ along the ring. Thomson assumed that a more realistic model would involve electrons in three-dimensional shells rather than in coplanar rings. This, he argued, would be stable only if the electrons rotated. Since he was unable to work out this general theory in a quantitative way, he tended to rely on formulas like (5.2) developed for his oversimplified model of coplanar electrons. The results were similar to that obtained by the magnetic-cork analogue model. In building up atoms, one adds outer electrons until the outer ring is on the verge of instability. For the next atom, electrons are added to inner rings until stability is achieved.

This was the only model of the atom that seemed to offer any hope of explaining chemical periodicity. Thus, atoms with fifty-nine through sixty-seven electrons all have outer rings of twenty electrons. The periodic table grouped atoms in layers of eight ranging from the noble gases and the most electropositive (valence = $+1$) to the most electronegative (valence = -1). Thomson's series of eight atoms with the same external ring seemed to offer a potential for explaining this basic feature of the periodic table. The correlation was, however, exceedingly thin. In Thomson's scheme it was necessary to assume that the chemical properties are explained by the *inner* electrons in spite of the outer rings shielding them from contact or combination with the valence electrons of other atoms. Furthermore, the fact that Thomson's scheme gave a series of eight atoms which differed successively by one electron could not really bear much explanatory weight, for in this scheme the number of electrons is proportional to the mass of an atom rather than to the atomic number. Oxygen, for example, was assumed to have sixty-five electrons. A succession of atomic configurations each differing from its predecessor by one electron need not be correlated with eight chemically different atoms.

The explanation of spectral frequencies presented an even more formidable problem. The Lorentz model of disturbed electrons undergoing simple harmonic oscillations led to the conclusion that the allowed frequencies should be fundamentals coupled to the sequence of overtones given by a standard Fourier expansion. The patterns observed in spectral series bore no discernible relation to such a Fourier series. Though Thomson did not attempt to calculate the frequencies of spectral lines on the basis of his model, Rayleigh did, using an idealized version of the Thomson model. He concluded that this model led only to linear-type series and could not explain the basic role of squared terms in the Balmer-type formulas.[22] The only way that seemed open for one who wished to explain spectral lines on the basis of disturbed electrons undergoing simple harmonic oscillations was to assume that the electromagnetic vibrations emitted by atoms are due to some complicated coupling of many oscillatory modes. If such were the case, then, as Bohr later put it,[23] inferring the structure of an atom through an analysis of spectral lines would be like trying to infer the basic laws of biology by analyzing the colors in a butterfly's wing.

Other atomic models were proposed by A. Haas, J. Stark, H. Nagaoka, and G. Schott.[24] They were no more successful than Thomson's. None explained chemical properties, observed spectral lines, or such newly discovered phenomena as the emission of alpha, beta, and gamma radiation. Following Lorentz's successful explanation of the normal Zeeman effect, it seemed reasonable to presume that some sort of electronic oscillations produced electromagnetic vibrations. This assumption, however, introduced a series of difficulties. The difficulty involved in relating any such electronic oscillations to specific heats, a difficulty noted in chapter 3, was accentuated rather than resolved by Einstein's use of the quantum hypothesis to explain deviations from the Dulong-Petit law. If the equipartition principle, which had withstood the objections brought against it, also applied to vibrating and rotating electrons, then the theoretical value for the specific heat of solids should be much higher than the value given by the Dulong-Petit law.

At the 1911 Solvay Conference there was a brief discussion of atomic models, chiefly the Haas model, and of the problems involved in attempting to relate any model to the quantum hypothesis. Einstein, Planck, Lorentz, and the others who constituted the leaders of the European scientific community thought it prudent to postpone a frontal attack on the atom in favor of piecemeal analysis and patient accumulation of more and more precise data. Only a brash young person, innocent of the true complexity of the problem, would attempt to develop a realistic model of the atom. Physics, fortunately, has regularly produced such people.

After completing his dissertation, Bohr went to England and first worked with J. J. Thomson in Cambridge. Their relations were cordial but distant. Bohr was still concerned with the electron theory of metals and with getting his dissertation published in English translation. Thomson preferred to work on positive rays rather than to reconsider the difficulties that Bohr was insisting were present in his earlier work on the electron theory of metals. In spite of such strains Bohr's stay in Cambridge was helpful. He attended some of Thomson's lectures and also acquired a more detailed familiarity with the Thomson model of the atom. Here he also studied P. Weiss's modification of Langevin's theory of paramagnetism. On the basis of careful measurements of the magnetic properties of certain metals Weiss had concluded that there were fundamental localized units of magnetism, *magnetons*, within the metal. A molecule should possess either no magnetic moment or an integral number of magnetons. The random orientation of these magnetons is modified by the presence of an external magnetic field. This, like Langevin's original account, cannot be explained on the basis of classical physics, or at least of the principles of classical physics assumed as basic in Bohr's dissertation. Yet, it has strong experimental support. Heilbron and Kuhn suggest the plausible hypothesis that Bohr's attempt to relate the Weiss theory of the magneton to his own speculations on the role of bound electrons in explaining magnetic effects might have led to the idea of putting special restrictions on allowed motions of orbiting electrons with an atom.[25]

In March 1912, Bohr transferred to Manchester, where he hoped to work with Rutherford on radioactivity while continuing his own work on the theory of metals. Rutherford had introduced his nuclear model of the atom a year before in an attempt to explain occasional backscattering of alpha particles from gold foil. Surprising as it may seem, very little attention was paid to Rutherford's model outside of Manchester. Even in Rutherford's own laboratory this model was not treated as a fundamental breakthrough.[26] Bohr, it seems, only became interested in the nuclear model three months after his arrival in Manchester, and then primarily because of Charles Darwin's use of it to calculate the scattering of alpha particles from atoms. Before considering Bohr's initial adaption of the Rutherford model, it is helpful to consider the two types of instability problems it encountered.

Radiative instability is consequent upon the loss of energy due to the emission of electromagnetic radiation. A negatively charged particle rotating about a positively charged nucleus should, according to the laws of electrodynamics, emit radiation, lose energy, and rapidly spiral into the nucleus. This difficulty did not seem to trouble Bohr. The fact that Rutherford's model could not explain how an atom produces radiation did not distinguish it, for better or worse, from any of its competitors. Since Bohr was already convinced that classical electromagnetic theory did not supply a basis for describing the motion of individual electrons, he would be neither surprised nor disturbed by the conclusion that a coupling of a simple descriptive account of electron motion to classical electromagnetic theory led to difficulties.

Mechanical stability was seen as a more pressing problem. If one assumes that electrons orbit a nucleus while exerting a repulsive force on each other, then it is necessary to show how such orbital motion is stable. The other members of the Manchester group did not focus on this problem. Bohr did. He was aware of the importance Thomson had attached to such stability considerations and quite familiar with the methods he used to assure the mechanical stability of each successive atom.

The draft manuscripts from this period, which have been analyzed by Rosenfeld and by Heilbron and Kuhn, suggest that Bohr tested different models of the atom and molecule. His most detailed calculations were for a model of the hydrogen molecule involving two electrons on a coplanar ring. In this case he could adapt Thomson's formula (5.2) as a basis for calculating electronic energy and stability conditions. Though he adapted Thomson's calculations, he rejected the Thomson model of the atom. Thus, on 19 June 1912, he wrote to Harald about a breakthrough he thought significant: "Perhaps I have found out a little about the structure of atoms. . . . If I should be right it wouldn't be a suggestion of the nature of a possibility (i.e., an impossibility, as J. J. Thomson's theory), but perhaps a little bit of reality.[27]

The nature of the breakthrough was indicated in a memorandum he wrote for Rutherford before leaving to get married in Copenhagen.[28] First, he showed that Thomson's calculations would not lead to stability for a central-force field. Nor, he concluded, did there seem to be any hope of achieving stability on a mechanical foundation. Accordingly, he introduced, as a nonmechanical hypothesis, the stability condition,

$$E = Kv. \qquad (5.3)$$

Here E (more commonly T) is the kinetic energy of the rotating electron and v its mechanical frequency. Bohr was following the technique of his dissertation, being very clear on what followed from classical mechanics and being quite explicit on any hypothesis that went beyond. In the present case he thought that the hypothesis (5.3) was "the only one which seems to offer a possibility of an explanation of the whole group of experimental results, which gather about and seems to confirm conceptions of the mechanism of radiation as the ones proposed by Planck and Einstein."[29] In spite of this reference, he did not relate his new constant K to Planck's constant h. This special hypothesis was the first distinct step along the path to quantized orbits. He also differed with particular models of atoms and molecules Thomson proposed, though he was not yet explicit on the principle of adding electrons to outer rings.

A further significant difference centered around a point that Bohr became aware of through a chance remark of his associate G. Hevesy that the number of radioactive elements considerably exceeded the slots allowed by the periodic table. The Thomson model, in which mass is the basic parameter, could not accommodate this idea. Bohr, through his familiarity with the work of his Manchester associates, was coming to accept the atomic number as basic and to equate, at least tentatively, the atomic number with an atom's place in the periodic table. Since atoms of the same atomic number would have the same place in the periodic table regardless of mass, there was room for a multiplicity of similar atoms with different masses. Though he did not publish it as a separate discovery, Bohr effectively developed the idea of isotopes independently of F. Soddy.

At the time when Bohr returned to Copenhagen in August for his wedding, he had not given any thought to spectral radiation, had no idea of orbital transitions, and, though he was convinced that quantum considerations played some role in explaining the stability of orbits, did not know what this role was. In spite of such perduring obscurities, he had already brought about a considerable clarification in the concept of the atom, its structure, and its activities. He had made a basic distinction between nuclear phenomena, such as radioactive decay, and atomic phenomena, such as chemical periodicity. In his developing model, the charge on the nucleus matched and determined the number of electrons while the mass of the nucleus determined the weight of the atom. The electrons were assumed to be in coplanar rings, the pancake model, with the outer valence electrons hopefully supplying a basis for explaining chemical properties.

When Bohr took up his new position as teaching assistant at Copenhagen, he tried to use his spare time to rewrite the memorial he had written for Rutherford in June and incorporate his new insights. The rewriting dragged on as it grew in complexity. Then two decisively new factors entered Bohr's conception of the atom, enabling him to revise and complete his study of the constitution of the atom. These new factors were, first, a series of papers by J. Nicholson on atoms and spectral lines and, second, Bohr's introduction to the Balmer series.

Bohr had some familiarity with the earlier work of John W. Nicholson, an astrophysicist, and in his private correspondence dismissed this work as "absolutely crazy."[30] In 1912 Nicholson wrote two papers which seemed to explain some lines found in stellar spectra through a new theory of atomic structure.[31] He postulated a class of primary atoms, simpler than any known on earth, which are responsible for some of the spectral lines found in starlight. A primary atom was postulated to consist of a single ring of electrons rotating about a nucleus. With a ring, the stability condition is that the vector sum of the central accelerations is zero. This condition ruled out one-electron atoms. The most primitive atoms were assumed to be coronium, containing a two-electron ring; hydrogen, with three electrons (not necessarily the same as terrestrial hydrogen); nebulium, with four electrons; and protofluorine, with a ring of five electrons. In each case the nucleus was assumed to have a balancing positive charge.

These models proved surprisingly successful in explaining some observed spectral lines. By balancing the centripetal and centrifugal forces acting on the electrons in neutral protofluorine, Nicholson obtained a formula for an energy divided by a frequency that was very nearly equal to $25\,h$. To interpret this he assumed that angular momentum is quantized in units of $h/2\pi$. Similar computations for singly and doubly ionized ions of protofluorine led to ratios of energy to frequency of $22\,h$ and $18\,h$. These

Nicholson interpreted as members of a harmonic sequence: 25, 22, 18, 13, 7, 0. The spectral lines determined on this basis accounted for fourteen previously observed, but unidentified as to their source, lines found in the solar corona with an error of less than 0.4 percent. Similarly, the model of the nebulium ($n = 4$) atom accounted for ten previously unexplained lines in nebular spectra. Nicholson's prediction of a previously unobserved line was found to be correct to within one part in ten thousand. Though Bohr did not take Nicholson's theory of primary atoms seriously, he concluded, in a Christmas card sent to Harald on 23 December 1912:

> Even though it doesn't belong on a Christmas card, one of us would like to say that he believes that Nicholson's theory is not incompatible with his own. For the latter's calculations should be valid for the final or classical state [klassicke Tilstand af Atomerne] of the atoms, while Nicholson seems to be concerned with the atoms while they radiate, i.e., while the electrons are about to lose their energy, before they have occupied their final position. The emission should then occur intermittently (there is much that seems to indicate that), and Nicholson should consider the atoms while their energy content still is so large that they emit light in the visible spectrum. Later, light is emitted in the ultraviolet, until all the energy that can be emitted is lost.[32]

Bohr's interpretation of Nicholson's results as fitting highly excited states of ordinary atoms rather than ground states of exotic atoms led to the idea that the emission of radiation was related to changes in electronic states. However, he still did not have a way of fixing states or of explaining the mechanism of radiation emission. Sometime in early February 1913, the final piece fitted into the puzzle. H. M. Hanson, a physicist who had just returned to Copenhagen after completing his studies in Göttingen, explained to Bohr how many spectral lines fitted into very simple formulas. Bohr looked through J. J. Stark's book *Prinzipien der Atomdynamik*, found Balmer's formula, and saw how all the fragments could fit together: "As soon as I saw Balmer's formula, the whole thing was immediately clear to me."[33]

This clarity is less than immediate to those attempting to explain the proper historical sequence in the development of Bohr's thought. Heilbron and Kuhn, whose account we are adapting, reconstruct a plausible step by step process of the way Bohr could have adapted Balmer's frequency formula to an energy formula by using the quantum expression $W = h\nu$ and then could have related this to his earlier stability condition, (5.3), now expressed in a form like

$$W = \tau K\nu, \qquad\qquad (5.4)$$

where τ could represent either an integer or some function of integers. To relate this to Nicholson's results, Bohr assumed that his own formula fitted the ground state of atoms while Nicholson's fitted the highly excited states found in hot solar coronas.[34]

Under this interpretation, radiated frequencies would be expressed by a formula relating the excited states, characterized by τ, to the ground state, characterized by Balmer's 2^2:

$$v = 2\pi^2 me^4/h^3(1/2^2 - 1/\tau^2). \qquad (5.5)$$

The physical interpretation to be accorded this formula played a crucial role in the further development of Bohr's argument. The familiar studies of sound production equate the mechanical frequency of an oscillator with the acoustical frequency of the sound it produces. In a similar way, Planck's original treatment of blackbody radiation assumed that the mechanical frequency of the oscillators was in an equilibrium relation with the frequency of the electromagnetic vibrations in the cavity. Under these circumstances there is no need to insist on the distinction between mechanical frequencies and the frequencies proper to electromagnetic vibrations. This is true in any model like Lorentz's where the oscillations of an electron produce electromagnetic vibrations with a corresponding number of oscillations per second.

In Bohr's developing interpretation, electromagnetic frequencies were correlated, not with mechanical frequencies (the number of rotations an electron makes per second), but with *changes* in electronic states. A simple way of relating the two hinged on the assumption that the atoms Nicholson treated existed in hot and very rare stellar atmospheres that would allow electrons in very excited orbits. In the transition from an orbit $\tau + 1$ to an orbit τ, where τ is assumed to be a large number, a simple extension of formula (5.5) yields

$$v = 2\pi me^4/h^3(1/\tau^2 - 1/(\tau + 1)^2). \qquad (5.6)$$

To simplify this, one may factor out the τ^2 in the parentheses and use the approximation, valid for $\tau >> 1$, $1/(1 + 1/\tau)^2 = 1 - 2/\tau$. This yields the familiar expression for the radiation frequency

$$v_r = 4\pi^2 me^4/\tau^3 h^3. \qquad (5.7)$$

This interpretation also suggests the first form of the correspondence principle: in the limit of large quantum numbers the equations of quantum theory (Bohr's) merge with the equations of classical mechanics (Bohr's interpretation of Nicholson's results) in spite of the difference in the way the two equations are interpreted. This is a topic that will be treated more extensively later.

With this relationship between mechanical and radiation frequency established, the fundamental aspects of the new model were now clear to Bohr. Yet, he could not expect this interpretation to be very clear to any of his contemporaries. As Heilbron and Kuhn put it, "Neither the atomic, nor the spectral, nor the quantum theory of the day could justify the necessary interpretation."[35] For this reason it may be helpful to review the growing complexity of Bohr's developing interpretation.

Bohr had reached his conclusions about the structure of the atom through a complex process of reasoning which centered around a quasi dialogue between mathematical formulas and the physical interpretations which could be read into or pulled out of them. In the course of this development he had achieved an unprecedented clarification in the conceptualization of the atom and its associated activities. As noted earlier, he had already distinguished atomic from nuclear phenomena, and explained the ring structure of the atom in such a way that the outside electrons explain both the disposition of atoms to enter into chemical combinations and the properties characterizing a particular atom's place in the periodic table. In making this distinction, he had come to recognize atomic number rather than mass as the basic parameter for atomic physics, a recognition that stimulated Moseley's epochal work relating atomic numbers to characteristic X-rays.[36] The association of atomic number with the charge on the nucleus and the number of electrons orbiting it allowed for a distinction between isotopic forms of the same atom (those with the same atomic number and different masses). Bohr's latest breakthrough led to the first coherent account of the atomic production of electromagnetic radiation. He had accepted the idea, first proposed by Conway in 1907, that the emission of radiation is due to the activity of one electron, rather than of all the electrons acting in concert. This, in turn, required a distinction between mechanical and radiation frequency. Mechanical frequency was related to the energy state of an atom, while radiation frequency was related to a change in energy state due to the transition of an electron from one orbit to another. Bohr had also found, at least in germinal form, a guiding principle for relating classical and quantum formulations of the same problem. However much the interpretations might differ, the classical and quantum formulas must merge in the limits of large quantum numbers.

When Bohr wrote his trilogy "On the Constitution of Atoms and Molecules," he still had some serious problems to contend with.[37] First, to make his conclusions intelligible and hopefully acceptable to his contemporaries he had to transform his own spiraling dialectic into something resembling the type of deductive pattern considered proper for scientific papers. Though Bohr made an effort to do this, he did not succeed in presenting his ideas in a completely consistent way. Second, Bohr was effectively relying on two different accounts of the process by which an atom emits radiation. His original idea was that when an atom absorbs energy, it loses an electron, or becomes ionized. In the subsequent recapture of an electron the atom emits radiation. The alternative account involved transitions from one energy state to another. Finally, there was the grand program of using the new atomic model to explain the periodic table. With this in mind Bohr postulated closed rings with numbers appropriate to the periodic table. His new theory, however, did not supply any theoretical justification for these closed rings, not even the sort of theoretical justification Thomson had developed in terms of stability considerations. For these reasons we will consider only the first paper in Bohr's trilogy and defer questions concerning chemical periodicity until we consider Bohr's later treatment of this topic.

The key ideas in the first paper of Bohr's trilogy have already been considered and are generally familiar. Rather than present one more summary, we will focus as a point of departure for understanding the direction of Bohr's later developments on the consistency problems he encountered in his initial attempt to present a coherent theory of the atom. After a brief discussion of the Rutherford and Thomson models of the atom, Bohr introduces quantum theory in his own characteristic fashion as something demanded by the inadequacy of classical electrodynamics in describing the behavior of systems of atomic size. This, he contends, should supply a basis for solving the stability problem the Rutherford model encounters.

After this introduction Bohr considers a single electron of charge $-e$ and mass m circulating around a heavy nucleus of charge E. If there is no energy radiated, the electron will describe stationary elliptical orbits of major axis $2a$ with a mechanical frequency W. If W is the energy required to remove the electron to infinity, then, on the basis of classical mechanics,

$$\omega = 2^{1/2}W^{3/2}/(\pi eEm^{1/2}); \ 2a = eE/W. \tag{5.8}$$

This does not take account of radiation. Classical electrodynamics would lead to continuous radiation with a decreasing orbit, a conclusion that does not accord with the stability of atomic size. The paper postulates accordingly that equation (5.8) describes a fixed orbit.

To relate this to radiation the paper next considers an electron at rest at a relatively large distance from the nucleus which gets captured and finally settles down to this fixed orbit: "Let us now assume that, during the binding of the electron, a homogeneous radiation is emitted of a frequency v, equal to half the frequency of revolution of the electron in its final orbit; then from Planck's theory, we might expect that the amount of energy emitted by the process considered is equal to $\tau h v$, where h is Planck's constant and τ is an entire number."[38] The apparently arbitrary assumption that the emitted radiation has an electromagnetic frequency equal to half the mechanical frequency of the electron in its stationary orbit is given without any justification. Its actual basis was Bohr's earlier adjustment of this formula (5.4) to the Balmer formula. With this assumption, however, Bohr was able to derive the size of the hydrogen atom, the mechanical frequency of the orbiting electron, the ionization potential, and the formulas for spectral series considered earlier. He could also give a theoretical value for the Rydberg constant that fitted the empirical value within the limits of experimental error.

After deriving these and some further results, the paper returns to the problem of justifying the assumptions used. It suspends as implausible the assumption that different stationary states correspond to the emission of different numbers, rather than different frequencies, of energy quanta—though this assumption was implicit in some of Bohr's earlier calculations. In place of the assumption that the electromagnetic frequency is half the mechanical frequency of the ground state, Bohr now assumes the more general expression for the energy, $W = f(\tau)h\nu$, where τ is an integer and ν the mechanical frequency. A derivation similar to the earlier one leads to the frequency formula

$$\nu = (\pi^2 m e^2 E^2/2h^3)(1/f^2(\tau_2) - 1/f^2(\tau_1)), \tag{5.9}$$

where $f(\tau_2)$ is for the energy state corresponding to the integer τ_2 and $f(\tau_1)$ relates similarly to the integer τ_1. To determine the form of this function Bohr uses the argument outlined earlier, that in the limit of large quantum numbers the classical formula equating mechanical and electromagnetic frequencies must be correct and that the electromagnetic frequency formula must have the same form as the Balmer formula. These correspondences obtain only if $f(\tau) = \tau/2$.

From a logical point of view Bohr had the choice of either assuming $K = \tau h/2$ in his original formulation and then deriving the Balmer-Rydberg equations, or assuming the empirical equations as established and deriving the value of K. In a sort of preview of what he was later to call "complementarity" Bohr chose both sides of the dilemma. The strength of Bohr's case clearly rested, not on the logical rigor of his deductions, but on the ability of his theory to give a unified coherent account of an otherwise disparate experimental data, empirical formulas, and speculative conjectures. The magnitude of Bohr's breakthrough is reflected in the autobiographical notes Einstein wrote some forty years later:

> That this insecure and contradictory foundation [of quantum theory as it existed in 1913] was sufficient to enable a man of Bohr's unique instinct and tact to discover the major laws of spectral lines and of the electron-shells of the atoms together with their significance for chemistry appeared to me like a

miracle—and appears to me like a miracle even today. This is the highest form of musicality in the sphere of thought.[39]

Before considering the later development of Bohr's thought it is of some interest, as a preparation to the later Bohr-Einstein debates, to compare the styles of reasoning manifested by these two men at the most creative periods in their careers. When Einstein wrote his 1905 trilogy on special relativity, the quantum hypothesis, and Brownian motion, he was still working in the Berne patent office. Largely self-taught, with no other physicists to talk to, with few technical books or current journals available even at the library of Berne's university, Einstein worked in the intellectual isolation he found most conducive to his unique and lonely genius.

The problems that most intrigued him were what he once called the "secrets of the Old One." He searched for the beautiful simplicity that, he was convinced, must lie hidden beneath the tangled phenomena of experience. Einstein repeatedly devised ingenious thought experiments which idealized and clarified the hidden core of a paradoxical situation. Only after he had worked out the mathematical implications of some new position would Einstein search for experimental data which might test its validity. The further and often grubbier details of development were left for lesser men.

Bohr never manifested Einstein's genius for intuitive creation. His talent lay in his ability to bring order out of chaos, to adjust models and equations, to adapt theories and interpret data until he had the best overall fit. Where Einstein tried to grasp a hidden essence by disregarding anything he felt to be irrelevant, Bohr insisted that nothing be left out. In a letter to G. von Hevesy, written 7 February 1913, Bohr listed the topics he was then working on and which he felt that an adequate theory of the atom must explain: atomic volume and its variation with valence, the periodicity of the system of elements, the conditions of atomic combination, excitation energies of characteristic X-rays, dispersion, magnetism, and radioactivity.[40] This letter was written just before his conversation with Hansen, a conversation that induced him to add to his list the further topics of spectral series, ionization energies, and the mechanism of absorbing and emitting radiation. In surveying the subsequent development of the Bohr-Sommerfeld theory of the atom, we will focus on Bohr's unceasing attempt to develop a coherent synthesis which accords even the most diverse elements their proper places.

The Bohr-Sommerfeld Atom

The development of Bohr's thought between 1913 and 1927 may be divided into three stages: the elaboration of the original Bohr theory (1913–15), the development of the Bohr-Sommerfeld theory (1916–23), and a critical reinterpretation of the nature of scientific explanation (1923–27).[41] In the first stage Bohr was still trying to make his theory consistent and show that it was superior to competing views. To extend his theory he tried to show how it could account, at least in a qualitative way, for the effect of electrical and magnetic fields on spectral lines.[42] Bohr treated this by attempting to describe the effects these fields would have on the motion of electrons. The electrical field, he assumed, would modify the energy levels, while a magnetic field would induce a Larmor precession which would be superimposed on the electron's circular or elliptical motion. An elliptical motion, he suggested, might explain spectral doublets.

By 1915 Bohr succeeded in organizing his theory around six basic assumptions.[43] The first two assumptions were (A) the existence of stationary states, and (B) the explanation of frequency in terms of the energy differences between stationary states ($A_1 - A_2 = h\nu$). The next two assumptions clearly express Bohr's early position on the use and limitation of classical concepts in a quantum context: "C. That the dynamical equilibrium of the systems in the stationary states is governed by the ordinary laws of mechanics, while these laws do not hold for the transition from one state to another," and "D. That the various possible stationary states of a system consisting of an electron rotating round a positive nucleus are determined by the relation $T = (1/2)nh\omega, \ldots$, where T is the mean value of the kinetic energy of the system, ω the frequency, and n a whole number."

Later, there appear the further assumptions (E) that the electrons are arranged in rings and move in circular orbits, and (F) that the system is stable if its energy is less than that of any neighboring configuration satisfying the same conditions.

Bohr's most distinctive and controversial assumption, that an atom has only discrete energy states, soon received striking experimental confirmation. James Franck and Gustav Hertz showed that electrons colliding with mercury atoms have a resonance reaction at 4.9 volts.[44] Below that energy there is no radiation. Above that energy level the spectral line of 2536 Å is emitted. Though Franck and Hertz originally interpreted this as an effect due to ionization, Bohr quickly pointed out that these experimental results are really explained by the Bohr theory of energy levels rather than through ionization.[45]

After completing the paper just cited, Bohr wrote a paper, "On the Application of Quantum Theory to Periodic System," which was to be published in the April 1916 *Philosophical Magazine*. When in March 1916 he received a copy of Sommerfeld's paper revising the Bohr theory of the atom, Bohr withdrew his own paper and began a revision which lasted some five years and was in fact never completed.[46] Part 1 was completed on 27 April 1918. Part 2 was published on 30 December 1918. Though part 3 was written in 1919, it was not published because of Bohr's intention of revising it. Finally, at Sommerfeld's suggestion, Bohr had it published in 1922 with an appendix indicating recent developments. Part 4 was never completed. This sustained effort brings out one of the two general problems that dominated Bohr's thought during this era, the attempt to develop a mathematical theory of electronic motion which would explain spectral lines and the effect of magnetic and electrical fields on these lines. We will consider this in some detail before treating the second major problem Bohr tackled, an atomic explanation of the periodic properties of the chemical elements.

Arnold Sommerfeld initiated the new directions. He also developed a way of doing atomic physics which partially complemented and partially competed with Bohr's style. At the 1911 Solvay Conference Sommerfeld had stressed the primacy of action, rather than energy, in quantum theory and proposed a general rule that in every purely molecular process the atom takes on or loses a quantity (or quantum) of action determined in a universal manner by the formula $\int_0^\tau H dt = h/2\pi$, where H is the Hamiltonian.[47] Unlike Bohr, however, Sommerfeld emphasized the point that his stress on the quantum of action would allow physicists to use the standard laws of mechanics and electrodynamics. After the Bohr trilogy Sommerfeld showed how his insistence on the primacy of action integrals supplied a basis for revising the Bohr theory.[48] He dropped Bohr's assumption of circular orbits and postulated that for each degree of freedom the motions allowed by the quantum theory are those corresponding to discrete changes in the action in units of h, in accord with the formula

$$\int p_i dq_i = n_i h. \tag{5.10}$$

By expressing equation (5.10) in polar coordinates, Sommerfeld developed a generalization of Bohr's formula that allowed an orbit, characterized by the principal quantum number n, to have integrally different degrees of ellipticity, characterized by a new azimuthal quantum number k (where we are using the symbols that eventually become standard). Later Sommerfeld introduced the magnetic quantum number m and the internal quantum number j. The physical significance to be accorded these numbers was initially somewhat confused.[49] In spite of such difficulties the general method supplied a promise of increasing success, chiefly because it made possible the introduction of such abstract and powerful methods of classical dynamics as Lagrange's equations, the Hamilton-Jacobi equation, and action-angle variables.

Bohr enthusiastically accepted Sommerfeld's revision and the further work done along these lines by P. Debye, P. Epstein, K. Schwarzschild, J. Burgers, and others.[50] Bohr's principal concern in his protracted revision of his unpublished paper on the quantum theory of periodic systems was to develop a coherent overall account. In this revision some of his earlier postulates describing the orbital motion of electrons ([C], [D], and [E] above) no longer played a foundational role. There were two abstract principles, however, which came to play an increasingly prominent role in Bohr's developing systematization, Ehrenfest's adiabatic principle, and Bohr's own correspondence principle. Since these principles later served Bohr as paradigm examples of formal principles, they deserve a more detailed consideration.

The term 'adiabatic' stems from thermodynamics and designates a process in which there is no heat change. The somewhat transformed extension used in atomic physics states that in slow changes of a periodic system the ratio of E, the kinetic energy, to v, the frequency, remains constant. More generally, if a multiply periodic system is changed in a reversible adiabatic way, allowed motions are transformed into allowed motions.[51] Bohr used this principle to consider the changes in circular orbits when they are transformed into elliptical orbits, or even into the more complicated orbits induced by electrical or magnetic fields.[52] In a Lagrangian formulation, the kinetic energy T is related to the generalized phase-space coordinates p_i and q_i by $2T = \Sigma_{i=1}^{n} p_i \dot{q}_1$. For any allowed periodic motion the average kinetic energy T divided by the frequency v should be invariant:

$$2\bar{T}/v = \int_0^P dt \sum_{i=1}^{n} p_i \, d\dot{q}_i = \sum_{i=1}^{n} \iint dp_i \, dq_i. \tag{5.11}$$

The right side of equation (5.11) expresses the trajectory of a point in a $2n$-dimensional phase-space representation of the system. Its value is independent of the choice of coordinates. Each summation, taken by $p_i q_i$ pairs, gives the projection of this trajectory onto a two-dimensional phase-space surface. Thus, for a simple one-electron case equation, (5.11) yields three equations expressing (for spherical coordinates) p_r, p_θ, and p_φ in terms of h multiplied by the three quantum numbers n', k, and m. The real force of this more general approach is manifested in its application to cases that could not be treated by the original Bohr method.

As a simple illustration of some notions needed for further discussion we will consider a two-dimensional harmonic oscillator with restoring forces along the x and y axes, with corresponding frequencies v_x and v_y. If v_x/v_y is a rational number, then the system is simply periodic. Each two-dimensional phase-space projection yields a closed Lissajous figure as the system develops in time. If the ratio of the two frequencies is not a rational fraction, then the two-dimensional phase-space figures never exactly retrace their paths. Such a system is known as a conditionally periodic system. It could be treated by an adaption of methods originally developed for astronomical problems, especially the Hamilton-Jacobi equation and the method of action-angle variables. While others were working out particular problems such as the Stark effect and the Zeeman effect, Bohr was more concerned with developing a coherent general theory which allowed a physical interpretation of the general mathematical formulas used.

It was in this new, more general formulation that Bohr's correspondence principle acquired extended significance. The earliest form of this principle simply involved a comparison of the quantum and classical frequency formulas. The quantum mechanical frequency for a transition from a state $n + \Delta n$ to a state n is

$$v_q = R(1/n^2 - 1/(n + \Delta n)^2), \tag{5.12}$$

where R is Rydberg's constant. The classical mechanical frequency for the electron in the nth orbit is

$$v_m = 2R/n^3. \tag{5.13}$$

The quantum frequency approaches the classical in the limit of $\Delta n = 1$ and $n \gg \Delta n$. This, however, is a purely mathematical comparison, since formulas (5.12) and (5.13) are about different things. Equation (5.13) expresses the mechanical frequency of an electron's vibrations (which we indicated by the subscript m). It relates to the electromagnetic frequencies of the emitted vibration only on the classical assumption that the electromagnetic frequencies are equal to the fundamental mechanical frequency and the overtones given by a Fourier expansion. This assumption is not compatible with the quantum assumption that the emitted radiation is the result of an orbital transition rather than a mechanical frequency.

In a Hamilton-Jacobi formulation, the action $I = nh$, where n is a quantum number proper to one degree of freedom. In a conditionally periodic system the projection of the trajectory of the phase-space point representing a system onto a two-dimensional subspace gives an approximately closed system such that I increases by one unit of h for each period. In symbols, $\Delta I = h\Delta n$. Since $hv = \Delta W$, this may be rewritten as

$$v_q = (\Delta W/\Delta I)\Delta n. \tag{5.14}$$

To express the classical mechanical frequency in a similar form we may substitute in equation (5.13) the value $W = -Rh/n^2$ to get

$$v_m = (1/n)(dW/dn) = dW/dI. \tag{5.15}$$

This mechanical frequency is correlated with a fundamental electromagnetic frequency and a series of overtones which are multiples of the fundamental, i.e., are equal to the fundamental times some number Δn. Accordingly, the general classical formula for frequency of electromagnetic vibrations produced by oscillating electrons may be written

$$v_c = (dW/dI)\Delta n. \tag{5.16}$$

A comparison of equations (5.14) and (5.16) indicates that the quantum frequencies approach the classical frequencies in the limit in which finite differences in units of h are relatively small enough so that difference ratios may be treated as differential ratios.

Classical theory, though incorrect, is complete in a way that the corresponding quantum theory is not. The quantum theory yields frequencies for transitions, only some of which occur. The classical theory not only yields frequencies, but also gives their polarizations and relative intensities. Bohr accordingly sought to extend the correspondence principle so that the classical results could be used as a guide in determining the selection rules, intensities, and polarizations for the quantum frequencies. To achieve this, Bohr expanded equations (5.14) and (5.15). By using angle variables ω_k, one may define a frequency for each action integral. In these variables (5.14) becomes

$$\nu_q = \sum_{k=1}^{s} \omega_k(n'_k - n''_k), \qquad (5.17)$$

where the n_k are quantum numbers characterizing each periodic component. Similarly, the mechanical frequencies corresponding to the displacement of particles in any given direction (in a phase space with coordinates $q_1, q_2, \ldots, q_s$) may be expressed as a Fourier expansion:

$$\xi = \Sigma C_{\tau_1 \ldots \tau_s} \cos 2\pi\{(\tau_1\omega_1 + \ldots + \tau_s\omega_s)t + c_{\tau_1 \ldots \tau_s}\}, \quad (5.18)$$

where the τ_s are integers and the ω_s are mean frequencies of oscillations for the different q's.

On the classical assumption that mechanical frequencies produce corresponding electromagnetic frequencies, equation (5.18) leads to an s-double infinite series of lines of frequencies equal to $\tau_i\omega_i + \ldots + \tau_s\omega_s$. These frequencies would agree with those given by (5.17) only in the limit of large n. In this case, the coefficients $C_{\tau_1 \ldots \tau_s}$ in (5.18) determine the intensity and polarization of the spectral lines with the corresponding frequencies $\tau_1\omega_1 + \ldots + \tau_s\omega_s$. Bohr extrapolated this classical-quantum correspondence from large to small values of n, assuming that there must exist an intimate connection between the probability of a given transition and the values of the corresponding Fourier coefficient in the classical expansion. Bohr's formulation gave no mathematical justification for this at all. Yet, it worked. Thus, Bohr could infer that if a certain $C_{\tau_1 \ldots \tau_s}$ is zero, then the corresponding quantum transition is forbidden. If $C_{\tau_1 \ldots \tau_s}$ is zero only for displacements in certain directions, then the corresponding quantum radiation should be plane polarized in a plane perpendicular to that direction.

The correspondence principle was not so much a formal principle as a method of extending classical concepts and arguments to quantum problems. Meyer-Abich, who has made the most detailed analysis of Bohr's use of this principle, aptly summarized its peculiar status: "The correspondence principle is a not yet fully determined form of the not yet fully determined analogy between the not yet fully determined quantum theory and the determined classical theory."[53] The justification of the type of argumentation was nebulous at best. Yet, it proved more successful than anyone could reasonably expect.[54] Why should it work so well? For Bohr, this gradually became a question of the relation between the formal and descriptive aspects of scientific explanations. The classical account of radiation was not, Bohr remained convinced, descriptively correct. Its explanation of how electronic vibrations produced electromagnetic radiation was unacceptable. Yet, classical theory had to have a certain formal correctness or it could not work as well as it did. From this time on, Bohr was less concerned with the mathematical formulation of theories than he was with the relative roles that descriptive accounts and formal principles play in scientific explanations.

A 1920 address to German physicists illustrates the way in which he related the two at that time:

> Although we must assume that ordinary mechanics can not be used to describe the transitions between the stationary states, nevertheless, it has been found possible to develop a consistent theory on the assumption that the motion in these states can be described by the use of the ordinary mechanics. Moreover, although the process of radiation can not be described on the basis of the ordinary theory of electrodynamics, according to which the nature of the radiation emitted by an atom is directly related to the harmonic components occurring in the motion of the system, there is found, nevertheless, to exist a far-reaching *correspondence* between the various types of possible transitions between the stationary states on the one hand and the various harmonic components of the motion on the other hand. This correspondence is of such a nature, that the present theory of spectra is in a certain sense to be regarded as a rational generalization of the ordinary theory of radiation.[55]

Bohr had come to insist on a sharp distinction between the things that can and those that cannot be described with classical concepts. The behavior of electrons in orbits can be described, a description that presupposes the particle nature of electrons and the direct applicability of mechanical concepts to such particles. Transitions between orbits and the origin of radiation cannot be described. However, the *propagation* of radiation in space and time can be described through Maxwell's electromagnetic theory. This, in fact, was the chief reason why Bohr still rejected Einstein's light-quantum hypothesis and his dualistic explanation of light.

The Challenge of the Periodic Table

In his original trilogy Bohr had attempted to explain the periodic table through the arrangements he postulated for electronic configurations. Partially because of the analogy between atoms and solar systems, but more because there seemed to be no other way of explaining stability, Bohr retained the idea of coplanar electrons. By about 1920 Bohr was becoming skeptical of this pancake model.[56] W. Kossel's work on X-rays[57] and even more the work of Born and Landé[58] led Bohr to think of cubical symmetry as a basis for explaining the structure of complex atoms. This, in turn, led him back to the problem of the periodic table of chemical elements.[59] Here he found a peculiar challenge. Irving Langmuir, building on the earlier work of G. N. Lewis, had developed a highly successful atomic explanation of the periodic table. This explanation relied on a descriptive account of atomic structure totally at variance with the Bohr theory and all of its fundamental postulates. The essential points of Langmuir's theory may be briefly summarized.[60]

The theory built on three basic assumptions. First, electrons are bound in an atom by both electrical and magnetic forces. Second, the valence forces which bind atoms into molecules act in directions which are nearly fixed with respect to each other, an idea which proved particularly useful in explaining carbon compounds. Finally, the stability and chemical inertness of the noble gases must be due to a particularly stable configuration of atoms.

The Lewis-Langmuir static model of atoms assumed that electrons are arranged around atoms in shells, each of which can be subdivided into cells. When all the cells proper to a given shell are filled, there is a noble gas configuration. The innermost shell can only have two electrons, which form a polar axis for further shells. The second shell can have eight electrons placed as if they were at the corners of a slightly distorted cube, a distortion due to the polar axis of the inner shell. In any explanation of the buildup of these shells, symmetry considerations assume a significant role. Thus, carbon is explained as having four electrons above the closed shell of the inner two electrons. If these four are arranged as if they formed the corners of a tetrahedron, then one has both the most stable configuration and also an atomic explanation of the directional aspects of valence forces in organic compounds. Using this basic schema and a set of eleven postulates, Langmuir was able to give a plausible account of all the elements in the periodic table, their systematic properties, the chemical valences of the elements, and the directional and polyvalent aspects of chemical binding. The challenge this chemical success offered to physicists was made quite explicit in Langmuir's opening paragraph:

> The problem of the structure of atoms has been attacked mainly by physicists who have given little consideration to the chemical properties which must ultimately be explained by a theory of atomic structure. The vast store of knowledge of chemical properties and relationships, such as is summarized by the Periodic Table, should serve as a better foundation for a theory of atomic structure than the relatively meager experimental data along purely physical lines.[61]

Though Bohr had been strongly influenced by Rutherford, he never endorsed Rutherford's contention that 'chemist' and 'damned fool' were synonymous. Now this chemical challenge presented a peculiar difficulty. Langmuir was relying on a descriptive account of atomic structure. Bohr had abandoned or minimized some of the descriptive assumptions of his original formulation. Norman Campbell, a philosopher as well as a physicist, suggested an epistemologically sophisticated way of defusing this challenge.[62] The Bohr theory, he argued, might work even though the description of electrons traveling in orbits is incorrect. He invoked Bohr's own correspondence principle to explain the success attributed to Bohr's descriptive postulates. In an answering letter Bohr emphatically rejected this way out:

> Insofar as it must be confessed that we do not possess a complete theory which enables us to describe in detail the mechanism of emission and absorption of radiation by atomic systems, I naturally agree that the principle of correspondence, like all other notions of the quantum theory, is of a somewhat formal character. But, on the other hand [the atomic explanation of spectral lines] appears to me to afford an argument in favour of the reality of the assumptions of the spectral theory of a kind scarcely compatible with Dr. Campbell's suggestion.[63]

Though, in characteristic fashion, Bohr never really explained what he meant by 'realistic' and 'formal', it is possible to reconstruct his position from the use he makes of these terms here and elsewhere. A realistic concept is presumed, without any special justification, to be one whose meaning stems from the object, quality, action, or whatever referred to when the concept is properly used. The meaning of a formal concept depends on its relation to other concepts and the conceptual framework (e.g., classical mechanics) in which it functions. Similarly, a realistic principle is one that is true (or false) because it correctly (or incorrectly) describes some process in nature. A formal principle has significance primarily because of the role it plays in a theory and relates to reality, not by any direct or immediate correspondence, but by the way in which the theory as a whole functions. What Bohr was then insisting on was that his theory's description of electronic motions must be considered realistic and that it supplied a better basis for explaining chemical properties and periodicities than Langmuir's account.

Bohr's explanation of the periodic table is the crowning achievement of the old atomic theory.[64] It is also one of the clearest manifestations of the type of multifaceted nonlinear reasoning in which Bohr excelled. He was not attempting to deduce the periodic table from his atomic theory. He was, rather, trying to develop a consistent atomic account that fitted the data coming from radically diverse sources: atomic theory, spectral lines, ionization energies, X-ray spectra, the correspondence principle, the periodic properties of the elements, chemical valences, and information on valent and covalent chemical binding. In blending these together, Bohr functioned more like a symphony director than a deductive logician.

Bohr's fundamental assumption was that each electron may be specified by three quantum numbers: the principal quantum number n specifying the orbit; the azimuthal quantum number k specifying the degree of ellipticity; and the inner quantum number j, which Bohr took to specify the orientation of the orbit of the outer electron relative to the axis of the core (nucleus plus inner electrons). Of these only k seemed to afford a secure basis for a descriptive account. Using these numbers and the information on binding energies available from X-rays, Bohr attempted to account for the buildup of atoms by the successive addition of electrons to the core of the preceding atom. The two guiding assumptions in the application of this *Aufbauprinzip* were that each configuration is in the lowest possible energy state and that the energy involved in binding the outermost electron (the *Leuchtelektron*) is manifested in the atom's optical spectra.

The first shell (a term Bohr adopted though he thought it descriptively inappropriate) can have two electrons. Hydrogen has one electron in a 1_1 orbit, where the designation n_k signifies the principal and azimuthal quantum numbers. The correspondence principle ruled out the possibility of adding another electron to the same 1_1 orbit. Bohr suggested two possibilities for helium. One, parahelium, has two electrons in equivalent 1_1 orbits in planes 120° apart. The other, orthohelium, has one electron in a 1_1 orbit and another in a 2_1 orbit. Though he and Kramers had never succeeded in calculating correct ionization energies for these configurations, Bohr was willing to assume that the parahelium configuration forms the inner core of higher atoms.

The second period, beginning with lithium, was assumed to be due to the addition of electrons above this core. Because of lithium's hydrogenic spectrum, the third electron was assumed to be in a 2_1 orbit that penetrates the core for a fraction of its period. More 2_1 electrons are added on one by one to explain beryllium, boron, and carbon. Bohr's treatment of carbon is a clear answer to Langmuir's contention that dynamic theories of the atom cannot explain the directional aspects of valence bonds. The most stable configuration for the orbits of the four outer carbon electrons would be in planes so inclined to each other that normals to these planes had directions relative to each other nearly the same as the lines from the nucleus to the vertices of a regular tetrahedron. This not only takes over Langmuir's idea of tetrahedral symmetry as a basis for explaining the directional aspects of carbon binding, but also includes his idea that the symmetry is imperfect due to the distorting influence of the two inner electrons. In other contexts this slight asymmetry in balancing core and outer angular momenta supplied a basis for the magnetic core–outer electron interaction used to explain the anomalous Zeeman effect.

The second part of this period was treated by assuming that the stable carbon configuration excludes any further electrons from the same orbits while the correspondence principle prohibits further 2_1 electrons. Accordingly, Bohr assumed that the seventh, eighth, ninth, and tenth electrons must be in 2_2 circular orbits whose diameters are considerably larger than those of the inner electrons. The four outer electrons of neon are assumed to rotate in planes whose angular momenta exhibit a distorted tetrahedral symmetry. This configuration, in turn, forms the core for the next period.

Bohr followed this step-by-step building process to explain all seven atomic shells and the corresponding chemical elements. The most striking achievement of this account was its ability to explain the chemical similarities of the rare earths (the same n_k outer electrons with differing inner configurations) and the eventual prediction that element 72 should not be a rare earth, as Dauvillier and Urbain had predicted, but should have chemical properties similar to titanium and zirconium. Two of Bohr's assistants, Hevesy and Coster, verified this in time for Bohr to use it in his Nobel Prize address on 11 December 1922.[65]

Reinterpreting Scientific Explanations

In treating electronic activity within the atom, Bohr had come to rely much more on formal principles and generalized mechanics than on a descriptive account of orbital motion. Yet, he still insisted that atomic theory could not be considered purely formal. His successful explanation of the periodic table seemed to vindicate the role he had assigned descriptive accounts. Bohr was well aware that this explanation still had shortcomings. Many of the details were tentative. The overall scheme employed had a rather loose coherence. Yet, such shortcomings notwithstanding, Bohr's mélange of theory and surmise, of established principles, interpreted facts, and conjectured symmetries, yielded the most complete and consistent explanation of the properties of atoms that had ever been given. Its success seemed an adequate justification of Bohr's faith in the possibility of using physical concepts and principles to give an objective description of the structure and behavior of unobservable atomic entities.

Within the next few years all the descriptive props in Bohr's atomic theory were undermined. The details of some of these developments will be considered in the next chapter when we examine Heisenberg's early work. Here we will concentrate on Bohr's reinterpretation of the foundations of atomic physics. As tentative new assumptions and ad hoc adjustments of old assumptions seemed to increase rather than resolve the incoherence of atomic theory, Bohr was more and more driven to reflect on the nature of scientific explanation. His questioning of the foundations of scientific explanation came to focus on three aspects which are usually not the concern of practicing scientists. These are the interrelation of the descriptive and formal aspects of scientific explanations, the nature of the concepts used to give descriptive accounts, and the role to be accorded statistical explanations in atomic physics. The present section will be concerned primarily with the first of these aspects. The last two will be treated in more detail when we consider the Copenhagen interpretation of quantum mechanics.

Since at least the time of Hertz and Boltzmann, physicists had been accustomed to speak of scientific theories as giving pictures of reality. The terminology had become familiar enough so that it could be used without analyzing what it meant for a theory to provide a picture. Though Bohr had used such terminology, he was gradually becoming aware of some difficulties it entailed when applied to atomic physics. Thus, in a 1920 address to the Royal Danish Academy of Science and Letters he claimed, "It is hardly possible to propose any picture which accounts, at the same time, for the interference phenomena and the photoelectric effect, without introducing profound changes in the viewpoints on the basis of which we have hitherto attempted to describe the natural phenomena."[66]

By 1922 Bohr had assigned a primary organizing role to three formal principles: the correspondence principle, the adiabatic principle, and the postulate of the invariance and permanence of quantum numbers. Descriptive accounts were seen as subordinate to such principles. However, Bohr did not yet make a very clear distinction between two rather different ways in which the need for such subordination could be interpreted. One is that the descriptive accounts of atomic processes need supplementation because the pictures they provide are incorrect or incomplete or fuzzy. Thus, in his Nobel Prize address, given in 1922, Bohr said, "We are therefore obliged to be modest in our demands and content ourselves with concepts which are formal in the sense that they do not provide a visual picture of the sort one is accustomed to require of the explanations with which natural philosophy deals."[67]

In remarks such as these Bohr is reflecting the idea that the pictures theories supply are some sort of iconic representations of hidden entities or processes. However, it is also possible to think of the pictures theories supply as conceptual representations to be analyzed in terms of the language in which they function. In the 1922 Göttingen lectures, which profoundly influenced Heisenberg and Pauli, Bohr discussed some paradoxical features of quantum concepts interpreted as pictures:

> The quantum theory, which may be regarded as a rational generalization of Planck's original theory, represents a sharp break with the conceptions of classical electrodynamics. This places us in a very peculiar situation. So far, only concepts developed in the classical theories, such as those of the electron and electric and magnetic forces, are available to us for describing the natural phenomena; however, we assume at the same time that the picture of the classical theories is invalid. Now, the question arises if there is any possibility at all of uniting the classical concepts with the quantum theory without contradiction. We are not in a position to settle this question; however, physicists hope that the ideas of both theories possess a certain reality.[68]

In late 1922 Bohr made a final attempt to fashion a coherent synthesis of atomic theory through another projected series of articles, a project that never saw completion. His introduction to the only article written in this series stresses the *prima facie* inconsistency involved in relying on classical concepts for descriptions while denying the validity of the classical conception of physical reality.[69] Bohr was clearly becoming aware that the use of classical concepts in quantum contexts could only be adequately explained against a general background of the nature of concepts and the functioning of conceptual frameworks. However, he seems to have done no systematic work on this aspect of the general problematic until 1927. This will be discussed in chapter 8.

In 1922 Bohr was still relying on the contrast between descriptive accounts and formal principles. In earlier accounts Bohr had insisted that the descriptive elements must be accorded some validity independent of their functioning in the overall theory. In his new synthesis the fundamental principles are exclusively formal, in Bohr's sense of that term. Thus, stationary states are now presented, not by describing the orbits, but by giving the classical equations which these states must satisfy:

$$dp_k/dt = -\partial E/\partial q_k, \ dq_k/dt = \partial E/\partial p_k \ (k = 1, \ldots, s), \qquad (5.19)$$

where s is the number of degrees of freedom. Though these equations are only approximate because they neglect such things as radiation reaction, nevertheless they specify the exactness with which such states can be fixed. Unlike stationary states, transitions between states do not admit of even an approximate description through classical concepts. For the stationary states themselves, descriptive accounts are subordinated to formal principles in the sense that these states are now specified, not by space-time descriptions of allowed orbits, but by the existence and permanence of quantum numbers.

The application of formal classical principles not only fixes the limits of allowable description, but also excludes any descriptions which involve the use of incompatible concepts. The most notable victim of this exclusion is Einstein's idea of light quanta: "We can even maintain that the picture, which lies at the foundation of the hypothesis of light-quanta, excludes in principle the possibility of a rational description of the conception of a frequency v, which plays a principal part in this theory."[70] Though he rejected Einstein's view, Bohr was not yet able to give an adequate alternative account of the production, absorption, or dispersion of radiation. Hence, he retreated into an even stronger emphasis on the formal nature of quantum explanations: "Our whole knowledge of the nature of radiation, which to a great extent plays a decisive role in the problems of atomic structure, of course rests solely on those phenomena, in the closer consideration of which the formal nature of the quantum theory stands out particularly clear."[71]

Bohr's growing skepticism on the realistic significance of quantum descriptions was occasioned by shortcomings or failures in the descriptive accounts he had relied on. Further developments which seemed to warrant a complete abandonment of this view may be presented in summary form. The Bohr paper we have been summarizing was received on 15 November 1922. Six months later, in May 1923, the *Physical Review* published A. H. Compton's paper on the scattering of X-rays from electrons.[72] The explanation of Compton scattering seemed to require the acceptance of light quanta with energy hv and momentum hv/c. Bohr had rejected the light-quantum hypothesis on the grounds that it involved a radical inconsistency. The energy of a light quantum is defined in terms of frequency, a wave concept; yet the light quanta are treated as particles. Now, forced to accept the Compton effect, Bohr had to devise some means of handling the conceptual inconsistencies resulting from the simultaneous use of wave and particle descriptions.

Another problem that perplexed Bohr in the 1921 paper previously summarized was dispersion, or the scattering from atoms of light of relatively long wavelength compared to the size of the atoms. If the Bohr picture of the orbital motion of electrons is essentially correct, then there should be a resonance reaction whenever the frequency of the incident light corresponds to the rotational frequency of the outside electron. Such a resonance was never observed, and no attempt to explain dispersion on the basis of the Bohr theory proved successful. What did prove surprisingly successful was an adaption of the Lorentz model picturing disturbed electrons as returning to an equilibrium position through oscillations which produce or modify electromagnetic vibrations. This model and Heisenberg's adaption of it will be given a detailed treatment in the next chapter. The point to be stressed here is that this model, interpreted as a realistic picture of the atom, contradicts Bohr's basic assumption of stationary states and quantum transitions. Yet, after R. Ladenburg calculated dispersion in this way, it was Bohr himself who isolated and effectively launched the virtual oscillator model implicit in Ladenburg's calculations.[73] If both the Bohr and the virtual oscillator models of the atom are interpreted formally, Bohr insisted, then they need not be considered contradictory. This was one of the clearest anticipations of Bohr's later doctrine of complementarity.

Descriptive accounts of electronic orbits had played an essential role in Bohr's account of the periodic table, particularly in symmetry and stability considerations and in explaining the directional properties of covalent binding. In 1924 E. C. Stoner revised this theory by substituting formal principles for descriptive accounts.[74] Landé had interpreted the inner quantum number, introduced by Sommerfeld, as characterizing the atom's total angular momentum. Stoner extended this by assigning a j quantum number to each electron and interpreting this as characterizing the angular momentum of individual electrons. From the explanation given for the anomalous Zeeman effect Stoner concluded that the number of possible electron states for the atom as a whole is always $2j$. All states are equally probable, though they only manifest themselves as separate states in the presence of an external magnetic field. Stoner extended this by suggesting that for each inner quantum level, or shell, the number of possible states is equal to twice the j number characterizing that level.

Stoner's argument led to the same number of electrons per atom as Bohr's and also to the same order in which electron shells are filled. However, the two accounts gave different results for the number of electrons in subshells. Thus, Bohr argued from symmetrical distributions of orbits to the assignment 4-4 for the second shell, 6-6-6 for the third, and 8-8-8-8 for the fourth. Stoner simply relied on the formal principle $N = 2j$ to get the uniform assignments 2-6-10-14.

When Coster wrote asking Bohr's opinion of Stoner's results, Bohr replied (on 10 December 1924) that he had from the first understood the formal beauty and simplicity of Stoner's classification, but could not accept it as final because it was not connected to a quantum theoretical analysis of electron orbits.[75] Bohr came to accept the Stoner account (developed independently by J. D. Main Smith) only after Pauli had developed his exclusion principle, which extended and rationalized Stoner's method of assigning quantum numbers.

The descriptive account of orbital motion no longer played a functional role in atomic explanations. Yet, a residue remained, the idea that the proper assignment of quantum numbers characterized the mechanical properties of orbital motions. Though Bohr had admitted difficulties in developing a consistent interpretation of the n and j numbers, he had felt secure in his interpretation of the k number:

> We shall not enter here on these problems but shall confine ourselves to the problems of the fixation of the two quantum numbers n and k, which to a first approximation describe the orbit of the outer electron in the stationary states, and whose determination is a matter of prime importance in the following discussion of the formation of the atom. In the determination of these numbers we at once encounter difficulties of a profound nature, which—as we shall see—are intimately connected with the question of the remarkable stability of atomic structure. I shall here only remark that the values of the quantum number n, given in the figure, undoubtedly can not be retained, neither for the S nor the P series. On the other hand, so far as the values employed for the quantum number k are concerned, it may be stated with certainty, that the interpretation of the properties of the orbits, which they indicate, is correct.[76]

This "certain" interpretation of the azimuthal quantum number as characterizing the ellipticity of electronic orbits led to apparent contradictions in the explanation of the anomalous Zeeman effect. The vector model developed by Landé and Heisenberg gave a correct account of spectral multiplicity only if one replaced the k number by the inexplicable value $(k(k-1))^{1/2}$. After Pauli proved that the accepted explanation of spectral multiplicities in terms of the magnetic interaction between the core and the optical electron was incorrect,[77] Heisenberg redeveloped the theory of the anomalous Zeeman effect and substituted a new and now standard quantum number $l (= k - 1)$ for the k number. In using this he was quite explicit in asserting that the l in the revised Landé formula for the g factor does not characterize mechanical properties of electron paths.[78]

Bohr was well aware of all these developments. Kramers, Landé, Pauli, and Heisenberg had done much of the research cited while working in Bohr's Institute. He was also well aware of the other difficulties atomic theory was encountering: the doublet riddle, the fact that Heisenberg's treatment of the anomalous Zeeman effect required both half integral quantum numbers and the assignment of quantum numbers incompatible with Bohr's *Aufbauprinzip*, and the repeated failure of the theory to give correct ionization values for the helium atom and the hydrogen molecule.[79] Bohr shared and to some degree shaped the consensus emerging around 1923 that atomic physicists should no longer continue attempting to put new patches on old bottles. Something new and radically different was needed. While others followed Max Born's lead in looking for a new quantum mechanics, Bohr sought a different basis for understanding the nature and function of atomic theory. This was presented in a paper which was in effect a preliminary draft of the Copenhagen interpretation. To interpret the paper correctly we must first consider two pieces of background material.

The Bohr-Kramers-Slater Paper

Some ideas introduced by John Slater which will be considered shortly seemed to supply a basis for a reinterpretation of ideas developed by Einstein and induced Bohr to reread in a critical way a paper Einstein had published in 1916.[80] A brief summary of its contents will indicate why this paper suddenly took on a new significance for Bohr. As indicated earlier, Einstein had never accepted as valid Planck's derivation of the radiation formula, though he accepted the formula itself. After completing the general theory of relativity in 1916, Einstein returned to quantum theory and developed a new derivation of Planck's formula, one that utilized the general features of the Bohr theory.

Einstein considered a gas of molecules in equilibrium with radiation and imposed the basic condition that the internal energy distribution of the molecules demanded by quantum theory should follow purely from the emission and absorption of radiation. From Bohr's theory Einstein took only the general assumption that molecules can exist only in discrete states $Z_1, Z_2, \ldots, Z_n$ with internal energies $\epsilon_1, \epsilon_2, \ldots, \epsilon_n$. For a gas at temperature T, the relative frequency W_n of such states as Z_n is given by the formula

$$W_n = p_n \exp(-\epsilon_n/kT), \qquad (5.20)$$

where p_n represents a statistical weight for a state. Einstein assumed that a molecule can pass from a state Z_m (with energy E_m) to a lower-energy state Z_n (with energy E_n) either through the spontaneous emission of radiation or through an emission (with corresponding absorptions) induced by a radiation density ρ of frequency ν. On the basis of these assumptions Einstein defined three different probabilities for emission and absorption of radiation:

spontaneous emission $\quad dW = A^n m\, dt,$

induced absorption $\qquad dW = B^m n\rho\, dt,$ $\qquad\qquad$ (5.21)

induced emission $\qquad dW = B^n m\rho\, dt.$

He then showed that for an equilibrium condition the assumption that spontaneous emission is balanced by induced absorption and emission led to Planck's radiation formula.

One distinctive feature of this paper was the support it lent to Einstein's light-quantum hypothesis. Maxwell's electromagnetic theory assumes that radiation is emitted spherically. This, Einstein argued, is incompatible with an equilibrium distribution of gas molecules. If, however, one assumes that a light quantum of energy $h\nu$ has a momentum of $h\nu/c$ and that the molecular recoil has an equal but opposite momentum, then a Maxwell-Boltzmann equilibrium distribution of molecular velocities can be explained.

In the next chapter we will consider the use made of these Einstein coefficients in explaining dispersion. Our present concern is with Bohr's adaption of Einstein's development. In spite of Einstein's arguments Bohr had steadfastly and repeatedly rejected the light-quantum hypothesis. In 1924, however, there were two new factors which induced him to reexamine Einstein's approach to radiation. The first was the Compton effect. Compton, at heart a conservative classical physicist, had resolutely resisted the light-quantum hypothesis until his own data seemed to leave him no alternative.[81] His explanation of the effect he discovered made an essential use of the still disputed idea that a light quantum of energy $h\nu$ has a momentum $h\nu/c$. The second factor was Slater's idea of a radiation field, an idea which will be considered shortly.

When Bohr reexamined Einstein's article, two of its distinctive features must have made a strong impression on him. First, though Einstein used the basic postulates of the Bohr theory, he did it in a way that made no use of the descriptive elements Bohr now found suspect. Second, Einstein, the acknowledged master of statistical reasoning, gave probability considerations a new, and apparently basic, role in atomic physics. Einstein obviously found himself uncomfortable with some of the implications of his probabilistic approach. His article concludes, "The weakness of the theory lies on the one hand in the fact that it does not get us any closer to making the connection with wave theory; on the other, that it leaves the duration and direction of the elementary processes to 'chance'."[82]

J. C. Slater, who went to Europe in 1924 as a postdoctoral fellow, introduced the idea of a virtual radiation field through which atoms communicate with each other.[83] In the position that Slater originally developed, the virtual radiation field is due to oscillators which are associated with atoms and which have the frequency of possible transitions. The virtual field proper to a given frequency produced by the atom itself accounted for spontaneous emission while the virtual field due to other atoms explained induced emission and absorption. The absorption of a frequency v_{12} in one atom was interpreted as a consequence of a prior transition from energy ϵ_2 to ϵ_1 in some other atom, where $\epsilon_2 - \epsilon_1 = hv_{12}$.

When Slater came to Copenhagen, Bohr and Kramers convinced him against, he later claimed,[84] his better judgment to drop the idea of conservation of energy and momentum in individual interactions. In this way it was possible to redevelop Einstein's earlier approach to radiation without assuming the existence of light quanta. The virtual radiation field postulated by Slater replaced the Planck radiation field used by Einstein. There was one other significant difference. The chance factor, which Einstein had reluctantly introduced and hoped to eliminate, was now assigned a basic role.

This paper, which utilized Slater's idea of a virtual radiation field but which was actually written by Bohr and Kramers, anticipated most of the basic contributions Bohr made to the Copenhagen interpretation. These interpretative features were set, however, in a pre–quantum mechanics framework, one which hinged on Bohr's formal-descriptive distinction. Thus, the paper begins by contrasting the wave and particle accounts of the interaction between radiation and matter and concludes:

> At the present state of science, it does not seem possible to avoid the formal character of the quantum theory which is shown by the fact that the interpretation of atomic phenomena does not involve a description of the mechanism of the discontinuous process, which in the quantum theory of spectra are designated as transitions between stationary states of the atom.[85]

The features of the Copenhagen interpretation anticipated in this paper were the following. First, the paper was essentially concerned with a conceptual clarification of the type of information quantum theory gives rather than with the development of a technical formalism. The paper contains only one equation, $hv = E_1 - E_2$.

Second, the authors accept the impossibility of giving an account of radiative processes that incorporates both a space-time description and a causal explanation. After discussing the role of the correspondence principle in estimating the possibilities of a transition, they conclude:

> In fact, together with other well-known paradoxes of the quantum theory, the latter difficulty has strengthened the doubt, expressed from various sides, whether the detailed interpretation of the interaction between matter and radiation can be given at all in terms of a causal description in space and time of the kind hitherto used for the interpretation of natural phenomena.[86]

Third, the interaction of the radiation field with matter is handled by using a virtual oscillator model of the atom in interaction with a virtual radiation field. This is the model of the atom, though not of the radiation field, that Heisenberg adapted to develop quantum mechanics. In the Bohr-Kramer-Slater (B-K-S) paper it led to a fourth significant feature, a probabilistic interpretation of the interaction between matter and radiation.

Finally, there is a groping after the idea that wave and particle pictures are complementary. Bohr's correspondence indicates his initial enthusiasm for this implication of Slater's idea of a virtual radiation field.[87] In the B-K-S paper the Compton effect is treated by considering both wave and particle aspects of radiation. The wave concept enters in the treatment of the scattering of radiation as a continuous process, the particle concept in considering momentum transfer from radiation to electrons as discontinuous rather than continuous. In a similar vein the authors postulated that the specification of the stationary states necessitates an indeterminism in the time an atom spends in a stationary state.

The essential features of the Copenhagen interpretation were present, but they were not yet brought together in a coherent fashion. The basic interpretative concept employed was the formal nature of quantum explanations. Here, and in a paper on the polarization of resonant fluorescent radiation,[88] Bohr justified the use of two apparently incompatible models of the atom, the Bohr model and the virtual oscillator model, on the grounds that both should be interpreted formally rather than as descriptive accounts. This, as Bohr saw it, necessitated the conclusion that the concepts of the theory could not be used in statements that pictured individual events. The correspondence was between the theory as a whole and the phenomena it explained. The best that could be done with individual statements was to interpret them as probabilistic accounts of individual events. This, in turn, entailed the conclusion that the principles of energy and momentum conservation, incorporated in the theory through its adaption of classical physics, had only a probabilistic significance with respect to individual interactions.

As Mara Beller has shown,[89] this was not the first attempt to reconcile the light-quantum hypothesis and the wave theory of light by abandoning the strict applicability of the law of energy conservation. Charles Darwin had developed such an idea and discussed it with Bohr. Einstein seems to have toyed with it, and Sommerfeld suggested it. These, however, had been suggestions or tentative hypotheses. In the B-K-S paper this suggestion had become an essential part of a radically new way of looking at the interaction of matter and radiation.

The flurry of excitement the B-K-S paper produced in the European physics community, in America, and even in the popular press has been well documented by M. J. Klein.[90] Most physicists, especially those engaged in experimental testing, saw the situation as a simple clash between two competing hypotheses: Einstein's light-quantum hypothesis and Bohr's abandonment of strict conservation laws for individual interactions. When the experimental results finally came in, they seemed to give unambiguous support to Einstein's position. The Bothe-Geiger experiment detected coincidences between Compton-recoil electrons and scattered X-rays, a result that seemed to require the admission of light quanta and also the validity of conservation laws for individual events.[91] Compton and Simon reached a very similar conclusion through their cloud chamber tests of the relationship between scattering angle and atomic recoil.[92]

Bohr rejected this simple dichotomy. He saw both the desperate expedient of the B-K-S paper and its experimental refutation as manifestations of the fundamental contradictions generated by attempting to provide a space-time description of the interaction of matter and radiation while preserving the fundamental laws of electrodynamics. Thus, in a letter to Heisenberg dated 1 April 1925—before the experimental refutations were available—Bohr complained that accepting as descriptively true Slater's idea of a coupling between distant atoms through a radiation field involved accepting a mystical view of natural forces: "The costs of these assumptions are in fact so high that they are not measurable in the customary space-time description."[93] His reaction to the experimental refutations was given in a postcript added in July 1925 to a paper on atomic collisions. The situation, as he interpreted it, was not one of a clear-cut choice between competing hypotheses. It was, rather, one of determining the limits within which one could apply to atomic processes the kind of space-time picture customarily used in the description of natural phenomena: "One must be prepared to find that the generalization of classical electrodynamic theory that we are striving after will require a sweeping revolution in the concepts on which the description of nature has been based up to now."[94] Similarly, in a talk which he gave in August 1925 before he was familiar with the breakthrough Heisenberg had made in quantum mechanics, Bohr said:

> From these results [the Compton effect and the Bothe-Geiger experiment] it seems to follow that, in the general problem of the quantum theory, one is faced not with a modification of the mechanical and electrodynamical theories describable in terms of the usual physical concepts, but with an essential failure of the pictures in space and time on which the description of natural phenomena has hitherto been based.[95]

Such reflections signal, I believe, the beginning of a third phase in the development of Bohr's interpretative framework. In the first phase, the goal of scientific explanation was thought to be a realistic description of physical structures and atomic events as they occur objectively. This early realism, which Bohr never developed in a systematic way, died the death of a thousand qualifications. By 1923 it was effectively replaced by an acceptance of something like a hypothetical-deductive account of scientific explanation. The theory as a whole was related to some domain of phenomena. The basic laws of the theory were interpreted formally, in terms of their function within the theory rather than in terms of some direct correspondence between basic laws and the class of phenomena they might be thought to describe. The peculiar complication here came from the correspondence principle interrelating theories which supplied incompatible descriptions of the same physical reality.

One consequence of this formalistic interpretation was that such basic laws as the laws of conservation of energy and momentum were not interpreted as supplying a true account of individual interactions. After it was conclusively established that energy and momentum were conserved in individual interactions, Bohr had a choice. He could either abandon the account of scientific explanation that he had painfully and rather reluctantly developed or reinterpret what it means for a theory to give a space-time description of the phenomena of nature. Bohr chose the second alternative and began serious reflection on a question that had long intrigued him: How do terms get meaning? For Bohr, however, this was not a general question or an abstract issue. His concern was with a clarification of the problematic terms used in atomic physics, particularly with the paradoxes generated by the extension of classical terms to quantum contexts. The biggest stumbling block was the lack of any account of atomic phenomena that could be considered descriptively adequate. Such accounts were just then being developed. Before returning to Bohr's developing views on the interpretation of scientific theories, we will consider the new breakthroughs initiated by Heisenberg, Pauli, de Broglie, and Schrödinger.

6. The Main Line: Matrix Mechanics

The Two Tracks and Their Backgrounds

Quantum mechanics and wave mechanics developed along parallel but separate tracks. The main line for the development of atomic physics linked Copenhagen, Munich, and later Göttingen in a shared endeavor of adjusting and revising the Bohr-Sommerfeld model to fit the data on spectral lines and the modifications they undergo in electrical and magnetic fields. There were, to be sure, significant differences in style and emphasis between the more mathematically oriented German centers and the rather philosophical emphasis stemming from Bohr's concern with the problems involved in developing a complete and consistent physical interpretation. These were essentially different styles of developing a shared program. In this program theoreticians and experimentalists worked closely together. Both saw a precise accord between mathematical results and experimental data as, if not a goal, at least a criterion of success. The *Zeitschrift für Physik*, founded in 1920, became the semiofficial journal for these main-line physicists.

In the late nineteenth century, thanks to the organizing efforts of Helmholtz and then Kirchoff, the University of Berlin emerged as the world's foremost center for research in physics, a position it maintained in the twentieth century. In the mid-twenties theoretical physics was represented by Planck, Einstein, and von Laue, while Nernst and Harber directed special institutes. All five were Nobel laureates. Yet, except for Einstein's theory of ideal gases, Berlin contributed almost nothing to the mid-twenties redevelopment of quantum theory. This was partially because Planck and Nernst were well past their productive peaks, while Einstein was more concerned with relativity and field theory. Yet, it also reflected a difference in emphasis and attitude that should be considered.[1]

Planck and Einstein had each passed through youthful periods when they were strongly influenced by the philosophy of Ernst Mach. From 1908 on, Planck led the opposition to positivism.[2] In Planck's view the goal that physics should ever strive for is a unified picture of nature based on absolute laws. Though the often abused term 'naive realist' would come closer to fitting Planck than any of the other figures we are considering, even Planck did not hold that the pictures given by scientific theories depict the world as it exists objectively. His argument, rather, was that a unified picture of reality is necessary for the coherence of physical science and the various theories it involves.

Einstein's relation to Mach's doctrine is much more complex and controverted, particularly concerning the influence of Mach's principle on the development of general relativity. Einstein's positive views will be treated in more detail in chapter 9. Here we simply wish to consider Einstein's contribution to a climate of opinion. The Einstein of the twenties, beginning his long and ultimately frustrating search for a unified field theory, was, like Planck, more concerned with cosmic absolutes than with the systematization of sense experiences. He too had come to reject positivism. In this intellectual milieu the thrust of the main-line atomists for precise correlations between mathematical conclusions and experimental results probably seemed too close to positivism. To foster and further it might give aid and comfort to the enemy. These internal disputes concerning positivist versus objectivist interpretations of physics were exacerbated by external attacks on the whole idea of scientific rationality.[3] Planck and Einstein came to see themselves as besieged defenders of causality and of the primacy of reasoned investigation.

The Einstein who peers at us from the pictures taken during his Berlin period is invariably dressed in a neatly pressed suit, complete with vest and necktie, and with hair carefully combed. He came closer to the image of the traditional academician then than at any earlier or later stage in his career. Yet, for those who, whether by choice or chance, developed new theories in isolation for the main line of atomic developments, Einstein still stood as a living proof that intellectual rebellion and isolated effort could succeed. Both de Broglie and Schrödinger were clearly inspired by Einstein's example and strengthened by his support.

Heisenberg's Early Development

The development of quantum mechanics, soon transformed into matrix mechanics, and the closely related developments of the Pauli exclusion principle and of electron spin are more dependent on the solution of technical problems in mathematical physics than any developments hitherto treated. Since this has been considered in some detail elsewhere, the technical aspects will merely be skimmed here.[4] The emphasis will be on the changes in the way Heisenberg practiced atomic physics and their motivation, changes that conditioned Heisenberg's decisive breakthrough to a new quantum theory and influenced the way in which he interpreted its significance. Pauli will be considered chiefly in his role as a dialectical foil to Heisenberg.

Prior to his paper initiating quantum mechanics, Heisenberg wrote a series of some sixteen technical papers, which can be classified under four general headings. The first is hydrodynamics, the topic of his dissertation.[5] The other three were on problems that were then current in atomic physics: the anomalous Zeeman effect, models of molecules, and the scattering of light from atoms.

The normal Zeeman effect, the splitting of spectral lines into three components in the presence of a magnetic field, had been explained by Lorentz in terms of a model of the atom which plays a somewhat surprising role in Heisenberg's development. In Lorentz's model each electron is considered to be quasi-elastically bound in the atom so that any displacement of it from equilibrium induces a force, proportional to the displacement, which returns it to its rest position through damped oscillations. Sommerfeld and Debye had independently worked out quantum explanations of the normal Zeeman effect which reproduced Lorentz's results.

The anomalous Zeeman effect, the complex splitting that multiplet spectral lines undergo in the presence of a magnetic field, proved to be a much more formidable problem. By 1921 it was known that atoms with a single electron outside a closed shell (alkalis) have doublet spectral lines, while atoms with two electrons above a closed shell (alkalines) have either singlet or triplet lines.[6] Higher multiplets were discovered in 1922.[7] In contemporary physics these splittings are explained through spin-orbit coupling. In the early twenties before spin was known, one had to use the Bohr model as the only available point of departure. Sommerfeld tackled this problem by various adaptions of Bohr orbits. When these failed, he fell back on what he called "number mystery." He introduced a new 'inner quantum number', j, and used it to characterize different sublevels proper to a given energy level. Though he thought that this number might be related to some hidden inner rotation, this assumption did not play a role in his formalism.

When an atom with multiplet lines is placed in a weak magnetic field, each multiplet splits into a number of components whose frequencies are rational fractions of the normal Zeeman effect. This, the anomalous Zeeman effect, is related to the normal Zeeman effect through the Paschen-Back effect. Paschen and Back showed that, when the magnetic field is increased gradually, the anomalous components merge continuously into normal Zeeman components.

For the normal Zeeman effect, the energy levels were given by the Sommerfeld-Debye formula

$$E = E_0 + mh\nu_L. \tag{6.1}$$

Here E_0 is the unperturbed energy; h is Planck's constant; m, the magnetic quantum number, was interpreted as the quantized projection of the azimuthal quantum number k onto the axis set by the magnetic field H. The term ν_L is the frequency of a Larmor precession $\nu_L = eH/4\pi mc$. In 1921 Alfred Landé presented a formula which had the same form as (6.1), but which fitted the anomalous Zeeman effect:[8]

$$E = E_i + gmh\nu_L. \tag{6.2}$$

Here E_i is the unperturbed energy level proper to one line in a multiplet. The term g is an unexplained proportionality constant adjusted to fit the types of splitting observed. Landé assumed that Sommerfeld's inner quantum number j characterizes the atom's total angular momentum. Since the magnetic quantum number m corresponds to projections of the angular momentum along the axis of the magnetic field, m should, he assumed, have the $2j + 1$ values: $m = 0, \pm 1, \pm 2, \ldots, \pm j$. While this assumption fitted singlets and triplets, it did not fit doublets. To accommodate these Landé assumed that m takes on the values: $m = + \frac{1}{2}, + \frac{3}{2}, \ldots, + (j - \frac{1}{2})$. This, like the values attributed to the g factor, was a purely ad hoc assignment with no theoretical justification.

Sommerfeld was unable to derive the Landé formula from the principles of the Bohr-Sommerfeld atomic theory. He did, however, develop a quasi derivation by relying on a different source. Woldemar Voigt had, in 1913, presented a classical mechanical model to explain the splitting of sodium lines. Three stationary electrons arranged in a triangle were represented as bound to each other and to a fixed center by forces such that any displacement produced simple harmonic vibrations. This was certainly incorrect as an atomic model. Yet, Sommerfeld adapted it by using the precedent of the Bohr-Sommerfeld atomic theory and interpreting each frequency as a difference between two energy levels.

Heisenberg, a first-year student in Sommerfeld's seminar, was assigned the project of trying to fit the Paschen-Back results into this general approach. Heisenberg found that by using half-integral quantum numbers, which Sommerfeld was unwilling to accept, he could adapt Sommerfeld's revision of Voigt's formula so that it accommodated all the empirical data. Heisenberg's new formula was[9]

$$E = \bar{E} + h\nu_L \left(m^* \pm \frac{1}{2}[(1 + (2m^*/k^*)\gamma + \gamma^2]^{\frac{1}{2}}\right). \tag{6.3}$$

In formula (6.3) an asterisk on a quantum number reduces it by $\frac{1}{2}$. Thus, $k^* = k - \frac{1}{2}$. $\bar{E}$ is the average value of the unperturbed doublet energy; and $\gamma = \Delta\nu/\nu_L$ where $\Delta\nu$ is the unperturbed doublet separation. Formula (6.3) fitted the normal Zeeman effect in the limit in which $\gamma = 0$. As $\gamma \to \infty$, formula (6.3) reproduced the Landé formula for the energy values proper to doublets. Since there is a continuous transition from one limit to the other as the value of γ varies, the formula also accommodated the Paschen-Back effect.

Heisenberg attempted to justify this formula through a new and, as it seemed to his contemporaries, quite arbitrary model of the atom, the core model. Since doublets were associated with alkali atoms while triplets were associated with alkaline atoms, Heisenberg began his construction with a noble gas configuration. Thus, the neon configuration is the core for sodium. The electron above this core was presumed to share a half unit of angular momentum with the core. On this sharing assumption, the electron has an angular momentum (in units of $h/2\pi$) of $k - \frac{1}{2}$, while the core has an angular momentum of $\frac{1}{2}$. If the axis of the core is at some angle θ to the electron's angular momentum, then the core precesses around this inner field. If a very strong external field is added, this inner field becomes negligible and one has the normal Zeeman effect. As this external field is weakened, there is a gradual transition to the anomalous Zeeman effect due to the combined internal and external fields. For the next atom, magnesium, there are two-valence electrons above the neon configuration core. Since this is a vector addition, different orientations could, presumably, explain the difference between singlet and triplet states.

The numerical results of this formula manipulation were surprisingly good. Heisenberg's was the only approach that seemed to fit the normal and anomalous Zeeman effects and also the Paschen-Back effect. Yet, the way in which this was done was arbitrary to the point of perverseness. As Cassidy (whose interpretation I am summarizing) put it, Heisenberg's model violated every principle in sight. The continuous transition from weak to strong fields did not fit space quantization. The building up of atoms through cores and electrons sharing angular momentum did not accord with the *Aufbauprinzip* Bohr had used in explaining the periodic table. The model also allowed violations of selection rules. Nor, if the model was used to describe the emission of radiation, did it accord with energy-momentum conservation. Initially, Heisenberg, still a very young student, was not fully aware of the massive incoherence the success of his approach introduced. When some of these difficulties were brought up, Heisenberg simply replied, "Der Erfolg heiligt die Mittel."[10] Success sanctifies the means.

In the face of recalcitrant problems the Bohr-Sommerfeld theory was being salvaged by an increasing number of ad hoc hypotheses, many of which were mutually inconsistent. There was a growing realization, which crystallized around mid–1923, that something new and different was needed. But how does one go from the old to the to-be-discovered new? The three principal figures in the development under consideration followed three rather different scenarios.

Bohr, as we have seen, devoted increasing attention to the development of a conceptual framework, or a basis for speaking about atomic phenomena, that would be epistemologically self-consistent and adequate both to the various types of empirical data and to the requirements imposed by his correspondence-principle approach. An adequate and self-consistent conceptualization should, he felt, be prior to and serve as a foundation for a new mathematical formalism. Heisenberg, the brilliant and very young iconoclast, was at the other extreme, effectively pursuing a policy of "anything goes." He was quite willing to introduce models which were both inherently implausible and also inconsistent with established physics, provided they supplied a basis for solving particular problems.

Pauli was impressed by Heisenberg's success, but somewhat repelled by his methods. In a letter to Bohr he commented on Heisenberg's methods: "Since he is so unphilosophical, he does not have a regard for a clear working out of the fundamental assumptions and of their relationships with the preceding theories."[11] Pauli advocated abandoning any reliance on models, since the pictures they supply are essentially misleading. His emphasis was on the search for new laws, both kinematic and dynamic, of general validity. This, as he realized, put him in something of a methodological bind. A superb calculator, characterized by a critical deductive approach to problems, Pauli did not want to use any methods or assumptions that could not be rationally justified. Yet, he realized that it would be necessary to make some sort of leap beyond established laws. Since Heisenberg's iconoclasm blazed the path to the new quantum mechanics, this chapter will pivot around Heisenberg's development.

Both in competition and in collaboration with Landé, Heisenberg did further work on the anomalous Zeeman effect. This phase of the problem came to completion in 1923 with the vector model and the Landé g-factor which handled even the newly discovered higher multiplet splitting.[12] For a multielectron atom in a magnetic field H, the classical expression for the interaction energy was multiplied by a factor g:

$$E_m = m \frac{h}{2\pi} \frac{eH}{2m_0 c} g \tag{6.4}$$

$$g = 1 + \frac{(J^2 - \frac{1}{4}) + R^2 - K^2}{2(J^2 - \frac{1}{4})}, \tag{6.5}$$

where m_0 is the mass of the electron, m is the magnetic quantum number, and the other terms in (6.4) have their familiar significance. Landé had defined his quantum numbers as follows: J was the total angular momentum, R the core angular momentum, and K the azimuthal angular momentum so that the half-integral quantum numbers were included. In contemporary notation (denoted by the subscript c), $J = J_c + \frac{1}{2}$, $R = S_c$ (total spin) $+ \frac{1}{2}$, and $K = L_c$ (total orbital quantum number) $+ \frac{1}{2}$.

Though this semiempirical formula fitted the data then available, two difficulties of a more basic sort remained. First, there was no theoretical explanation of the *g*-factor and of the novel features it incorporated, half-integral quantum numbers and the ad hoc assumption that the core's contribution to the magnetic interaction was double the expected contribution. Second, the assignment of quantum numbers to the cores was not consistent with the numbers Bohr had assigned in using the *Aufbauprinzip* to explain the periodic table. Heisenberg, echoing evaluations by Bohr, Born, and Pauli, concluded that a profound modification of quantum theory was required.[13] He found a temporary stopgap solution through a formula which allowed the assignment of two different quantum numbers to the core.[14] This fitted the *g*-formula without contradicting the *Aufbauprinzip*. It had, however, no theoretical justification.

After completing his doctoral studies at Munich, Heisenberg followed Pauli's precedent and became Max Born's assistant at Göttingen. Born and Heisenberg wrote a series of papers concerned with models of molecules, especially helium. Here again, the use of the core model as a basis for approximate calculations had a limited success. Born, however, was more conscious of the problems the calculation raised than of the piecemeal solutions achieved. In their first joint paper Born and Heisenberg, following Bohr's precedent, explicitly assumed that the electrons traveled in definite orbits whose phase relations are, in principle at least, determinable.[15] Yet, they found no way in which this could lead to a unique solution. Born's feelings are indicated in a letter he wrote to Einstein on 7 April 1923:

> We have been looking at perturbation theory (Poincaré's) to determine whether it is possible to obtain the observed term values from Bohr's models by exact calculations. But it is *quite* certainly *not* the case, as was demonstrated with helium, where we found any number of multiple periodic orbits (to a sufficient approximation). I had Heisenberg there during the winter (as Sommerfeld was in America); he is easily as gifted as Pauli, but has a more pleasing personality. He also plays the piano quite well. Apart from the work on helium, we examined together some questions of principle in connection with Bohr's aomic theory—particularly with regard to the phase relations in atomic models. . . . Then I am going to put this subject into cold storage until the question of homeopolar binding forces between atoms had been solved from Bohr's point of view. Unfortunately every attempt to clarify the concept fails. I am fairly sure that in reality it must all be very different from what we think now.[16]

In concluding their first article, they brought up a difficulty of a different sort. A thought experiment based on the model they employed seemed to lead to consequences incompatible with Ehrenfest's adiabatic principle, one of the pillars of the old quantum theory. Born concluded that a new quantum mechanics was needed.[17]

In the fall of 1924 Heisenberg went to Copenhagen as a fellow in Bohr's Institute. There, as he repeatedly testified, he was strongly influenced by Bohr's stress on developing a clear consistent interpretation of the physical significance of the concepts and method used.[18] He was also caught up with the then current Copenhagen concern with radiation. Bohr and his followers had consistently rejected Einstein's light-quantum hypothesis until the discovery of the Compton effect precipitated a rethinking of the interaction of radiation and matter.[19] The immediate problem Heisenberg tackled, in collaboration with H. A. Kramers, Bohr's assistant, was dispersion. This term was then used to refer to the scattering from atoms of light of a wavelength which is long relative to the size of the atoms. If the optical electron really orbits the atom the way the core model pictures it, then there should be a resonance reaction when the frequency of the incident light corresponds to the mechanical frequency of the electron's rotation. Such a resonance reaction, if observed, would constitute strong evidence for the reality of electron orbits. It was never observed.

Somewhat earlier, R. Ladenburg had a rather surprising success in developing a formula for dispersion based on an adaption of the Lorentz model.[20] He treated the electrons in a collection of atoms as oscillators, a certain fraction of which are disturbed by the incident radiation and then return to equilibrium positions through damped oscillations. He more or less quantized this essentially classical treatment by using, for the fraction of oscillators involved, the coefficients Einstein had introduced for transitions of atomic states.

A reliance on damped oscillations in place of quantum jumps would contradict Bohr's fundamental quantum postulates, if interpreted realistically. But, Bohr argued,[21] there is no inconsistency if the quantum theory is interpreted formally. Then the correspondence principle allows one to treat the total reaction of a number of atoms as if they were the oscillators proper to the Lorentz theory of radiation. Thus, in spite of an apparent contradiction with Bohr's postulates, the virtual oscillator model had won papal approval—in his letters to Pauli, Heisenberg sometimes referred to Bohr in papal terms. For Bohr, however, the virtual oscillator model was not treated as a model of an individual atom, but simply as a means of extending classical methods via the correspondence principle.

A. Smekal had given a competing account, relying on the concept of light quanta.[22] If a molecule was velocity v and energy state E_κ collides with a light quantum of energy $h\nu$ and passes to a state of velocity v' and energy E_ℓ, then the law of energy conservation yields

$$mv^2/2 + E_\kappa + h\nu = mv'^2 + E_\ell + h\nu'. \tag{6.6}$$

Since the velocity change is relatively negligible,

$$\nu' = \nu + (E_\ell - E_\kappa)/h. \tag{6.7}$$

Kramers extended Ladenberg's formula to include transitions to and from excited states as well as the ground state. His dispersion formula is

$$P = E\sum_i A_i^a \tau_i^a e^2 / [4\pi^2 m(\nu_i^{a2} - \nu^2)] \tag{6.8}$$
$$- E\sum_j A_j^e \tau_j^e e^2 / [4\pi^2 m(\nu_e^2 - \nu^2)].$$

Here E is the intensity of the incident electromagnetic plane wave, ν its frequency, P the induced polarization, $A_i^a(A_j^e)$ the Einstein coefficients for the probability of a particular absorption (emission) per unit time, τ the decay time characterizing the same transition, and m the mass of the electron.[23] Though this seemed to accommodate the data on dispersion, it had no theoretical explanation and did not include the new frequencies predicted by Smekal on the basis of the light-quantum hypothesis. To achieve these ends Kramers secured the assistance of the newly arrived *Wunderkind*. The joint paper Heisenberg and Kramers wrote handled dispersion through a semiclassical treatment of radiation.[24] They began with a generalized classical expansion for the electrical moment of an atom exposed to a plane monochromatic train of light waves of frequency ν, split the expansion into parts corresponding to coherent and incoherent scattering, and then went from a classical to a quantum formulation by the trick Born had introduced a replacing differential equations by difference equations. The first part of their general expansion then yielded Kramers's dispersion formula (6.8), while the second part effectively reproduced Smekal's results without introducing light quanta.

The Kramers formula, equation (6.8), effectively treated atomic electrons as a doubly infinite set of virtual oscillators. Should this be interpreted purely formally via the correspondence principle, or did this model have physical significance for individual electrons? Heisenberg faced this question in a paper he wrote on the polarization of resonant florescent radiation, a topic which had briefly emerged from technological obscurity. Fluorescence, a luminescence stimulated by radiation, differs from phosphorescence in that it does not continue more than a minute fraction of a second after the stimulating radiation is extinguished. In 1923 Wood and Ellet discovered that even a weak magnetic field induces a strong polarization in the resonance D line of sodium vapor, a line that exhibits an anomalous Zeeman effect.[25] The relative polarization of the different components in this splitting were soon given a qualitative explanation in terms of the

Lorentz model. Heisenberg was inevitably intrigued by a problem that interrelated both the virtual oscillator model, which had been used in dispersion theory, and the anomalous Zeeman effect, the problem he had treated by using the core model.

Heisenberg treated this problem in a paper that differed radically from his other papers in that it contained almost no mathematics. It exemplified the type of physicalistic reasoning he claimed to have learned from Bohr. Through a series of thought experiments and a consideration of the pertinent data he tried to determine the polarization to be expected for different relative orientations of source, detector, electrical, and magnetic fields. Though the core model served as his point of departure, he found in some cases that only the virtual oscillator model led to reasonable predictions. He concluded that "the standard virtual oscillators employed in quantum theory for radiation obey laws according to which the closest analogy between the classical theory and the quantum theory is preserved."[26]

From this time on, Heisenberg took the virtual oscillator model seriously as a basis for calculating individual atomic processes rather than as a mere extension of classical physics justified by the correspondence principle. This was a decisive step beyond the methods of atomic calculation developed in the Bohr tradition and toward the new quantum mechanics.[27] This does not mean that Heisenberg simply dropped the core model. It had proved too helpful to be completely wrong. However, it was no longer possible to consider its ascription of orbital motion to be a description of what really happens in the atom. This had not worked in explaining the helium atom, in predicting a resonance in dispersion, and in confronting aspects of the problem of the polarization of resonant florescent radiation. Such negative conclusions were reinforced by Bohr's insistence that the quantum theory should be considered formal rather than realistic and by the influence of Pauli, who merits special consideration.

Pauli, Master of Criticism

Wolfgang Pauli and Werner Heisenberg had remarkably parallel early careers. Pauli was a year ahead of Heisenberg at Munich, and, as in Heisenberg's case, his outstanding potential was immediately spotted and cultivated by Sommerfeld. Before specializing in atomic physics, he wrote, at Sommerfeld's suggestion, a *Handbuch* survey of the special and general theories of relativity which was immediately recognized as one of the classics in the field. He was then nineteen years old. Pauli became Born's assistant in Göttingen a year before Heisenberg and also preceded Heisenberg in becoming a research fellow at Bohr's Institute. Yet, their temperaments, life-styles, and ways of doing physics were quite different. Heisenberg was a brilliant innovator with a remarkably fast mind, who seems to have had an almost effortless success in sports, in music, and in learning languages as well as in physics and mathematics.[28] He was quite willing to introduce novel, even wild, assumptions and then judge them in terms of the consequences that could be calculated from them, rather than in terms of their inherent reasonableness. Pauli considered the young Heisenberg to be very unphilosophical.

Pauli's father, also a scientist, had been a friend and professional associate of Ernst Mach.[29] Their joint influence supplemented Pauli's own native skepticism, making him, in Heisenberg's term, the master of criticism. According to Heisenberg, Pauli had considered the concept of electron orbits to be horribly mystical even in his days as a student at Munich.[30] The rationale behind this evaluation was given in Pauli's first published paper, where he rejected the idea of an electrical field at a point as meaningless. He insisted, "One should accordingly hold fast to the idea that in physics only quantities which are observable in principle should be introduced."[31]

During the period we have been considering Pauli was Heisenberg's closest correspondent and severest critic. Pauli was quite unhappy with the core model for a number of reasons in addition to his general suspicion of models, especially mechanical ones. The core model seemed arbitrary in its assignment of half-integral quantum numbers. It was also inconsistent with established principles of physics. Bohr's *Aufbauprinzip* was based on the invariance of quantum numbers. As one adds electrons in reconstructing the periodic table, the latest electron becomes the optical electron outside the core, the electronic configuration of the preceding atom. This might involve a redistribution of orbits. Yet, Bohr believed, it did not involve any reassignment of quantum numbers. The system of magnetic coupling that Heisenberg postulated and that he and Landé developed involved changing, by half a unit, the j quantum number of an electronic configuration when it becomes the core of the next atom in the periodic table. Such inconsistencies did not disturb Heisenberg: *der Erfolg heiligt die Mittel* (success sanctifies the means). They did disturb Pauli. Yet, Pauli was unable to deduce the Landé g factor from general principles, even when he suspended suspect mechanical principles. The best he could do was to develop a phenomenological formula to reproduce the Landé-Heisenberg results without explicitly relying on either the vector model or Heisenberg's half-integral quantum numbers.[32]

This confusion was soon compounded by the problem that Forman has dubbed the "Doublet Riddle."[33] Physicists had generally accepted the idea of two different types of spectral doublets. Though the spectral patterns are similar, the underlying physical causes were thought to be different. The first kind of doublets, observed in X-ray spectra, were thought to be due to the relativistic change of mass an electron experiences as a function of its changing velocity. Though X-ray spectra still presented some unexplained features, there was one strong supporting argument for this interpretation. The relativistic calculation indicated that this doublet splitting should depend on the fourth power of the atomic charge. Experimental data seemed to support this Z^4 dependence.

The second type of doublet was that proper to optical electrons in multielectron atoms. This was interpreted as an effect of the interaction between the optical electron and the core. For this case the key parameter was the azimuthal quantum number k, which characterized the orbit's ellipticity, rather than Z^4. The physical accounts were different; the spectral structures were remarkably similar. Landé attempted to unify the two by extending the core-model account to explain X-ray doublets as well as optical doublets. The attempt failed, because it could not reproduce the Z^4 dependence of the X-ray doublets.

As a purely formal exercise, Landé then attempted to extend the X-ray–type calculations to optical doublets. To his dismay, this seemed to work quite well. This very success altered the nature of the crisis in atomic physics. There had been a growing consensus that the Bohr-Sommerfeld theory was not adequate for multielectron atoms, not even for helium, the simplest multielectron atom. This, however, did not necessarily indicate that the basic principles of the Bohr-Sommerfeld theory were wrong. One could retain the principles and pin the blame for failure on a lack of understanding of the coupling proper to multielectron atoms. Then the magnetic interaction postulated by core-model theorists could be understood as a first step in attempting to understand this coupling. Similarly, the peculiar tension (*Zwang*) that Bohr had postulated might be another step.

The Bohr-Sommerfeld theory did not explain multielectron atoms because the electronic coupling was not properly understood. But it did explain hydrogen and hydrogen-like atoms. This, its previously undoubted success, now seemed suspect. If the splitting in optical doublets could be explained by relativistic effects rather than through orbital eccentricity, characterized by the azimuthal quantum number, then the Bohr-Sommerfeld method of specifying orbits through the assignment of quantum numbers might require reinterpretation. As Forman has shown, this now obscure doublet riddle was the key factor precipitating a feeling of crisis in the atomic community.

Pauli's appraisal of the current situation is vividly revealed in a letter he wrote to Bohr in 21 February 1924:

> The atomic physicists in Germany are now divided in two classes. The first calculate a definite problem first with half-integral values of the quantum numbers and, if this does not accord with experience, then they calculate it with integral

quantum numbers. The second group calculate first with integral numbers and, if it does not work, they calculate with half-integral numbers. However, both groups of atomic physicists share a common property: no a priori arguments at all can be drawn from their theories concerning which quantum numbers and which atoms one calculates with half-integral values and which with integral values. . . . I myself have developed no taste at all for this sort of theoretical physics and withdraw from it back to my heat conduction in solid bodies.[34]

In this letter Pauli explicitly exempted Heisenberg from membership in either group on the grounds that he is cleverer, especially in doubting the reality of electron orbits. Yet, this particular exemption hardly sufficed to exclude Heisenberg from criticism. In a letter he sent Bohr some six months after Pauli sent his letter, Heisenberg complained, "Pauli grumbles about everything, especially about my atomic physics."[35]

Early in December 1924, Pauli abruptly ended his short-lived retirement from active work in atomic physics. While writing a survey article on quantum theory, Pauli realized that relativity theory, still his field of predilection, showed that the core model must be incorrect. The reasoning and the conclusion Pauli draw from it were sketched in a letter to Bohr and then developed in more detail in an article.[36]

If the standard description of orbital motions is correct, then the innermost electrons of heavy atoms should be moving at an appreciable fraction of the velocity of light. According to the core model, these contribute to the core's magnetic moment and this, in turn, is responsible for the splitting of electron lines manifested in the anomalous Zeeman effect. It follows that the magnitude of the splitting should manifest a systematic variation with atomic mass. It does not. Pauli accordingly assumed, on the basis of symmetry considerations, that a closed core must have zero angular momentum and hence also a magnetic moment of zero. If this is true, then the magnetic effects attributed to the core must, at least in the case of alkali atoms, be due to the optical electron itself. Pauli made the novel assumption that the electron has a classically nondescribable "doublevaluedness." To accommodate this doublevaluedness Pauli associated two new quantum numbers with the electron, k_1 and k_2, which have projections m_1 and m_2 on a magnetic field.

This shift in emphasis from the core to the optical electron might seem to eliminate the need for the peculiar tension that Bohr had suggested to explain coupling in multielectron atoms. Pauli, however, was not yet clear on that. In the letter to Bohr previously cited, Pauli expressed his endorsement of Stoner's formal reinterpretation of Bohr's account of the number of electrons in each atomic shell. Pauli had considered Bohr's assignment to be rather arbitrary, and he did not think that the correspondence principle sufficed to explain why shells close at the numbers they do. In his later account of the discovery of the exclusion principle Pauli claimed that one sentence in Stoner's article particularly impressed him: "For a given value of the principal quantum number the number of energy levels of a single electron in the alkali metal spectra in the external magnetic field is the same as the number of electrons in the closed shell of the rare gases which correspond to this principal quantum number."[37]

The way in which this sentence might have suggested the Pauli exclusion principle is hardly obvious. It helps to see how this principle emerged from a complex of more particular problems. Pauli wanted an account of multielectron atoms that would accommodate in a unified way both the building up of the periodic table and the multiplicity of spectral lines. This stress on consistency is something of a leitmotiv of the paper. Doublet splittings, whether optical or X-ray, are handled in a uniform way. Splittings in alkalis and alkalines are treated by the same rules. One assignment of quantum numbers to a core fits both the explanation of the periodic table and the treatment of spectral multiplicities without compromising the integrity of the *Aufbauprinzip*.

The means of achieving this overall consistency is through the assignment of quantum numbers and the interpretation accorded them. In a strong magnetic field, which serves to remove degeneracy, each electron is specified by four quantum numbers, n, k_1, m_1, and m_2. Here n is the principal quantum number, k_1 the old azimuthal quantum number, and m_1 the component of k_1 along a magnetic field with the assumption that the total angular momentum is equal to the maximum value of m_1. The second magnetic quantum number m_2 is the projection of the new auxiliary quantum number k_2 along the magnetic field. Instead of attempting to give a physical interpretation to this quantum number specifying a classically indescribable doublevaluedness, Pauli simply specifies that in the case of a doublet k_2 has the values $k_1 - 1$ and k_1. This gives m_2 the values $m_1 + \frac{1}{2}$ and $m_1 - \frac{1}{2}$. By the adiabatic principle this assignment of quantum numbers should retain its significance as the strong magnetic field is gradually reduced to a weak magnetic field and then shut off.

The problem this assignment of quantum numbers engenders is the way in which it is to be related to the correspondence principle. One cannot simply dispense with the correspondence principle, since it still supplies the only basis for selection rules and polarizations. However, it is now difficult to accord this principle its customary descriptive significance. In Pauli's new account the core of both alkaki and alkaline atoms is the same noble gas configuration, which has a net angular momentum and magnetic moment of zero. The two differ only in the number of optical electrons and in the states assigned these electrons. The optical electron bears the burden of explaining multiplicity. If the correspondence principle is interpreted as extending classical modes of explanation, then the plane of the optical electron must be thought of as precessing around the axis proper to the central force field. The imposition of a weak magnetic field adds a further precession around an axis through the nucleus in the direction of the field. This way of interpreting the situation, however, reintroduces the difficulties associated with the doublet riddle and with the type of core-optical electron interaction that Pauli has excluded.

Pauli's way around this impasse was to sidestep considerations of the correspondence principle and rely on rules governing the proper assignment of quantum numbers. His new rule is: There can never be two or more equivalent electrons in an atom for which in strong fields the values of all quantum numbers n, k_1, k_2, m_1 (or, equivalently, n, k_1, m_1, m_2) are the same. If an electron is present in the atom for which these quantum numbers (in an external field) have definite values, this state is 'occupied'.

Pauli was unable to give a deductive justification of this principle. However, he could show that it supplied a consistent means, in accord with the *Aufbauprinzip*, of explaining the closing of shells, Stoner's assignment of the number of electrons in each subshell, and the spectral multiplicities proper to alkali and alkaline atoms. Though Pauli's results differed, in a few cases, from the Landé-Heisenberg branching rule, the available experimental data could not adjudicate the divergence.

Pauli had repeatedly criticized Heisenberg for using formulas which had little or no theoretical justification. Suddenly, the situation seemed reversed. Pauli was attributing a basic significance to a rule which had no deductive justification and which seemed to have a purely formal significance. It did not describe electronic activity. It simply specified how quantum numbers could be assigned. Heisenberg, who saw Bohr's copy of Pauli's paper, replied with a hastily written postcard:

> Today I have read your new work and it is certain that I am the very man who enjoys it the most, not only because you drove the swindle to a previously unsuspected swindling height, and through this have easily broken all earlier records concerning which you revile me (by introducing individual electrons with four degrees of freedom), but especially I am elated that even you (et tu Brute) with bowed head have returned to the land of the formalistic Philistines. But, do not be sad, you will be received with open arms.[38]

The crisis in atomic physics seemed to be coming to a climax. Bohr invited Pauli to Copenhagen. There, in March 1925, Heisenberg, Pauli, and Bohr argued through new developments and conflicting interpretations. The substance of their discussions is not recorded. However, the general tenor may be gleamed from some published writing. When Pauli sent Bohr his exclusion principle paper, he included a letter indicating how his new and still unjustified principle might relate to the work that preceded it:

> The interpretation, which serves as my point of departure, is undoubtedly nonsense. . . However, I believe that what I am doing here is *no greater* nonsense than the former interpretation of complex structure. My nonsense is joined to the now customary nonsense. It is just for this reason that I believe that this nonsense must be established as necessary at the present state of the problem. The physicist who should ever succeed in adding together both of these nonsenses, he it is who will reach the truth.[39]

Pauli could not interpret his principle as the needed breakthrough. It supplied a method for specifying states, but could not give polarizations, intensities, or selection rules proper to spectral lines. These still had to be handled through the correspondence principle, the old nonsense. Pauli clearly saw the need for fusing these two approaches, but did not know how to accomplish it.

Heisenberg, problem-solver extraordinaire, seems initially to have accepted Pauli's new quantum number and the exclusion principle as two more tools to add to his bag of tricks. Before leaving Copenhagen, he wrote one more paper on the anomalous Zeeman effect.[40] Here he accepted the doublevaluedness Pauli had introduced, but noted that it might be possible to explain it either through an extra quantum number, as Pauli urged, or by attributing a doublevaluedness to the core-optical electron interaction. Unlike Pauli, Heisenberg was still straddling the fence on the doublet riddle.

Regardless of the difficulties they might entail, Heisenberg had no intention of abandoning the models that had proved so helpful in solving problems. After considering different types of anomalous Zeeman patterns, Heisenberg turned to the more general problem of how the motion of electrons is to be explained. Sommerfeld's azimuthal quantum number k had been interpreted as specifying the eccentricity of an electron's orbit. Such a mechanical interpretation led to the relativistic difficulties Pauli had noted and also to the unsatisfactory ideas of penetrating orbits. Heisenberg replaced the azimuthal quantum number k by the new orbital quantum number $\ell = k - 1$, and insisted that this should not be interpreted as characterizing any mechanical property, such as orbital ellipticity.

How could one handle the kinematics and dynamics of electrons if mechanical specification of orbital motion is disallowed? The virtual oscillator model, Heisenberg argued, provides an alternative route. Since it is based on classical physics, it provides a clear mechanical model. But, since this classical model relates to the atom only in a formal way via the correspondence principle, it cannot be interpreted as giving a descriptive account of the kinematics of the electron. In concluding his paper, Heisenberg sketched the program suggested by his new approach. Adapting the work he had done with Kramers, he assumed that for radiation of frequency v the allowed atomic frequencies must have the general form $v + \omega_{n,j}$ or $v + \omega_{n,i} \pm 2\omega_k$, where ω_k is the fundamental vibration and the set $\omega_{n,j}$ represents its overtones. These frequencies, as the virtual oscillator model requires, are simply the Fourier components given by a spectral expansion.

The basic idea seemed simple in principle. One should calculate the frequencies proper to the optical electron and then, to include its interaction with the core, calculate all possible combinations of frequencies. Yet, the execution of this program was, Heisenberg admitted, more than he was then able to handle. Some simplification was needed.

Fabricating Quantum Mechanics

Heisenberg's paper initiating quantum mechanics reversed the procedures and suppressed the methods he actually used. Here I will try to summarize his actual development while skimming the technical details, which are given elsewhere.[41] When Heisenberg found even the hydrogen atom too difficult to solve by his proposed program, he turned to the anharmonic oscillator as a problem that was mathematically tractable, though physically unrealistic.

The easiest problem might seem to be the simple harmonic oscillator specified by the formula $\ddot{x} + \omega^2 x = 0$. This, however, only allows radiation frequencies equal to the mechanical frequency of oscillation. This assumption, applied to atoms, had proved a failure in the development of dispersion theory. The simplest way to overcome this limitation is to add a correction term, dependent on x^2, yielding the formula for an anharmonic oscillator:

$$\ddot{x} + \omega^2 x + \lambda x^2 = 0, \tag{6.9}$$

which has the Fourier expansion,

$$x = a_0 + a_1 \cos \omega t + a_2 \cos 2\omega t + \ldots \tag{6.10}$$

Though this anharmonic oscillator obeys a different force law than the hydrogen atom, the two expansions involve terms similar enough that one might suggest the methods for handling the other. In the hydrogen atom an electron at a distance a from the nucleus is bound to it by a force $F = -e^2/a^2$. For the anharmonic oscillator pictured in figure 6.1, the diagram Heisen-

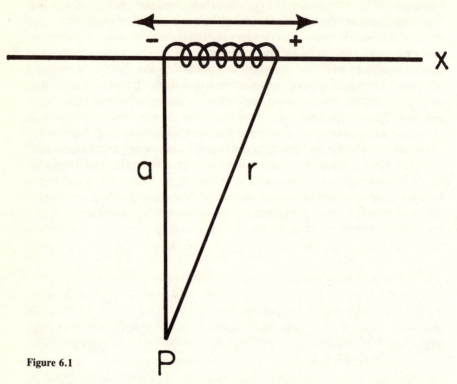

Figure 6.1

berg used in a letter of 8 June 1925,[42] the dipole charge can vibrate back and forth along the x direction. The force exerted on it by a positive charge at P would be

$$K = (e^2/a^2)[-1 + 1/(1 + x^2/a^2)]. \tag{6.11}$$

The denominator in the last term can be given a direct Fourier expansion,

$$1/(1 + x^2/a^2) = b_0 + b_1 \cos \omega t + b_2 \cos 2\omega t + \ldots \tag{6.12}$$

One could also expand this denominator algebraically in powers of x and then substitute for each x term its expansion derived from equation (6.10). For the two expansions to be identical, i.e., equal for any value of $\cos \omega t$, the coefficients of each $\cos n\omega t$ in the two expansions must be equal term by term. This requirement leads to a series of equations relating the a coefficients of equation (6.10) to the b coefficients of equation (6.12).

So far we have simply discussed mathematical tricks. The more pertinent question is, How can the terms in this expansion be given a physical interpretation in quantum terms? In the Lorentz, or virtual oscillator, model, the key parameter is not the distance a, but the displacement x. The distance a formerly specified an orbital radius. This was the type of mechanical specification Heisenberg had come to renounce. The distance x could represent a displacement from whatever equilibrium position or motion the electron had. This could be quantized through an extension of the correspondence principle. The quantum correlative of the distance a from the nucleus is the principal quantum number n specifying the orbit, while virtual transitions supply the quantum analog for oscillations about an equilibrium position. More concretely, if a b coefficient in the expansion (6.9) is interpreted as a transition from a state specified by quantum number n to another state specified by $n - 1$, then the a coefficients that supply an expansion of b should be interpreted as a summation of the transitions that lead from the same initial to the same final state by means of intermediate virtual transitions. The relation of this to the Fourier coefficients may most easily be seen by abstracting one fragment from the equations obtained by identifying the coefficients in the two expansions,

$$b(n,\ n - 2) = (1/a^2)a_1\ (n,\ n - 1)a_1\ (n - 1,\ n - 2). \qquad (6.13)$$

The left side of equation (6.13) is interpreted as a transition leading by one jump from a state n to a state $n - 2$. This is equivalent to the two virtual transitions on the right, one leading from the state n to the state $n - 1$, the next from the state $n - 1$ to the state $n - 2$. Because of this physical interpretation, it is necessary to keep the product on the right of equation (6.13) in the order given. Or,

$$a_1 (n,\ n - 1)a_1\ (n - 1,\ n - 2) \qquad (6.14)$$
$$\neq a_1\ (n - 1,\ n - 2)a_1\ (n,\ n - 1).$$

This inequality, now familiar from matrix multiplication, was not posited on mathematical grounds. Heisenberg has repeatedly testified that he had no familiarity with matrices at the time his quantum mechanics paper was written. It was justified exclusively by means of its interpretation through the virtual oscillator model. The electron cannot go from the state $n - 1$ to the state $n - 2$ until after it has gone from the state n to the state $n - 1$. Therefore, the left, rather than the right, side of equation (6.14) represents the correct ordering.

Only someone deeply immersed in the difficulties of the old quantum theory could see this trick expansion applied to an artificial problem as a potential breakthrough. In classical radiation theory the solution of the equations of motion yield not only the allowed frequencies, but also their relative intensities. The Bohr-Sommerfeld quantum theory did not supply any direct means for the calculation of relative intensities. Instead, one had to get them indirectly by using classical physics and the correspondence principle, as Sommerfeld and Heisenberg had done,[43] or use extrapolations from experimental data. Heisenberg's new method with its interrelation of coefficients allowed for a direct calculation of relative intensities. This new method, if correct, clearly went beyond the old.

For Heisenberg himself, the decisive test of the correctness of his new method came from a different anharmonic oscillator, one with the formula

$$\ddot{x} + \omega_0^2 x + \lambda x^3 = 0. \qquad (6.15)$$

Since the anharmonic correction depends on x^3, an odd power, the even terms in the expansion drop out. With this technical simplification Heisenberg was able to calculate both the energy and the frequency in expansion to order λ^2, i.e., through second-order correction terms.

It was also possible to solve this same problem in a way that did not depend on Heisenberg's new method. Following a method Born had developed with Heisenberg's assistance,[44] one sets up a Hamilton-Jacobi equation for a simple harmonic oscillator and then treats the anharmonic term by perturbation methods, a long, complicated procedure involving a series of interlocked Hamilton-Jacobi equations.[45] When Heisenberg finally succeed in solving these and determining all the energy terms to order λ^2, he found that both expansions were identical term by term. Furthermore the frequencies he calculated fitted the basic quantum law $E_n - E_m = h\nu_{nm}$. The new method had passed its initial test.

Heisenberg was convinced. But could he convince others? After leaving Helgoland, a little island in the North Sea where he made his decisive breakthrough, Heisenberg visited Pauli in Hamburg. When he returned to Göttingen, he wrote Pauli a series of letters commenting on aspects of the *Handbuch* article Pauli was completing and seeking Pauli's reaction to the tentative justification he was then developing for his new method. Pauli, who disliked any reliance on models and was especially outraged at the virtualization of physics, could not be expected to accept a new method based on the virtual oscillator model and on transitions which, as virtual, are unobservable in principle. If he, the master of criticism, could be brought to accept Heisenberg's new method, then less critical physicists might be expected to follow his lead. When Heisenberg finally completed his manuscript and sent it to Pauli on 9 July, he included a note asking Pauli to return the manuscript in two or three days:

> for I must either complete it during the final days of my stay
> here or burn it. My own view concerning this scribbling, about
> which I am not at all happy, is this: I am convinced of the
> negative heuristic part, but I consider the positive part to be
> excessively formal and insufficient. But perhaps people who
> know more than I can make something reasonable out of this.
> So please read especially the introduction.[46]

The introduction summarized reasons for rejecting the old quantum theory and then suggested that it seemed reasonable to establish a theoretical quantum mechanics analogous to classical mechanics, but in which only relations between observable quantities appear. Pauli would surely be placated by seeing his own principle given such prominence. The new justification conformed to this by suppressing any explicit reference to virtual oscillators. It was essentially an extension of the correspondence principle using classical formulations of the kinematics and dynamics of a particle's motion as a guide in setting up the corresponding quantum mechanical formulations. These, supplemented by one new quantum condition, constituted the new quantum mechanics.

In classical physics one can treat a system either through a straightforward space-time description that gives position as a function of time $x(t)$ or through a Fourier expansion of $x(t)$. For the new kinematics Heisenberg simply postulated that, since position is unobservable, one should work only with Fourier expansions of position functions. He did not explain why the Fourier expansion of a position function should be physically meaningful, if the position function itself is not physically meaningful.

For radiation due to a transition from one energy level to another he simply postulated that the quantum analogue of classical combining frequencies should have a form

$$v(n, n - \alpha) + v(n - \alpha, n - \alpha - \beta) = v(n, n - \alpha - \beta). \qquad (6.16)$$

This equation is actually quite counterintuitive on the basis of the then currently accepted principles of quantum theory. The left-hand side represents a frequency due to a transition from a state n to a state $n - \alpha$ plus a second frequency due to a transition of a state $n - \alpha$ to a state $n - \alpha - \beta$. The right-hand side represents a third frequency due to a transition from a state n to the state $n - \alpha - \beta$. The two are simply not the same. Any of Heisenberg's readers who accepted the light-quantum hypothesis would find equation (6.16) even more counterintuitive. The equation is intelligible only if it refers to virtual, rather than real, transitions. Then it does not claim that two frequencies combine to form a different third frequency, but that a virtual transition from state n to state $n - \alpha$ followed by another virtual transition from state $n - \alpha$ to state $n - \alpha - \beta$ is energetically equivalent to one transition from the initial to the final state.

Heisenberg's new dynamics was a conceptually simple development fleshed out by mathematical prestidigitation. Newton's second law takes the general form

$$\ddot{x} + f(x) = 0. \qquad (6.17)$$

In the old quantum theory one integrated equation (6.17) and imposed the quantum condition that the total action around a closed path equals an integral multiple of Planck's constant

$$\oint m\dot{x}\,dx = nh.\qquad(6.18)$$

Heisenberg accepted (6.17), but with the proviso that one replace $f(x)$, referring to an unobservable orbit, by its Fourier expansion. Equation (6.18), which explicitly depends on orbits, was not allowed. What Heisenberg did in effect was to substitute a Fourier expansion in equation (6.18), differentiate with respect to n (a bastard move which was given some sort of legitimacy through the correspondence principle), and obtain a new quantum condition

$$h = 4\pi^2 m \sum_{\alpha=0}^{\infty} \left[\, |a(n,\ n+\alpha)|^2\ \omega(n,\ n+\alpha)\right.\qquad(6.19)$$
$$\left. -\ |a(n,\ n-\alpha)|^2\ \omega(n,\ n-\alpha)\right].$$

Equations (6.17) and (6.19) constitute the basis of the new quantum mechanics. They are sufficient, when soluble, to give a complete determination of frequencies, energy values, and transition probabilities. To show the efficacy of the new formalism Heisenberg used it to solve the problems he had solved through the virtual oscillator model. The mathematics remained the same. All references to the virtual oscillator model, however, had vanished.

Around 11 July 1925, Heisenberg gave Born the final version of the paper we have been considering. Born was particularly intrigued by the novel noncommutative multiplication rules. After some false starts he remembered that matrices, then a somewhat esoteric topic, obeyed a similar rule. He sought Pauli's help in working out the details. Pauli refused, on the grounds that Born's mathematical formalism would spoil Heisenberg's physical ideas. Born then secured the assistance of his student Pascual Jordan. Together they redeveloped Heisenberg's new mechanics in terms of matrices.[47] Heisenberg was enthusiastic. He quickly mastered the new mathematics and collaborated with Born and Jordan on a new paper (the "three-man paper") giving a systematic overview of quantum mechanics in matrix form.[48] It contained the basic elements that have since become a normative part of matrix mechanics; diagonalization of the Hamiltonian, canonical transformations, commutation relations, perturbation theory, eigenvalues and eigenvectors.

The crisis and confusion of the old quantum theory was beginning to precipitate into separate tractable problems. The new mechanics supplied the means for calculating energy levels, spectral frequencies, polarizations, and intensities. The Pauli exclusion principle clarified the method of state specification and the attendant problems concerning the *Aufbauprinzip* and the doublet riddle. Pauli had attributed his new fourth quantum number to a classically indescribable doublevaluedness. Other, less austere physicists attempted more descriptive interpretations. Kronig suggested that it might be due to a rotational motion, a suggestion Pauli rejected.[49] Slightly later, and independently, Goudsmit and Uhlenbeck introduced the idea of the spinning electron.[50] Both Pauli and Heisenberg rejected it: Pauli on the grounds that it was too mechanical, Heisenberg on the grounds that it led to results that were off by a factor of 2. Yet, the rejection was not total. In a letter of 5 February 1926 Pauli informed Bohr that neither he nor Heisenberg had succeeded in calculating the hydrogen fine structure with either a point electron or the Goudsmit electron.[51] Two weeks later Bohr sent Pauli a copy of Thomas's relativistic calculation, which solved the factor of 2 in the Goudsmit-Uhlenbeck work.[52] Though Pauli originally rejected this correction, he soon came to accept it.

The new mechanics seemed to many to be extremely abstract and forbiddingly austere, couched in a type of mathematics that few physicists were familiar with. Yet, after Pauli used this mechanics to solve the still outstanding problem of the behavior of the electron in crossed electrical and magnetic fields and to give a systematic treatment of the hydrogen atom, few could doubt its efficacy.[53] Some unsolved problems remained, such as fine-structure splitting. Yet, these now appeared as questions of technical detail rather than as indications of an underlying incoherence. There seemed to be no reason to believe that they would not eventually yield to the new methods. To most atomic physicists the task at hand was one of giving the new formalism a more secure mathematical formulation and using it to solve the problems left unresolved in the old quantum theory. The method of interpreting the new breakthrough seemed to be adequately settled by the doctrine that Heisenberg presented in the introduction to the three-man paper—from this time on, the doctrine was considered Heisenberg's—that theoretical formulations in atomic physics must be exclusively geared to observables.[54]

This evaluation encountered three strong dissenting opinions. Shortly after his paper on quantum mechanics appeared, Heisenberg was invited to give an address at the University of Berlin. Afterward Einstein invited Heisenberg to his apartment and questioned him about his doctrine of observables. Against Einstein's protests Heisenberg insisted that his doctrine of observables simply followed the precedent Einstein himself had set in his treatment of space and time in the special theory of relativity. Einstein's answer, as Heisenberg later remembered it, was: "Possibly I did use this type of philosophy, but it is nonsense all the same. Perhaps I can put it more diplomatically by saying that it may be heuristically useful to keep in mind what one has actually observed. But, on principle, it is quite wrong to try founding a theory on observable magnitudes alone. In reality the very opposite happens. It is the theory that decides what we can observe."[55] Though Heisenberg did not then accept Einstein's doctrine, he claimed that a remembrance of Einstein's statement "It is the theory that decides what we can observe" played something of a guiding role in his development of the indeterminancy principle.[56]

The two other strong dissenting opinions came from Erwin Schrödinger, who developed an alternative interpretation, and Niels Bohr, who insisted that the two apparently conflicting interpretations were really complementary. We will consider the other track leading to the development of wave mechanics before returning to these conflicts.

7. The Branch Line: Wave Mechanics

Wave mechanics resulted from the work of two men who, both by location and temperament, were not a part of the main line we have been considering. Though Louis de Broglie introduced the novel and highly successful idea of matter waves, his conception of the way quantum theory and atomic physics fit together ultimately had very little influence. Elsewhere I have reconstructed de Broglie's original development and argued that it rests on such a fundamental conceptual confusion that the consequent success of the de Broglie formula $\lambda = h/p$ must be considered one of the most fortuitous results in the history of physics.[1] For these reasons the present account will give but a brief summary of de Broglie's ideas. Schrödinger's work and the problems it generated will be treated much more fully. His interpretation of wave mechanics helped to stimulate the development of the Copenhagen interpretation and supplied its most developed and formidable competition. The development, alteration, and demise of the Schrödinger interpretation constitute a significant and revealing case study in the problem of interpreting scientific theories. It merits and will be accorded a detailed study.

De Broglie's Wave-Particle Duality

Louis de Broglie first became interested in quantum theory through a study of the papers of the 1911 Solvay Conference, which his older brother Maurice was preparing for publication.[2] Since the two brothers worked together on X-ray research, Maurice's specialty, Louis was familiar with X-ray aspects of wave-particle duality, as well as with Einstein's light-quantum hypothesis.[3] In 1922, just before the discovery of the Compton effect, he wrote a paper attempting to resolve the problem that had puzzled Einstein, the fact that the light-quantum hypothesis seemed to fit Wien's rather than Planck's radiation law.[4] By a mathematical expansion of Planck's law, de Broglie showed that it could be interpreted as a collection of Wien's laws applying to monatomic light atoms, diatomic light molecules, and ultimately n-atomic light molecules, where n has no upper limit. This was the most extreme corpuscular view of light that had yet been presented.

On his thirty-first birthday, in 1923, de Broglie conceived the idea of extending the wave-particle duality to electrons. To understand the actual historical development it is important to disregard the latter textbook summaries. These begin with the de Broglie wave formula $\lambda = h/p$ and argue that the Bohr orbits are those that exactly accommodate an integral number of wavelengths. The original argument de Broglie presented was based on the special theory of relativity and considerations of frequency rather than wavelength.[5] By combining the basic quantum postulate $E = h\nu$ with the relativistic idea of the equivalence of mass and energy $E = m_0c^2$, de Broglie obtained for a body of mass m_0

$$h\nu_0 = m_0c^2 \text{ or } \nu_0 = m_0c^2/h. \tag{7.1}$$

The immediate difficulty de Broglie encountered was that equation (7.1) seemed to admit of two different Lorentz transformations. An observer, for whom the electron is moving with a velocity $v = \beta c$, could argue that frequency is inversely proportional to time and use a standard Lorentz transformation for time to get

$$\nu_1 = \nu_0 (1 - \beta^2)^{\frac{1}{2}}. \tag{7.2}$$

On the other hand the stationary observer could use the same argument that led to equation (7.1) but apply it to the energy he observes, $E = E_0/(1 - \beta^2)^{\frac{1}{2}}$. This yields the result

$$\nu = m_0c^2/h(1 - \beta^2)^{\frac{1}{2}}. \tag{7.3}$$

A comparison of equations (7.2) and (7.3) yields

$$\nu_1 = \nu(1 - \beta^2). \tag{7.4}$$

What significance do these two new frequencies have? I believe that most atomic physicists, who already looked on de Broglie as something of a crank theoretician,[6] would have answered that the frequencies have no significance. The quantum formula $E = h\nu$ applies to differences in atomic energy levels, not to rest masses. De Broglie, however, effectively followed the principle: the formula knows best. In the remainder of his article he showed that the frequency ν_1, now associated with a relativistic phase wave, is in phase with an electron's motion along an orbit, provided that motion is also treated relativistically, only in the case of Bohr orbits. This would certainly seem to indicate that the peculiar new frequencies introduced must have some physical significance.

In the doctoral thesis "Recherches sur la theorie des quanta" which he defended on 25 November 1924, de Broglie attempted to give an overall coherence to the novel ideas he had been developing.[7] He introduced an interesting analogy to explain the interrelation of the three frequencies we have been considering. Consider a large circular horizontal plane from which are suspended a set of equal weights on identical springs so distributed that they are dense in the middle and thin out rapidly toward the periphery. If all are oscillating with the same amplitude and phase, then the centers of gravity of these weights form a plane which rises and falls with respect to the fixed plane from which they are suspended. The uniform frequency of this pulsation is analogous to the internal frequency of the electron, ν.

An observer watching this plane of pulsating weights pass by him with a velocity $v = \beta c$ could focus his attention either on a particular oscillating weight or on the pulsating plane of the centers of gravity. A particular oscillating weight would effectively constitute a clock. The stationary observer would measure its frequency as $\nu_1 = \nu_0 (1 - \beta^2)^{1/2}$. This is the intrinsic frequency as measured from a relatively moving platform.

The pulsating plane constitutes a more interesting phenomenon. The stationary observer would see this dephased. Thus, if he looked at some particular oscillator at its point of maximum descent, the neighboring oscillators would not be at their points of maximum descent. If the stationary observer tried to trace this point of maximum descent, he would have a relativistic phase point which would seem to move along the oscillators with a velocity $v_p = c/\beta$. A frequency could be associated with the velocity by determining the time interval between the points of maximum descent for a fixed point in the *observer's* frame of reference. This frequency would be $\nu = \nu_0/(1 - \beta^2)^{1/2}$, the frequency associated with the relativistic phase wave. To obviate confusion I will always refer to this by the term "relativistic phase wave." Since these relations between ν_0, ν_1, and ν flow exclusively from the Lorentz transformation, they should obtain independently of how ν_0 is defined, whether for vibrating springs or for the intrinsic frequency associated with an electron's rest energy.

In his thesis, as in his earlier and later writings on the same subject, de Broglie identified the relation between particle velocity and relativistic phase velocity with the relation between the group velocity of a wave packet and the velocity of the individual waves. This identification is not argued; it is simply presented. In the article cited in note 1, I have argued that this identification is seriously wrong and have attempted to present some reasons why this conceptual confusion in so historic a work had not previously been brought to public attention. It was nevertheless an amazingly fruitful error; serendipity striking in the core of atomic physics. An elementary exposition of the relation between waves and groups may help to bring out both the nature of the misidentification and the grounds of its fruitfulness.

A stone dropped into a pool sends out circular waves. If one watches a single wave in the spreading group of waves, it will be seen to begin in the rear, pass through the center, and gradually disappear in the front. Stokes first explained this by treating the group as a superposition of two infinite trains differing slightly in wavelength and velocity. Lord Rayleigh showed the significance of this for optical problems. In a vacuum the wave and group velocities of light are the same, but in a dispersive medium different wavelengths travel at different frequencies, leading to a group velocity that is less than the velocity of an individual wave. The relationship between the wave velocity V the group velocity U and the wavelength λ is

$$U = V - \lambda dV/d\lambda. \tag{7.5}$$

If all wavelengths travel at the same velocity, then $dV/d\lambda = 0$ and $U = V$.

This is a phenomenon which is independent of the theory of relativity. Consider an observer traveling with the peak of the group, or the moving point of maximum reinforcement. Immediately below him he always sees the point where all the waves are in phase. Yet, if he tries to watch an individual wave, he will see it rise in the rear, crest under him, and die out in front. This relationship between wave and group velocities is operative in the group's center-of-mass system and is significant only when waves of different frequencies travel at different velocities. As a consequence there is dispersion. Using a Lorentz transformation to relate this to a different observer adds nothing significant.

Now let the same observer ride a de Broglie electron or plane with oscillating springs. In the center-of-mass system there is no distinction to be drawn between particle velocity and phase velocity. Nor is there any spread in frequencies and velocities. There is simply one system moving at constant velocity v and oscillating with one frequency v_0. In the rest system there is nothing analogous to the relation between wave and group velocity. Consequently, there is no dispersion in the rest system. A Lorentz transformation to any other system does not introduce dispersion. De Broglie's identification of the relationship of particle and relativistic phase velocity with the relationship between wave and group velocity is simply wrong. It rests on a conflation of two radically different concepts.

Yet, it suggested some remarkable conclusions, which were soon verified. It led to a prediction of the diffraction of matter waves, it seemed to explain why only the Bohr orbits were allowed, and it led to a novel explanation of the statistical equilibrium of a gas. Only the final point will be considered here. A gas molecule moving with a velocity v has associated with it a relativistic phase wave of velocity c^2/v. If interpreted literally, this would mean that a relativistic phase wave travels an average of about 9 km between collisions. This extremely high velocity allows for the existence of stationary phase waves in a gaseous mass in equilibrium. For a particle moving back and forth and with relativistic corrections neglected, the wavelength for stationary phase waves is

$$\lambda = h/p \tag{7.6}$$

This is the only time the celebrated de Broglie formula appears in his thesis. It appears only as an approximate expression for the length of stationary phase waves in a gas in equilibrium.

Paul Langevin, one of de Broglie's examiners, sent a copy of this thesis to Einstein, who was then immersed in the problems of ideal gases. A young Indian physicist, Satyandra Nath Bose, had developed a new derivation of Planck's radiation formula and sent this deriviation to Einstein. Einstein, recognizing the wider significance of this new approach, translated the article from English into German and had it published.[8] Then he adapted Bose's technique of counting occupation cells in phase space, rather than the light quanta themselves, to develop a quantum theory of ideal gases. After writing his first paper on a new theory of ideal gases. Einstein read de Broglie's thesis and then wrote a second paper developing a formal analogy between the behavior of radiation and the behavior he predicted for a gas.[9]

To achieve these new results Einstein worked out an analogue to the *Gedanken* experiment of the 1909 paper, which we summarized earlier. Consider a gas of volume V connected by a partition to another container of infinite volume containing the same gas. Assume further that this partition allows molecules within an energy range ΔE to pass freely, but completely blocks all other molecules. If z_ν is the number of cells which belong to the infinitesimal energy range ΔE and n_ν is the average number of molecules in this region, then the result Einstein obtained for the mean square fluctuation $\langle(\Delta_\nu)^2\rangle$ of these molecules can be expressed by the formula

$$\langle(\Delta\nu)^2\rangle = n_\nu + n_\nu{}^2/z_\nu. \qquad (7.7)$$

The corresponding formula for radiation fluctuation, equation (4.33), can be rewritten by noting that the radiation equivalent to z_ν is the number of cells in a unit volume of phase space $8\pi\nu^2/c^3$. With this substitution equation (4.33) becomes

$$(\Delta\nu)^2 = (h\nu\rho + \rho^2/z_\nu)A\tau \, d\nu/c. \qquad (7.8)$$

The first term in equation (7.8) was previously interpreted as the Wien term (or the term corresponding to the assumption of independent light quanta), while the second term was interpreted as a Rayleigh-Jeans term (or the term corresponding to the assumption of light waves). By analogy Einstein interpreted the first term in equation (7.7) as the term that would obtain if the molecules were completely independent. The second term in equation (7.7), the analogue of the light-wave term, cannot be given an interpretation in the Maxwell-Boltzmann theory of molecular behavior. The analogy between the two equations suggests that it should be interpreted as a wave phenomenon proper to molecules. It was in this context that Einstein invoked de Broglie's idea that material particles should have wave properties. Then Einstein wrote back to Langevin saying that de Broglie had lifted a corner of the great veil. This endorsement was responsible for the original acceptance of de Broglie's novel ideas. Before considering the new developments this stimulated, we should conclude this section with a look at the interpretative problems de Broglie faced.

At the conclusion of his thesis de Broglie described his new theory as a form whose physical content was not entirely specified. His immediate concern was to find a content, an account of what the electron is, that fitted the mathematical formalism he had developed. His original idea, one that accorded with his equating of the electron's rest-mass energy with a quantum frequency, was that the electron is essentially energy spread through all space, but with a sharp condensation in a region of very small dimensions. His analogy of oscillating weights, packed densely in the center of the plane and thinning out rapidly as one moves from the center toward the periphery, supplied a suggestive model. He soon rejected this model, however, probably on the grounds that it could not accommodate the electron's particle properties, and sought to develop a new theory of the electron.

As long as de Broglie considered relativistic phase waves the heart of his theory and the idea of group velocity a mere auxiliary device, dispersion did not seem to be a problem. He rejected Schrödinger's idea of electrons as wave packets and his way of treating waves in configuration rather than real space,[10] and sought better interpretations through more rigorous relativistic developments. He was one of many who developed the Klein-Gordon equation simultaneously and independently (along with de Donder, Klein, Gordon, Fock, Kudar, and Schrödinger).[11] When this encountered difficulties, he tried a new formulation, interpreting the motion of material particles as geodesics in a five-dimensional universe.[12] A slightly later attempt involved the type of dual solution of Schrödinger's equation that Bohm redeveloped in 1952.[13] This involved mathematical difficulties which de Broglie was unable to handle. Accordingly, at the 1927 Solvay Conference he gave a less technical version, the pilot-wave hypothesis. Pauli's criticism,[14] coupled to the interpretative problems that de Broglie was now acutely aware of, led him to capitulate to Copenhagen and defend the probabilistic interpretation of wave mechanics from 1928 to 1952, when he returned to a modified form of his dual solution. His final position was that his original interpretation was correct.[15]

Schrödinger's Philosophical Perspective

To understand the world as both one and many, one must grasp a coincidence of opposites and affirm a learned ignorance as the highest form of knowledge. This theme, expressed in Nicholas of Cusa's semipantheistic philosophy and embodied in his efforts as an ecclesiastical diplomat to reconcile the Eastern and Western branches of the Church, could be adapted to characterize the thought and activity of Erwin Schrödinger. He was a brilliant theoretical physicist who claimed that the physical world is not real; an avowed follower of Mach's sensationalism who affirmed the ancient doctrine of the Upanishads that mind and matter are one; a contemplative philosopher who was a fiercely competitive physicist; a published poet who, when a citizen of Ireland, entered his name as a candidate for the highest political position in Austria; a loner who sought to be a leader. In considering his contribution to the problematic we are examining, we must have some awareness of the complex intellectual background from which it emerged and through which it was interpreted.[16]

Most of the other physicists we have been considering backed into philosophical issues from puzzles and paradoxes their scientific work engendered. From the inception of his career Schrödinger cultivated a professional interest in both physics and philosophy. He claimed that he intended to concentrate on philosophical research in the post he expected to get at Czernowitz. When, at the end of World War I, Czernowitz no longer belonged to Austria, the offer was withdrawn and Schrödinger had to stick to theoretical physics.[17] For this reason we will briefly consider Schrödinger's philosophy before examining his physics.

Though Mach had retired from the University of Vienna, his influence was still strong when Schrödinger entered in 1906. From Mach, Schrödinger accepted not only the primacy of sensation as the basis of knowledge, but also a doctrine of neutral monism. The term comes from William James. The doctrine was developed in somewhat similar ways by James, Mach, and Bertrand Russell. In opposition to a subject-object dualism, which Schrödinger regularly attributes to Kant,[18] neutral monism affirms that the basic stuff of reality is neutral and that both the physical and psychic realms are constructions of the mind based on this neutral material. These are the elements that ground sensation.

Sensations are personal and subjective; science aims at interpersonal objectivity. The empiricist tradition has come to grips with this tension in various ways. Schrödinger accepted the conclusion that the common content of consciousness cannot be adequately explained by the doctrine that different people experience the same object. The Vedantic vision, affirming a transcendent unity of all consciousness and its ultimate identity with reality—I am the whole world—expresses, Schrödinger thought, a profound doctrine in an allegorical form.[19] To relate this ancient doctrine to modern science Schrödinger developed an interpretation of evolution as increasing complexification and of conscious awareness as something emerging in individuals like the tips of waves from a deep and common ocean. In both ontogenesis and phylogenesis, consciousness is the instructor supervising the education of living tissue.[20]

As Schrödinger admitted,[21] this metaphorical and metaphysical aspect of his philosophy was rather detached from the work he did in science. One other aspect of his philosophy had a closer connection with his scientific work. From Boltzmann, his lifelong hero, he accepted the idea that the development of physics involves the mental construction of pictures of physical reality. For Schrödinger, such pictures cannot be taken as representations of physical reality as it exists objectively; he considered physical reality a collective construct. The real function of such pictures is to supply a ground for intuition and a test for consistency. Before he came to accept the indeterminacy principle, Schrödinger thought of space-time continuity, comprehensibility, and objectivity (the cognizing subject does not include himself as part of the picture) as criteria which such pictures must meet.[22]

Even after he accepted indeterminacy Schrödinger did not completely forsake the ideal of a complete and consistent picture as a proper goal for scientific striving. Thus, in a lecture on the image of matter he said:

> There is a widespread hypothesis that an objective image of reality in any previously believed interpretation cannot exist. Only the optimists among us (and I consider myself one of them) consider this a philosophical eccentricity, a desperate measure in the face of a great crisis. We hope that the vacillation of concepts and opinions signifies only an intense process of transformation, which will finally lead to something better than the confused series of formulas that today surround our subject.[23]

Boltzmann's doctrine of the role of pictures in science was a secondary contribution to Schrödinger's thought. The primary contribution, which Schrödinger absorbed through Boltzmann's students Fritz Hasenöhrl and Franz Exner, was the role of statistics as a tool and foundation for physics. In his inaugural address as Planck's successor at the University of Berlin, Schrödinger said of Boltzmann's way of doing physics: "His line of thought may be called my first love in science. No other has ever thus enraptured me or will ever do so again."[24] We will consider something of Schrödinger's work in statistics later. What we wish to consider now is the role he ascribed to statistical methods as opposed to determinism.

In his inaugural lecture at Zurich in 1922, Schrödinger furthered an idea which Exner had developed, that the fundamental laws of physics are statistical in nature:

> It is quite possible that Nature's laws are of a thoroughly statistical character. The demand for an absolute law in the background of the statistical law—a demand which at the present day almost everybody considers imperative—*goes beyond*

the reach of experience. Such a dual foundation for the orderly course of events in Nature is in itself improbable. *The burden falls on those who champion absolute causality, and not on those who question it.* For a doubtful attitude in this respect is today the more natural . . . I prefer to believe that, once we have discarded our rooted prediliction for absolute Causality, we shall succeed in overcoming these difficulties, rather than expect atomic theory to substantiate the dogma of Causality. [Italics in original.][25]

When Bohr, Kramers, and Slater developed the view that the conservation of energy is merely a statistical law on the atomic level, Schrödinger came out as one of their strongest supporters.[26]

Even when Schrödinger, in his pre–wave mechanics period, accorded pictures a primary role, he did not insist on logical coherence as a sine qua non. Thus, in the defense of the Vedantic view he wrote in autumn 1925, he claimed:

> We intellectuals of today are not accustomed to admit a pictorial analogy as a philosophical insight; we insist on logical deduction. But, as against this, it may perhaps be possible for logical thinking to disclose at least this much: that to grasp the basis of phenomena through logical thought may in all probability be impossible, since logical thought is itself a part of phenomena, and wholly involved in them; and we may ask ourselves whether, in that case, we are obliged to deny ourselves that use of an allegorical picture of the situation, merely on the grounds that its fitness cannot be strictly proved.[27]

If Schrödinger had fathered the statistical interpretation of wave mechanics, it would be quite easy for later historians of science to show that this represented a legitimate and almost inevitable development of his earlier perspective. Yet, Schrödinger strongly opposed the statistical interpretation of wave mechanics. Before entertaining any conjectures on this apparent philosophical reversal, we should first consider Schrödinger's early work in physics.

Schrödinger's Early Work in Physics

Throughout the course of his scientific career, Schrödinger made contributions to many branches of physics. Before his pioneering work on wave mechanics he had contributed some forty-six papers treating electricity, magnetism, classical dynamics, roentgen rays, atomic processes, special relativity, general relativity, probability theory, thermodynamics, a theory of colors, and ideal gases. One gets the distinct impression of a man conscious of his talent, casting about for a problem where he could make a decisive contribution. Before considering his work on gas theory, we will briefly indicate the relation between Schrödinger's work and Einstein's, on one hand, and the main-line atomists, on the other.

In 1918 Schrödinger wrote a brief article attempting to clarify a problematic feature of general relativity, the energy component in a gravitational field. This provoked Einstein to write two short comments on different aspects of Schrödinger's suggestion.[28] Though Einstein's second note expressed some disagreement with one aspect of Schrödinger's proposal, he clearly regarded Schrödinger as an ally, one of the few who supported general relativity before the decisive 1919 eclipse expedition of Eddington. Though this did not directly relate to atomism, Schrödinger's later work in calculating planetary orbits familiarized him with some of the mathematical tools needed for the treatment of orbital motion.[29] Schrödinger did find one potentially interesting connection between general relativity, especially the gauge factor Herman Weyl had introduced, and atomic physics. He suggested that the quantum conditions express the requisite phase relations which a length associated with an electron should have if its orbital motion is treated as the type of parallel displacement considered in general relativity.[30] Thus, prior to de Broglie's work, Schrödinger tried to relate the general theory of relativity to atomic physics by associating a phase relation with an electron's motion.

Shortly before the developments that initiated the path to wave mechanics, Schrödinger wrote an article extending Heisenberg's core model.[31] If this model is correct, then the optical electron should induce a polarization in the core. Schrödinger's paper gives a detailed treatment of the polarization induced in different types of cores and compares this with available experimental evidence. It also brings out the modifications this induces in the frequencies due to quantum jumps associated with penetrating orbits (another idea Schrödinger introduced), especially those associated with transitions from f- to 3-d orbits. Schrödinger later claimed that he found the idea of quantum jumps abhorrent from the beginning. Yet, in his early papers on atomic physics he made detailed calculations of orbits, orbital transitions, and the interactions between penetrating electrons and the atomic core.

Schrödinger's work on the quantum theory of gases supplied the setting from which his wave mechanics emerged. His papers on this topic constituted his contribution to an ongoing dialogue with Planck and Einstein.[32] Nernst had introduced the idea that a gas should become degenerate at sufficiently low temperatures. This conclusion, flowing from Nernst's theorem, required some readjustments in other aspects of gas theory. The key adjustment was given by Planck, who pointed out that the standard formula for the entropy of a gas must be divided by $N!$ to keep S (the entropy) an extensive quantity and to accord with degeneracy. Planck justified this division by $N!$ on the grounds that, if it were not included, one would be counting the same states too many times.

Schrödinger accepted Planck's division by $N!$ but not the reasons Planck gave for it.[33] Schrödinger's argument was that, in the condensed state, the atoms are held fast and accordingly are identifiable in principle. In this case the permutation number $N!$ becomes physically meaningful.

Einstein's first paper on gas theory seemed to undercut these difficulties.[34] His new statistics led to a definition of entropy that was both extensive and in accord with Nernst's heat theorem. Einstein's second gas theory paper, considered earlier, introduced de Broglie waves and predicted a peculiar quantum mechanical condensation. Planck devised an alternative and more conservative theory, treating the gas as a quantum system with an ordered set of states, where the molecules in each state have energy values in the same range.[35] He was in effect treating the gas as a unit. By using the same partition function for phase-space cells and introducing an adjustable parameter, Planck developed an equation of state which could fit either Maxwell-Boltzmann or Bose-Einstein statistics.

When, in July 1925, Planck presented the paper we have been considering to the Prussian Academy, Schrödinger also presented a paper, "Observations on the Statistical Entropy Definition of an Ideal Gas."[36] What Schrödinger did was similar to Planck's method of treating the gas as a whole, but not his method of counting distributions. Instead of basing statistics on counting individual molecules, Schrödinger suggested summarizing over energy levels of the entire gas. Since all rearrangements describe the same distribution, the troublesome permutation number $N!$ is automatically taken care of. This approach led to an entropy definition which Schrödinger thought appropriate to Bose-Einstein statistics at low temperatures.

Schrödinger was aware that his fusion of Planck's method of treating a gas as a whole and of Einstein's statistics represented more of an ad hoc patchwork than a coherent integration. Furthermore, it did not explain the most peculiar feature of the Bose-Einstein statistics. Classical statistics presupposes independent molecules. The new statistics, which leads to condensation at sufficiently low temperatures, implied some mysterious interaction between molecules. Schrödinger's tentative synthesis did nothing to explain the nature of this interaction.

Schrödinger's final paper on gas theory suggested a new way of treating atoms and molecules, a way which prepared the conceptual underpinning for the Schrödinger wave equation.[37] At Debye's suggestion, Schrödinger gave a seminar in the fall of 1925 on de Broglie's thesis. Though he had been familiar with de Broglie's idea of associating a wavelength with an electron from Einstein's adaptation of this idea, Schrödinger had not yet subjected de Broglie's thesis to a critical examination. His subsequent reaction is indicated in two letters.

The first letter, one he sent Einstein on 3 November 1925, related de Broglie's new idea to Schrödinger's own earlier paper associating a phase with an electron: "The de Broglie interpretation of the quantum rules seems to me to be related in some ways to my note in the *Zs. f. Phys.* 12, 13, 1922, where a remarkable property of the Weyl 'gauge factor' $\exp(-\int \phi_i dx_i)$ along each quasi-period is shown."[38] Two weeks later, on 15 November 1925, Schrödinger sent Landé a letter in which he mentioned the difficulties he was having with de Broglie's doctrines: "I have been occupied a great deal these days with Louis de Broglie's ingenious thesis. It is extraordinarily suggestive, but has nevertheless, very great difficulties. I have tried in vain to make for myself a picture of the phase wave of the electron in the Kepler path, considering as 'rays' the neighboring Kepler ellipses of the same energy. This gives horrible caustics or the like for the wavefronts, however."[39] A caustic is an envelope curve which gives the boundaries of an initially plane wave after reflection or refraction.

Very great difficulties notwithstanding, de Broglie's idea of matter waves suggested a new way of handling the gas problem. Einstein had stressed the formal parallel between Planck's treatment of radiation in a cavity and his own treatment of molecules in a gas. He also endorsed Bose's idea of counting cells in phase space as a means of getting a statistics which could fit either light quanta or molecules. Schrödinger retained the parallel, but not the new way of developing statistics. Jeans's development of the Rayleigh-Jeans law treated the radiation within a cavity as a whole and attributed to it certain frequencies, $\nu_1, \nu_2, \ldots, \nu_s$. P. Debye had shown how this method could be adapted to give Planck's rather than Rayleigh-Jean's law.[40] One imposed the constraints that the energy in any one mode, e.g., the sth, is either $h\nu_s, 2h\nu_s, \ldots, nh\nu_s$ and that the sum of all the energy in all the modes equals the total gas energy.

By the radiation-gas parallel one should be able to adapt this method, treat the gas as a whole, and derive Bose-Einstein statistics by a natural way of counting. The gas considered as a whole should have various modes of vibration. One could allocate the energy $n_s \epsilon_s$ to the sth mode on the assumption that it contains n molecules, each on the energy level ϵ_s, and also add the constraint $\Sigma_s n_s = n$. Then, through an application of the Darwin-Fowler method, Schrödinger obtained the Bose-Einstein statistics.

The pivotal assumption here is that the allowed degrees of freedom of gas, considered as a whole, have energies $h\nu_s, 2h\nu_s, \ldots, nh\nu_s$. Planck had obtained this in his original blackbody derivation by quantizing, not the radiation itself, but the fictitious oscillators in equilibrium with the radiation. Schrödinger was now attributing simple harmonic oscillator energy levels to a gas. This bears a peculiar parallel to the development of matrix mechanics. Bohr had introduced the virtual oscillator model to treat, not molecules themselves, but the radiation emitted by a set of atoms as if it were emitted by a set of simple harmonic oscillators. In developing quantum mechanics, Heisenberg adapted this model to treat the energy states of individual atoms. Though Schrödinger was not treating radiation, he was treating molecular energy levels, which are intimately related to radiation frequencies, as if they were the levels proper to a set of harmonic oscillators.

For Schrödinger, however, these were not virtual oscillators. In section 3 of the paper we are considering, he justified his new way of treating molecules by adapting the ideas of de Broglie. Though he briefly summarized de Broglie's relativistic treatment, he actually used only the nonrelativistic form. The key identification was pinpointed in a later summary Schrödinger gave.[41] If one considers simple waves trapped within a box of volume V, then the number of vibrations with wavelength greater than λ is $(4\pi/3)V/\lambda^3$. If these are thought of as de Broglie waves, then one can substitute for λ the momentum value given by de Broglie's formula, $\lambda = h/p$, and consider the states with momentum between p and $p + dp$ to get the value $(4\pi/3)(V/h^3)p^2\,dp$. This value was already familiar as the number of quantum states a single molecule is allowed in phase space. Now it reappears as the number of stationary vibrations within a certain momentum range. Particles had been replaced by vibrations.

Schrödinger used de Broglie's formula, but not his interpretation. In spite of the relativistic formula cited, Schrödinger was using the nonrelativistic relation between wave velocity and group velocity. Also, the analogy Schrödinger was using between the gas in a box and the radiation in a 'blackbody box' would seem to suggest stationary waves, like the fixed nodes in a vibrating string, rather than moving waves. What physical significance is to be attached to this?

In this final gas article Schrödinger simply offered a suggestion and noted the difficulties it involved. The suggestion was that a moving molecule could be considered as nothing but a signal, the wave crest of a wave system. The overriding analogy between radiation and molecules had something of a muting effect on this suggestion. Planck's treatment of blackbody radiation had not quantized radiation itself, but the fictitious oscillators in equilibrium with this radiation. A direct quantification of radiation involved the light-quantum hypothesis. Though both Einstein and de Broglie had used this hypothesis in the works Schrödinger was adapting, Schrödinger himself did not accept it. In the opening paragraph of the paper, Schrödinger claimed that the basic point of this article was to reproduce Einstein's results through natural statistics by treating ether oscillators rather than gas molecules.

These ether oscillators played a role analagous to Planck's fictitious Hertz-ian oscillators.

Schrödinger's new interpretation also introduced a basic difficulty, dispersion. Though this is not a problem for de Broglie's relativistic phase waves, it is a problem for any interpretation of a molecule as a signal in a wave system. The natural analogy is to identify a molecule with a wave packet and its velocity with the group velocity of this packet. But, as Schrödinger explicitly noted, such wave packets disperse and hence do not preserve the singularity thought proper to molecules. His article concludes, "If one can avoid this consequence through a quantum theoretical modification of the classical wave law then it appears that a path is prepared for the solution of the light-quantum dilemma."

The Development of Wave Mechanics

This final paper on gas theory was submitted to the *Physikalische Zeitschrift* on 15 December 1925. Five weeks later, on 26 January 1926, Schrödinger submitted to the *Annalen der Physik* the first of his four monumental papers on quantization as an eigenvalue problem (henceforth referred to as Q1, Q2, etc.).[42] Thanks to recent research it now seems possible to give a plausible reconstruction of what happened during this, the most creative period in Schrödinger's career. He had become convinced by his work on gas theory that the representation of particles by waves had some validity. Yet, he was well aware of the difficulties this model involved. In the discussion following Schrödinger's second seminar on de Broglie's thesis, Debye offered the suggestion that the proper way to treat wave motion is to look for a wave equation.[43] Apparently Schrödinger followed this suggestion and first attempted to develop a relativistic wave equation, probably the initial form of the equation now known as the Klein-Gordon equation.[44] On 27 December 1925, Schrödinger wrote a letter to Willy Wein, then the editor of the *Annalen der Physik*, expressing his optimism about his new approach. In this letter, reflecting work that had been lost, he had a relativistic formulation for the frequencies. In terms of observable results, however, the relativistic parts dropped out. For two frequencies, v_n and v_m, his theory gave

$$v_n = mc^2/h - R/n^2, \quad v_m = mc^2/h - R/m^2, \tag{7.9}$$

$$v_n = v_m = R\,(1/n^2 - 1/m^2), \tag{7.10}$$

where R is Rydberg's constant.[45]

Sometime after this, he dropped the relativistic formulation and developed the nonrelativistic formulation that proved successful. This reconstruction leaves a very short time for the writing of his first quantization paper. Yet, it seems correct. In the first half of 1926 Schrödinger turned out a series of fundamental papers in theoretical physics at a rate that has few precedents in the history of science. The only parallels that come to mind are Newton in 1666, Maxwell in the early 1860s, Einstein in 1905, Bohr in 1913, and Heisenburg during precisely the same period when Schrödinger peaked.

The formal development of Schrödinger's wave mechanics is familiar, not only from the basic historical surveys, but even more from the fact that his methods and solutions have become an integral part of modern physics. Here I wish to concentrate on the interpretative problems Schrödinger encountered rather than present one more summary account of how the Schrödinger equation is set up and solved. The fulfillment of this wish, however, presents some peculiar problems.

Schrödinger found himself in a situation similar to that Heisenberg had encountered some six months earlier. A nonrelativistic adaption of a simple harmonic oscillator model to a gas suggested a new mathematical method for treating problems in atomic physics. Initial calculations indicated that the new method worked quite well. Yet, the model of matter-waves that suggested this formalism was underdeveloped, not yet coherent, and not related to the mathematical formalism in any clear and unambiguous way. The strategy Schrödinger initially followed in this impasse was quite similar to Heisenberg's. For all its limitations, the hypothetical-deductive model of scientific explanation can supply a bridge over troubled waters. Schrödinger simply presented his new equation without giving any real justification, worked out particular solutions, and showed how they could be interpreted in terms of measurable phenomena. In the first two quantization papers the basic formalism was developed in such a way that its intelligibility did not depend on an acceptance of the wave interpretation.

Heisenberg had suppressed his dependence on the virtual oscillator model and, before his struggles with Bohr, had argued that a mathematical formalism related to observable phenomena in a rule-governed way supplied a sufficient basis for the interpretation of a scientific theory. Schrödinger did not hold such a view. As his private correspondence manifests, he was interested in developing, rather than suppressing, the new matter-wave model. Yet, in Q1 and Q2 he was careful not to accord any foundational role to such models. For these reasons we will first present the public interpretation accorded the new formalism in Q1 and Q2 and then consider Schrödinger's private efforts to fill in the gaps in his public performance.

Schrödinger's first quantization paper (Q1) attempted to show, for the simple case of the nonrelativistic and unperturbed hydrogen atom, that the usual quantization procedure can be replaced by another postulate in which integers are not assumed but follow in a natural way, as, for example, in the number of nodes of a vibrating string. This hint of a wave analogy, together with a brief allusion to vibrations at the conclusion of the article, supplies the only textual basis for interpreting Schrödinger's new equation as a wave equation. The emphasis in the first communications is on the fact that the new method of quantizing yields integers in a natural way.

The usual form of the quantum conditions is connected with the Hamilton-Jacobi partial differential equation,

$$H(q, \partial S/\partial q) = E. \tag{7.11}$$

A solution for S is sought which is a sum of functions, each of a single one of the independent variables q. Schrödinger modified this by introducing a new unknown ψ through the substitution

$$S = K \log \psi. \tag{7.12}$$

Here K is a constant which, like S, has the dimensions of action (or of h). The basic advantage resulting from this substitution is that ψ will now appear as a product of the functions of the individual coordinates rather than a sum. With this substitution, equation (2) takes the form

$$H(q, K/\psi, \partial\psi/\partial q) = E. \tag{7.13}$$

Before developing any particular solutions for (7.13), Q1 stipulates the conditions that ψ must fulfill.[46] It must be real over the whole of configuration space, unique-valued, finite, continuous, and twice differentiable.

These conditions of adequacy are simply presented with no justification. Yet, they deserve some comment. The most notable feature about them is that they are conditions imposed on ψ considered as a mathematical function. The physical interpretation to be accorded ψ is only indirectly operative through the stipulation that ψ must be real. The reason for this stipulation is undoubtedly the reason given in later parts of the series: if the ψ function corresponds to something physically real, such as a matter-wave, then it should be real rather than imaginary or complex. The force of this reason is somewhat blunted, however, by the fact that the ψ function is in configuration space (a product space with three coordinates for each particle) rather than ordinary space. Though being real is listed as a basic criterion for the acceptability of the ψ function, it is, in fact, never fulfilled in this or any other article in the quantization series.

The rest of Q1 has a rather chiaroscuro quality. The solution of the Schrödinger equation, to use the now standard term, is developed with such elegance and generality that this treatment has become a stable part of physics ever since. Yet, the justification of this virtuoso performance is so obscure that Schrödinger, in the introduction to his next article, referred to it as an unintelligible transformation for an incomprehensible transition.[47]

Q2 is primarily concerned with making this transition more intelligible. This peculiar order of development would seem to reflect the fact that Schrödinger had more confidence in the conclusions flowing from his equation than in any justification he was then able to give for its introduction.

After the general specification of requirements that ψ must fulfill, Q1 is chiefly concerned with solving equation (7.13) for the hydrogen atom. The method of solution employed need only be outlined here. By applying the variational method to equation (7.13), Schrödinger developed two equations. The second of these, concerning the behavior of an integral of $\psi \partial \psi / \partial_{normal}$, is handled by requiring that physically significant quantities vanish in a suitable way at infinite distances. Though this might seem to be suggested by the wave interpretation of ψ, Schrödinger relies on mathematical rather than physical reasons.[48] The first, and more familiar, equation is

$$\nabla^2 \psi + 2m/K^2(E + e^2/r)\psi = 0, \tag{7.14}$$

where K must, for numerical agreement, have the value $h/2$.

Solving equation (7.14) was far from routine. Schrödinger's was the first solution of a partial differential equation exhibiting both a continuous and a discrete eigenvalue spectrum.[49] One significant point about this solution was noted by Schrödinger himself;[50] the solution was developed in a way that was neutral with respect to interpretations of atomic structure. Though he would have preferred to relate ψ to some vibratory process within the atom, something analogous to beats in music, he relied instead on the fact that the correct numerical relations came out in a natural way without the imposition of arbitrary conditions.

Schrödinger's second quantization paper related more directly to the work of de Broglie.[51] The paper began with a clarification of the basis of Q1, showing how Schrödinger's variation principle corresponds to Fermat's principle for wave propagation, though Schrödinger's principle is in configuration space. Similarly, the Hamilton-Jacobi equation Schrödinger used in Q1 could be interpreted as expressing Huygens's principle for wave propagation. The justification that Q2 presented can be described as a nonrelativistic reinterpretation of de Broglie's results. In place of de Broglie's relativistic phase waves, however, Schrödinger concluded to a system of wave surfaces which form a progressive but stationary wave motion in configuration space.[52]

The problems configuration space presents will be considered later. Of more immediate interest is the comparison drawn in Q2:[53] de Broglie's theory (rightly reinterpreted) stands to Schrödinger's theory as geometric optics stands to wave optics. Geometric optics represents a valid approximation when one is studying phenomena, such as reflection and refraction, for which wave effects are so negligibly small that light may be treated as a set of rays. Similarly, de Broglie's geometric wave mechanics (or ray mechanics) is an approximation no longer valid when one is treating phenomena of a size comparable to the wavelengths involved. In using such a reinterpretation of de Broglie's work, Schrödinger faced a basic problem of showing how his mathematical formulation of wave surfaces in configuration space related to de Broglie's theory of matter waves.

What Schrödinger took from de Broglie was actually the wave velocity–group velocity formulation rather than the relativistic phase wave formulation. Though the way Schrödinger handled this is quite involved mathematically, the basic method can be summarized in a fairly qualitative way. One postulates a group of waves and then sets up and solves the Hamilton-Jacobi equation to get a point of agreeing phase for a whole aggregate of wave groups. This singularity in configuration space takes the place of de Broglie's particle riding a wave, and its motion defines the geometric locus of points of agreeing phase. Schrödinger was unable to prove mathematically that a superposition of such wave disturbances really produces a notable disturbance only in a very small region—or that there is no spreading of the singularity. Accordingly, he postulated a lack of spreading and presented this as a physical, rather than a mathematical, hypothesis subject to the test of experimental trial.[54] The conclusion Schrödinger drew from this protracted comparison was that in dealing with small-range phenomena, where the ray-mechanics approximation is invalid, one must abandon any images of electrons moving along definite paths and rely on a wave equation. To back up this deliberate disregard of visualizable models, Schrödinger cited the work of Heisenberg, Born, and Jordan and noted that their strivings seemed to manifest the same tendency. Though he had not yet found the link connecting his development with theirs, Schrödinger expressed the hope that his own method would eventually lead to a more intuitive understanding of the microscopic mechanical processes.[55]

After this rather abstract discussion, Schrödinger set up and solved his wave equation for some basic and now familiar problems: the Planck-Hertz oscillator, the rigid rotator with fixed axis and with free axis, and the nonrigid rotator. In solving these problems, he concentrated on achieving mathematical results while sidestepping the problem of the precise physical interpretation to be accorded these results. As he put it, "The question of how the energy is really distributed among the proper vibrations, which has not been taken into account here up till now, will, of course, have to be faced some time."[56]

Schrödinger's First Interpretation of Wave Mechanics
Schrödinger intended his four quantization papers as a series covering the whole foundation of wave mechanics. Yet, he interrupted this series to write two further papers: a short note concerned with the physical inter-

pretation of the new mathematical formalism and a much longer one on the relation of his new wave mechanics to Heisenberg's matrix mechanics (Schrödinger regularly referred to this as Heisenberg's, rather than that of Heisenberg, Born, and Jordan). When he returned to his quantization series, he developed a new and somewhat different interpretation of his formalism. This accordingly seems to be the appropriate place to consider his first interpretation.

In a summary account of his new breakthrough which he sent to the *Physical Review* on 3 September 1926, Schrödinger claimed, "The point of view taken here, which was first published in a series of German papers, is rather that material points consist of, or are nothing but, wave-systems."[57] This is the essence of his first interpretation. Though some further details were supplied in letters, it seems that Schrödinger never worked out this interpretation in anything like complete detail.[58] What we will attempt to do here is to explain the basic model Schrödinger was developing and then indicate the difficulties it encountered.

In the citation just given as well as in the earlier paper on Einstein's gas theory, Schrödinger used the general term 'wave system' rather than the more specific 'wave packet' used in Q2. A wave system is a system confined within a volume as the gas in a box or, in an intuitive picture, an electron in an atom. Since any dispersion that occurs should also be confined within the same volume, dispersion for a wave system might seem to present a technical complication to be solved in due time rather than an obstacle to progress.

A wave packet is a bundle of waves moving through a dispersive medium. Schrödinger clearly wanted to develop the idea that an electron moving freely through space or even in an outer atomic orbit could be treated as if it were really nothing but a wave packet. This presents a basic and obvious difficulty. A freely moving electron displays the perduring particulate properties manifested, for example, in the Wilson cloud chamber photographs that were just becoming public.[59] A wave packet disperses. Schrödinger was well aware of this difficulty and had, in fact, discussed it briefly in his final paper on gas theory. Schrödinger was not alone in noting this difficulty. Since the 1911 Solvay Conference and his 1912 appointment as director of the Teyler Institute in Haarlem, H. A. Lorentz had effectively assumed a position as referee of the European physics community, the man whose competence and objectivity everyone respected. On 27 May 1926, he sent Schrödinger a long letter commenting favorably on his new breakthrough, but pointing out in some detail the difficulties with dispersion.[60] In his reply to Lorentz, Schrödinger admitted that solving this problem was an urgent requirement.[61]

Yet, it seems that Schrödinger also thought of this as a technical problem which might be overcome by a more elaborate mathematical construction of wave packets. In his earlier gas theory paper he indicated the hope

that this might be accomplished by the Debye–von Laue method of constructing wave packets from waves which differed infinitesimally both in
frequency and in wave normal. The first of the two articles he sandwiched in
between Q2 and Q3 was a short note demonstrating that, at least in one case,
a group of proper vibrations behaves like a particle.[62] The case chosen is a
somewhat artificial one, a one-dimensional simple harmonic oscillator with
frequencies confined to a specially selected range. The results Schrödinger
developed can be described as the motion of a tall narrow Gaussian hump,
corrugated with vibrations, which moves like a classical particle in that the
wave packet does not spread out in the course of time. Thus, in March of
1926, there seemed to be reasonable grounds for hoping that the difficulty of
dispersion might be overcome by more sophisticated mathematical
methods.

There were also further difficulties which are clearer in retrospect than
they were in the hectic days of rapid problem solving. A basic one was that
Schrödinger's mathematical formalism did not really fit the wave packet
interpretation he was attempting to impose on it. This difficulty has two
separate aspects. The first is that the Schrödinger ψ function, which should
be related to the amplitude of a wave packet, was a function in configuration—rather than real space. In the one-particle case, illustrated by the
problems Schrödinger had solved in Q1, the distinction is a mere technical-
ity. In the many-particle case, it is crucial. Thus, for a two-particle system
which is not reduced to an equivalent one-particle system, ψ is a function in a
six-dimensional space. For an n-particle system it is a function in a $3n$-
dimensional space. Real matter waves, if such there be, should exist in a real
three-dimensional space. The second gap between the formalism and the
interpretation stemmed from the fact that a wave packet, to be at all
coherent, must be made up of a large number of waves with slightly different
frequencies. The Schrödinger equation for a bound system does not yield
such frequencies. It simply yields eigenfrequencies, which would not in
general constitute a wave packet.

The wave model of matter for all its difficulties also had some intuitively
appealing features. It seemed to supply a basis for replacing Bohr's quantum
jumps and the idea of discontinuous processes by some sort of connection
between continuous oscillations in the atom and frequencies in the emitted
radiation. The importance this intuitive appeal had for Schrödinger himself
comes through very forcefully in his 6 June 1926 letter to Lorentz:

> I was so extremely happy, first of all, to have arrived at a pic-
> ture in which at least something or other really takes place
> with that frequency which we observe in the emitted light that,
> with the rushing breath of a hunted fugitive, I fell upon this
> something in the form in which it immediately offered itself,

namely as the amplitudes periodically rising and falling with the beat frequencies. . . . The frequency discrepancy in the Bohr model, on the other hand, seems to me (and has indeed seemed to me since 1914) to be something so monstrous, that I should like to characterize the excitation of light in this way as really almost inconceivable.[63]

This interpretation for all its intuitive appeal did not adequately come to grips with all the difficulties Lorentz had raised. One of the most basic and solidly established properties of atomic radiation is that the emitted frequency corresponds to the difference of two energy levels, $\nu_{nm} = (E_n - E_m)/h$. In Schrödinger's interpretation, which accorded frequency rather than energy a primary role, the radiation frequency was thought to correspond to the difference of two atomic eigenfrequencies which were simultaneously excited. The idea was that just as radio waves (very high frequency waves which are not heard) can carry sound waves which are heard, so atoms are like resonators whose unseen eigenfrequencies have beat differences which are seen as visible radiation. The analogy breaks down rather quickly. Sound-carrying radio waves represent a superposition of two quite dissimilar waves rather than a difference of two similar waves. This interpretation leads to false results for higher excited states. For multielectron atoms or, in Schrödinger's interpretation, for atoms which have many simultaneous eigenvibrations, the differences between these eigenvibrations should produce detectable radiation even in the ground state.

A Struggle for Primacy

Schrödinger was encountering formidable difficulties both with his new mathematical formalism and with the physical interpretation he was attempting to develop. Yet, he continued to make unprecedented progress, turning out about an article a month, each of which represented a contribution to theoretical physics of perduring value. The second article which Schrödinger produced between Q2 and Q3 established the mathematical equivalence of matrix mechanics and wave mechanics.[64] The idea was in the air. The mathematical equivalency was established independently by Pauli, who did not publish his results though he circulated them privately,[65] and by C. Eckart in America.[66] Though Schrödinger's paper must have been written in haste, it is characterized by his usual balance of generality in scope and precision in detail. For any orthonormal set of functions $u(x)[\rho(x)]^{1/2}$, where the density function $\rho(x)$ is defined by $\int \rho(x) U_i(x) U_x(x) = \delta_{ik}$ and x is an abbreviation for the product of configuration space coordinates $q_1, q_2, \ldots, q_n$, one may set up matrices for any two general functions,

$$F^{kl} = \int \rho(x) u_\kappa(x)[F, u_l(x)] \, dx, \qquad (7.15)$$

$$G^{lm} = \int \rho(x) u_l(x)[G, u_m(x)] \, dx. \qquad (7.16)$$

The general function $[F, u_l(x)]$ usually reduces to $Fu_l(x)$. In modern notation the density function is usually absorbed by normalizing the wave function so that the matrix component has the simpler form $F^{kl} = \int \bar{\psi}_k F \psi_l \, dx$, where F is an operator. Schrödinger showed that matrices, defined by equations (7.15) and (7.16), conformed to the standard law of matrix multiplication

$$(FG)^{kn} = \Sigma_l \, F^{kl} \, G^{ln}, \qquad (7.17)$$

that they fitted the Hamiltonian equations proper to matrix mechanics

$$(dq_l/dt)^{ik} = (\partial H/\partial p)^{ik}; \; (dp_l/dt)^{ik} = (\partial H/\partial q_l)^{ik}, \qquad (7.18)$$

and that for the orthonormal functions in equations (7.15) and (7.16) one can use functions which are solutions of the Schrödinger equation.

This demonstrated mathematical equivalency immediately enriched both approaches. To make the two formulations agree Schrödinger had to introduce the operator substitutions $p_k \rightarrow (ih/2\pi)\partial/\partial q_k$. This was soon generalized into the basic method for setting up a Schrödinger equation for an arbitrary system. Further, since Schrödinger could now incorporate the results of matrix mechanics, he automatically had a method of handling transition probabilities and line intensities, items not covered in Q1 and Q2. Matrix mechanicians, in turn, could now incorporate Schrödinger's simpler and more familiar mathematical methods of setting up and solving problems.

Mathematically, the two theories were demonstrated to be intertranslatable. If a scientific theory is just a mathematical formalism related to observable phenomena by correspondence rules (the mode of interpretation Heisenberg was stressing), then Heisenberg, rather than Schrödinger, had made the decisive breakthrough. He is the new Newton, laying down the basic laws of new mechanics. If, on the other hand, the physical interpretation accorded the mathematical formalism is an essential part of the theory and if Schrödinger's interpretation is different from and superior to Heisenberg's, then there would be good grounds for regarding Schrödinger as the new master builder.

The traditional image of the research scientist as the detached, disinterested searcher after truth was rather publicly shattered by James Watson's vivid depiction of the contest to unravel the DNA molecule and the feverish activity inspired by the desire to be first with a fundamental discovery.[67] Subsequent reflection has indicated that such intense competition for priority or primacy is often more a normal pattern than an aberation. An illustrative example is the protracted and often bitter competition between Roger Guillemin and Andrew Schally that finally led to their sharing the 1977 Nobel Prize for medicine.[68]

From March of 1926 through June of 1927 Schrödinger and Heisenberg clearly, and I believe quite self-consciously, struggled for dominance in quantum physics. Though such struggles are generally not a part of the issues we have been considering, this one complicated the problem of how scientific theories are to be interpreted in some subtle but significant ways. For this reason we will interrupt the chronological development of the physics to consider the way in which this competition highlighted the physical significance to be accorded the mathematical formalism.

Since the two theories were now accepted as mathematically equivalent, the only really significant difference between them consisted in the physical interpretation Schrödinger was attempting to develop for his formalism, as opposed to the minimalist interpretative position Heisenberg favored. This seems to have been immediately clear to both men. Before establishing the intertranslatability of the two methods, Schrödinger indicated in his initial reaction to the Heisenberg formalism:

> I did not at all suspect any relation to Heisenberg's theory at the beginning. I naturally knew about his theory, but was discouraged if not repelled [*abgestrossen*] by what appeared to me as very difficult methods of transcendent algebra and by the lack of intuitiveness [*Anschaulichkeit*].[69]

This is a bit peculiar when interpreted as a straight scientific criticism. Schrödinger's intertranslatability paper, hastily written while Schrödinger was in the throes of his own theory construction, clearly shows that he had little difficulty in mastering and reinterpreting Heisenberg's "very difficult methods." The lack of intuitiveness in matrix mechanics clearly bothered Schrödinger. Yet, it was somewhat odd to criticize Heisenberg for constructing his theory in a nonintuitive way after Schrödinger had characterized the initial development of his own theory as an unintelligible transformation for an incomprehensible transition.

In his 6 June letter to Lorentz, Schrödinger discussed in detail the objections Lorentz had brought against his interpretation and then gratuitously added some further considerations: "In conclusion may I emhasize several serious difficulties of a fundamental nature in the matrix mechanics (without any connection with your letter), which have gradually become clear to me and in which I see an advantage in the wave mechanics, quite apart from its intuitive clarity."[70] The points stressed were symmetrization of the Hamiltonian, which the matrix mechanicians did in an ad hoc and rather limited way but which wave mechanics easily handled in a general fashion, and the fact that wave mechanics always yields completely determined eigenvalues while matrix mechanics can do so only by supplementing detailed calculations with ad hoc assumptions. These proved to be technical shortcomings rather than fundamental difficulties. What is clear is that Schrödinger was hoping to win Lorentz, the ultimate referee for the physics community, over to his persuasion.

Heisenberg's reaction to Schrödinger's physical interpretation is bluntly revealed in a letter he wrote to Pauli from Copenhagen on 8 June 1926: "The more I ponder the physical part of the Schrödinger theory, the more disgusting I find it."[71] In the same letter he called the Schrödinger interpretation of the electron "crap" (*Mist*) and claims that the theory's most significant achievement is the calculation of matrix elements rather than some to-be-discovered relation to de Broglie's theory.

This personal competition was reinforced by something of an institutional rivalry. On one side was the main-line development of atomic physics, the Copenhagen-Göttingen-Munich school, led by Bohr, which emphasized atomic spectroscopy as *the* data supplying the definitive tests for any theory of the atom. On the other side was the Berlin school, led by Einstein and Planck, which emphasized the statistical approach to gases and light and slighted the tedious and often grubby bookkeeping involved in calculating the frequencies, intensities, and polarizations of spectral lines under the influence of various sorts of fields. Before his gas theory papers and his extended correspondence with Planck and Einstein, Schrödinger, the loner, was not affiliated with either school. Now, though he preserved his intellectual independence, he was clearly lining up with the Berlin school. After Schrödinger's spectacular success, the Berliners were delighted to have on their side someone who could beat the spectroscopic calculators at their own game. It probably suprised no one when, in 1927, Schrödinger succeeded to Planck's chair at the University of Berlin. This competitive atmosphere supplies a background for the further developments we will consider.

Schrödinger's Second Interpretation

In the article demonstrating the mathematical equivalence of the two formalisms, Schrödinger also argued for the superiority of the intuitive interpretation associated with wave mechanics. Since it gives a clear significance to the amplitude of the atom's electrical oscillation, it makes the polarization and intensity of emitted light intelligible on the basis of the Maxwell-Lorentz theory. To develop this, Schrödinger introduced a new conjecture, that the spatial distribution of electrical charge within an atom is given by the real part of $\psi \partial \bar{\psi} / \partial t$. On this basis Schrödinger was able to calculate the dipole moment for radiation, automatically including the contribution of the Heisenberg matrix element. This, as Schrödinger interpreted it, served as the real grounds for making Heisenberg's results intelligible:

> The Heisenberg matrix elements q_i^{km} come into the coefficients in such a manner that their cooperating influence on the intensity and polarization of the part of the radiation concerned is completely intelligible on the grounds of classical electrodynamics.[72]

The spatial distribution of the wave function should explain charge distribution within the atom as well as the intensities and polarization of spectral lines. Unfortunately, as Schrödinger realized, this interpretation labored under three serious difficulties. First, only the real part of ψ was used, while the imaginary part was arbitrarily discarded. Second, if the real part of this factor represents charge distribution, then its integration over space should yield the total charge, a conserved quantity. It gave zero. Finally, the perduring difficulty that ψ is a configuration- rather than a real-space wave function still impeded a straightforward realistic interpretation of ψ.

The third communication in the quantization series was received on 10 May 1926.[73] In it Schrödinger treated time-independent perturbation theory. By adapting some methods Lord Rayleigh had developed in acoustics and utilizing the mathematical tools which had just become available through the monumental work of Courant and Hilbert, *Methoden der mathematischen Physik*, Schrödinger was able to give a development of time-independent perturbation theory that was both general and rigorous. To illustrate the scope of this method he gave a very detailed treatment of the Stark effect including line intensities, selection rules, and polarization rules; compared his results with the best data available; and even predicted the results of some special cases that had not yet been tested. Next to the derivation of the hydrogen spectrum this was the most resounding success of the new theory and conclusively established the efficacy of the general formalism that Schrödinger had developed.

The basic outline of this paper and the Stark effect calculations were done before Schrödinger realized the need to work out a physical interpretation different from Heisenberg's. The perturbation theory he developed did not depend on differences between the two interpretations. But Schrödinger found one key point that did seem to differentiate the two interpretations. In matrix mechanics the generalized position coordinate is represented by a square matrix. Schrödinger could use his ψ function, together with the density function, to determine an arbitrary component $q^{rr'}$ of this matrix:

$$q^{rr'} = \frac{\int q\rho(x)\psi_r(x)\psi_{r'}(x')\,dx}{\{\int \rho(x)[\psi_r(x)]^2\,dx'\,\int \rho(x)[\psi_{r'}(x)]^2\,dx\}^{1/2}}. \tag{7.19}$$

The interpretation of equation (7.19) supplied Schrödinger with the opportunity to distinguish between his and Heisenberg's interpretations of quantum theory. For Heisenberg, as Schrödinger interpreted his position, the square of (7.19) should be a measure of the probability of a transition from the rth to the r'th electronic state. More precisely, it is a measure of the radiation resulting from this transition that is polarized in the q direction. Where Heisenberg's interpretation utilized the idea of quantum jumps and transition probabilities, Schrödinger was able to give this matrix element a simple physical interpretation that relied on continuity rather than quantum jumps. According to Schrödinger's interpretation, $q^{rr'}$ is a particular component of the amplitude of the periodically oscillating electrical moment of the atom. Consequently, the matrix can be interpreted as a kind of Fourier analysis, but one that uses the actual frequencies of emission rather than the unobserved harmonic overtones given by a straightforward Fourier analysis. Schrödinger brought out the physical significance of this:

> However, the idea of wave mechanics is not that of a sudden transition from one state of vibration to another, but according to it, the partial moment concerned—as I will briefly name it—arises from the simultaneous existence of the two proper vibrations, and lasts just as long as both are excited together.[74]

This formed the nucleus of Schrödinger's second interpretation of the ψ function. In the 6 June letter to Lorentz cited earlier, Schrödinger stressed the point that instead of considering the Bohr energy levels basic, one should consider frequency basic and treat an individual orbit, at least for large quantum numbers, as made up of a superposition of very many proper oscillations.[75] Two weeks later, Schrödinger submitted his final paper in the quantization series, Q4.[76] In this paper Schrödinger presented the time-dependent Schrödinger equation, though not in the full generality Dirac later gave it; time-dependent perturbation theory; and a relativistic wave equation equivalent to the Klein-Gordon equation. These achievements, significant as they are, are subordinated to the special emphasis characterizing this paper, the attempt to work out the promised new interpretation of wave mechanics in a way that comes to grips with the difficulties we have been considering. In developing this interpretation, Schrödinger was particularly concerned with showing how his theory yields explanations that are different from and superior to those given by matrix mechanics. These are the aspects of Q4 that we will consider here.

The basic interpretative problem confronting Schrödinger was the one present from the beginning, to explain the physical significance of the ψ function. It is convenient to consider this problem in two states, the ψ

function for a free electron and the ψ function for a bound electron. Though Schrödinger was favorably disposed to the idea that free electrons may be considered as wave packets, he still would not make this a basis for the quantization series, but relegated it to his later paper on the Compton effect. In Q4 he focused on the physical significance to be accorded the ψ function within the atom, especially the multielectron atom.

The basis of Schrödinger's interpretation is a modification of the idea, previously developed for the one-electron case, that $e\psi\bar\psi$ rather than $e\psi\partial\bar\psi/\partial t$ represents the charge density. This is extended to the multielectron case by a five-step procedure:

  a) selecting some particular particle;
  b) keeping the triplet of coordinates that describe its position in ordinary mechanics fixed (this triplet [*x, y, z*] assigns a location in real space);
  c) integrating $e\psi\bar\psi$ over all the rest of the coordinates of the system (this is an integration in a configuration space of $3n - 3$ dimensions for an *n*-particle system);
  d) multiplying by *e*;
  e) and then doing the same for every other particle.

After giving these rules, Schrödinger summarized the interpretation they supported:

> This rule is equivalent to the following conception, which allows the true meaning of ψ to stand out more clearly. $\psi\bar\psi$ is the kind of weight-function in the system's configuration space. The wave-mechanical configuration of the system is a super-position of many, strictly speaking of all point mechanical configurations kinematically possible. [Italics in original.][7]

This interpretation seemed to solve the problems presented by configuration space. The product $e\psi\bar\psi$, when summed for the atom as a whole, is physically real and corresponds to the spatial distribution of electrical density. However, one can only get at this total result by a series of partial calculations. These intermediate steps require a detour through a physically unreal configuration space. Unfortunately, this solution, though suggestive, could not be considered final, for the Schrödinger equation still led to a complex, rather than a real, ψ function. Even though his new interpretation did not require it, Schrödinger still thought that the true wave function would be real.

The best that Schrödinger could do with this difficulty was to suggest a plausible hypothesis. A complex wave function can always be written in terms of sines and cosines by the formula $e^{i\phi} = \cos \phi + i \sin \phi$. It may be that one is really dealing with a real function (cos φ) and its time derivative (− sin

ϕ). If the total wave function of the atomic system can only be approximated through the five steps summarized earlier, then it seems reasonable to suppose that there is a real wave equation, probably of fourth order, to which Schrödinger's present equations are just first approximations. This true wave equation should have functions that are real rather than complex.

This interpretation remained tentative and incomplete. Yet, it supplied the only basis Schrödinger could develop for showing the superiority of his system to Heisenberg's. To bring this out, Schrödinger tried, in effect, to beat Heisenberg at his own game. The justification of the Ladenburg-Kramers dispersion formula had played a crucial role in the development that led Heisenberg to quantum mechanics. Schrödinger now presented an interpretation which, he believed, could make the formal results obtained by Kramers and Heisenberg physically intelligible. Earlier we considered the way Schrödinger used the ψ function, and the electromagnetic interpretation he accorded it, to explain the dipole moment of an atom subject to electromagnetic radiation. Now these considerations were extended to the case where two free vibrations are excited, u_k and u_l. In addition to the sum of these two terms, Schrödinger expected the interaction terms between the forced vibrations of u_k and the free vibrations of u_l and vice versa. The frequency of these interaction terms is not v, the frequency of the incident radiation, but the new frequencies Smekal had introduced,

$$|v + (E_k - E_l)/h|. \qquad (7.20)$$

The Kramers-Heisenberg formula, developed prior to quantum mechanics, had this term, as did the Born-Heisenberg-Jordan paper on quantum mechanics. But the Heisenberg approach, Schrödinger insisted, could explain neither the physical significance of this term nor the conditions for its experimental verification.

The conditions for experimental verification are, he argued, uniquely given by Schrödinger's electromagnetic interpretation of the ψ function. For this postulated, as yet unobserved, radiation to occur, both states (u_k and u_l) must be strongly excited in the same individual atom. For such radiation to be detectable, such simultaneous excitations must occur in many atoms. The interpretation of this difference term in the frequency as due to the simultaneous excitation of different states or vibrational levels is peculiar to Schrödinger's interpretation of wave mechanics. Matrix mechanics, which abandons space-time descriptions, has no way of even expressing this. Schrödinger made this difference quite explicit:

> As far as I can see, the above-mentioned dispersion theory of Heisenberg, Born and Jordan does *not* allow of such reflections as we have just made, in spite of the great formal similarity to the present one. For it only considers *one* way in which

the atom reacts to incident radiation. It conceives the atom as a timeless entity, and up till now is not able to express in its language the undoubted fact that the atom can be in *different* states at different times, and thus, as has been proved, reacts in different ways to incident radiation. [Italics in original.][78]

Conflicting Interpretations

With the completion of Q4 the most productive period in Schrödinger's career came to a close. The basic formulas and methods of wave mechanics had been so securely established that they became a permanent part of subsequent physics. The way this formalism is to be interpreted, however, remained a subject of lively debate. Until the summer of 1926, the community of atomic physicists had essentially two options: the forbiddingly difficult formalism of matrix mechanics, with its austere prohibition of space-time descriptions of an electron's orbit; or Schrödinger's much more manageable differential equation and the intuitively appealing interpretation associated with it.[79] Even such main-line physicists as Sommerfeld were showing signs of switching to Schrödinger's position.

In Q4 Schrödinger had shown that if $e\psi\bar{\psi}$ is interpreted as the spatial distribution of charge density, then conservation of electrical charge and current obtains, regardless of any problems associated with spreading wave packets. A new interpretation of $\psi\bar{\psi}$ as representing a probability came from the work of Max Born, though not in as straightforward a way as indicated in the standard historical accounts. Prior to Heisenberg's initiation of quantum mechanics, Born, with the assistance of Jordan, had been attempting to develop a quantum theory of collision processes. They wished especially to account for the peculiar Ramsauer-Meyer effect. On classical grounds one would expect that the faster an electron is moving, the less it is deflected by collisions with atoms. Experiments, however, indicated that slow electrons are deflected less frequently and with smaller deflection angles. Franck, the leading experimental physicist at Göttingen, suggested that a low-energy electron might give up less of its energy in high-frequency radiation than a fast electron.

To give this suggestion a theoretical development Born and Jordan found it necessary to identify stationary states of a freely moving electron and then calculate the probabilities of transitions between these states. Then came Heisenberg's paper, the Born-Jordan transformation of quantum mechanics into matrix mechanics, and the three-man paper. Born tried to include collision problems in the matrix theory and encountered some technical difficulties. Periodic motion, such as the motion of bound electrons, fitted nicely into a matrix formulation. Aperiodic motion, such as the motion of electrons in atomic collisions, did not. Such motion required continuous matrices, an obscure and poorly explored topic.

At the end of October 1925, Born went to America to lecture at MIT. While there he worked with the young Norbert Wiener. To get around the problems posed by continuous matrices they tried to replace matrices by operators, a more general notion. Though this was not very successful, it suggested methods of adapting Born's earlier treatment, based on stationary states of freely moving particles, to quantum mechanics.

When Born returned to Göttingen in the spring of 1926, he studied Schrödinger's papers on wave mechanics and accepted the idea, which Pauli had communicated to the Göttingen group, of the mathematical equivalence of wave and matrix mechanics. Since the Schrödinger wave equation yielded both discrete and continuous solutions, it seemed to offer the best prospects for treating collision problems. In a short paper written in June 1926, Born suggested that it should be possible to formulate collision problems by using the asymptotic values of the electron's initial and final states as boundary conditions in the solution of a wave equation representing both the electron and the atom. In a wave-mechanics formulation the asymptotic solution is a superposition of plane waves. Born commented that, if one wished to interpret this result in terms of particles rather than waves, then the coefficient should be interpreted as a probability that a particle coming in from one direction should be thrown in the direction proper to the plane wave associated with the coefficient. In a note added to the proof copy around 25 June, Born revised this and claimed that the probability should be proportional to the square of the coefficient.[80]

In his subsequent fundamental paper on collision processes Born generalized this.[81] According to Wessels, whose interpretation I am summarizing, the crucial feature behind the generalization was the realization that any wave function, not merely those representing collisions, could be represented as a superposition of eigenfunctions. Thus, one has a summation for the discrete case and an integral for the continuous case:

$$\psi(q) = \sum_n c_n \psi_n(q), \tag{7.21a}$$

$$\psi(x) = (\tfrac{1}{2}\pi) \int c(k) e^{ikx} dx, \tag{7.21b}$$

where $k = 2\pi/\lambda = 2\pi p/h$. In the discrete case, Born proposed that $|c_n|^2$ be interpreted as the probability that a system represented by $\psi(q)$ is in the state with coefficient c_n. If $\psi(q)$ is thought of as an ensemble, then c_n^2 gives the frequency of the state $\psi_n(q)$ in the ensemble. Similarly, in the continuous case, $|c(k)|^2$ gives the frequency of the plane wave e^{ikx} in the ensemble $\psi(x)$.

In Born's paper, relative probabilities, or statistical frequencies, were associated with the expansion coefficients rather than with the wave functions themselves. This was done only for two cases: a superposition of energy eigenfunctions for a bound electron and a superposition of momentum eigenfunctions for a free electron. Adapting an earlier suggestion of Einstein's, Born indicated that the ψ function could be thought of as a kind of ghost field guiding the electron.

Jordan, and especially Pauli, extended Born's ideas. Pauli was the first to interpret the wave function itself, rather than merely its coefficients, as yielding a probability, and doing so for position as well as for momentum and energy. The now standard "Born interpretation" that $|\psi(q_1 \ldots q_f)|^2 dq_1 \ldots dq_f$ represents the probability of finding a system that is in the indicated quantum state in the volume element $dq_1 \ldots dq_f$ first appeared in a footnote to an article on gas degeneracy and paramagnetism Pauli sent to the *Zeitschrift für Physik* in mid-December 1926.[82]

The scientific community for the most part was not particularly concerned with the intricacies of interpretation that worried Schrödinger and Heisenberg. Primary emphasis was given to solving the problems that the old Bohr-Sommerfeld theory had left unresolved.[83] Here, however, we will concentrate on the resolution of interpretative problems. The scientific community quickly came to accept the probability interpretation of the ψ function developed by Born and Pauli, but not the general interpretation of quantum situations developed by Born.[84] This acceptance effectively shifted the burden of proof onto Schrödinger's shoulders. Since Born's minimalistic interpretation seemed to suffice for most applications and fitted both matrix and wave mechanics, any more extensive interpretation required justification.

Schrödinger's efforts to refine and justify such an interpretation involved two interrelated problems. The first was to find and develop some points where his interpretation would lead to observable differences. Since Born's interpretation of $\psi\bar{\psi}$ as a probability distribution paralleled Schrödinger's interpretation of $e\psi\bar{\psi}$ as a spatial distribution of charge density, the difference could not come from any consequences of the ψ function. There was, however, a significant difference in the way in which Schrödinger interpreted the emission of radiation. In his view the emission of a frequency ν_{mn} depended on the simultaneous excitation of the two eigenfrequencies E_m/h and E_n/h. In the standard eigenfunction expansion, the intensity of the radiation would be expected to be proportional to $c_m^2 c_n^2$, the coefficients for both eigenfrequencies. The matrix mechanicians still held to the Bohrian idea that the intensity is simply a function of the higher-energy state, and hence of the coefficient c_m^2. This difference was not yet testable.[85]

The second issue was more a question of the goal of scientific explanation than of testable conclusions. Schrödinger's electromagnetic interpretation had difficulties he was well aware of. So too did his interpretation of electrons as wave packets. Lorentz had replied to Schrödinger's 6 June letter with a long letter dated 19 June, two days before Q4 was submitted. This letter contained a detailed calculation, requiring some twelve pages, showing that a wave packet in a high Bohr orbit would not remain intact.[86] Lorentz concluded the letter with the hope that Schrödinger might still find some easy way to retain the wave packet interpretation in spite of the setback Lorentz's calculation represented. This hope manifests a goal which both Lorentz and Schrödinger thought proper to scientific explanation. It should be possible in principle to give a spatiotemporal description of what goes on within the atom. As long as this criterion is retained, difficulties with particular models may lead to modifications of the models used or to a search for new models, but they do not induce an abandonment of the role of such models or the hope of a spatiotemporal description.

On this point there was a clear conflict between on the one hand Schrödinger and the other physicists, especially of the Berlin school, who thought of spatiotemporal descriptions as a legitimate and necessary goal, even in atomic physics, and on the other hand the main-line atomic physicists. Bohr's quantum postulates of discrete energy states and discontinuous jumps between such states prohibited any spatiotemporal description of the process responsible for the radiation. Later, as we have seen, Bohr had concluded that any account of the motion of electrons within the atom must be considered formal rather than realistic. The new matrix mechanics effectively pushed this mode of interpretation to an extreme. Spatiotemporal accounts of electronic activities within the atom were simply and totally abandoned. The remainder of this chapter will be concerned with the historical resolution of these two problems: Schrödinger's account of the process of the emission of radiation; and the criterion that induced him to concentrate on such an account, the requirement that a complete scientific explanation necessarily includes a spatiotemporal description of the system treated.

Before considering this historical development, I would like to raise, without resolving it, a psychological problem concerning Schrödinger's motivation. There is no doubt of Schrödinger's opposition, then and later, to quantum jumps and indescribable processes, and his insistence on the need for spatiotemporal descriptions, or Boltzmannian pictures, as a goal for science. The fact that Planck, Einstein, and Lorentz clung to such classical ideals of scientific explanation is not surprising; that Schrödinger also did is surprising. He was the one who had taught that statistical laws are basic and who had enthusiastically accepted the idea that in atomic processes causality be considered a statistical rather than an absolute law. Schrödinger did accept the Hertz-Boltzmann idea of the role of pictures in scientific explanation. Yet, Schrödinger, *qua* Vedantic philosopher, also insisted that no pictures supplied by science represent physical reality as it exists objectively, for physical reality is just a collective construct. One cannot help but wonder why Schrödinger was so insistent on retaining semiclassical ideals of scientific explanation.

In our earlier debate on this issue, Linda Wessels explained Schrödinger's insistence in terms of his prior philosophical commitments and supplied impressive documentation to support this position.[87] To me, it still seems more likely that a competition for primacy between Heisenberg and Schrödinger lent a unique significance to the issue of how the new formalism is to be interpreted physically, the only remaining difference of significant import between wave mechanics and the prior matrix mechanics. I suspsect that if matrix mechanics had not already been developed as a competing system, Schrödinger would have reacted quite differently to Born's probabilistic reinterpretation of his ψ function.

Both intellectually and emotionally Schrödinger seems to have had more unresolved tensions than most of the creative thinkers we have been considering. It is quite likely that Schrödinger himself would not have been able to evaluate the relative force of his tangled intellectual and emotional motives. Regardless of the motivation that inspired it, Schrödinger's attempt to develop a consistent and adequate physical interpretation of the new formalism stimulated Bohr and Heisenberg into developing the Copenhagen interpretation, and it also provided that interpretation with its most serious competition. For these reasons Schrödinger's efforts, though ultimately self-defeating, merit serious consideration.

On 23 July 1926, Schrödinger gave a lecture on wave mechanics in Munich, which Heisenberg attended. It seems to have been the first time the two met. By that time Heisenberg's earlier dismissal of Schrödinger's physical theory as crap had changed considerably. Heisenberg had accepted Schrödinger's demonstration of the equivalence of the two formalisms and had recognized the advantages that Schrödinger's methods presented for calculation. Heisenberg, in fact, had already begun his calculations of the helium atom energy levels, calculations that eventually made extensive use of the Schrödinger wave functions.

In the discussion following Schrödinger's talk, Heisenberg raised the objection that Schrödinger's interpretation did not even seem to fit Planck's original blackbody radiation derivation. Willy Wien cut him off with the assurance that Schrödinger would figure out such details sooner or later. Schrödinger himself was much less certain.[88] Two letters Heisenberg wrote after this encounter reflect the new modifications in the problematic we have been considering. One to Bohr, stressing the difficulties of the Schrödinger interpretation, was probably responsible for the invitation Schrödinger received to come to Copenhagen and discuss the interpretation of quantum theory.[89]

Heisenberg's 28 July letter to Pauli is more revealing.[90] He reported that he found Schrödinger personally quite nice and thought his physics quite important. However, Heisenberg noted, it is not very difficult to construct a physical theory if one simply throws overboard the photoelectric effect, the Franck collision process, the Stern-Gerlach effect, etc. He then went on to a technical point. Contrary to the opinion Heisenberg and Pauli shared, Schrödinger did not think transition coefficients sufficed for calculations. Heisenberg accordingly requested that Pauli, who was already familiar with the calculations involved, calculate the damping coefficients for the Balmer and Lyman series and compare them with Wien's measurements. Though Wien's measurements proved ambiguous and Pauli was not the one who carried through the requisite calculations, this calculation of radiation coefficients did eventually provide, as Heisenberg anticipated, a decisive test of Schrödinger's interpretation.

On 1 October 1926, Schrödinger arrived in Copenhagen for discussions with Bohr and Heisenberg on the interpretation of the new quantum theory. The anecdotal aspects of this meeting are known, chiefly from Heisenberg's later accounts, and have become part of the established lore of modern physics: Bohr's intensive and unremitting argumentation; the collapse of the already exhausted Schrödinger; Schrödinger's confinement, with Magrethe Bohr tending to Schrödinger's physical needs while her husband sat at the other side of the bed and continued to argue physics. For our purposes, however, the content of their disagreement is of more concern than its emotional aspects. Bohr accepted the mathematical formalism that Schrödinger had developed, agreed with Schrödinger that the wave properties of matter had to be included in any acceptable interpretation, but rejected the physical interpretation Schrödinger had developed on the grounds that it did not account for the corpuscular aspects of matter and could not even explain Planck's treatment of blackbody radiation. This, like the old quantum theory that sprang from it, required discrete energy states and quantum jumps between states. Ultimately the only reply the beleaguered Schrödinger could give was, "If all this damned quantum jumping were really here to stay, I should be sorry I ever got involved with quantum theory."[91]

Heisenberg's account reflects Bohr's side of the argument. Schrödinger's own reaction is best seen from a letter he wrote to Bohr on 23 October 1926,[92] two weeks after leaving Copenhagen, in which he explained why he could not accept the doctrine of complementarity which Bohr was then developing and which we will treat in the next chapter.

Here there apparently exists a contradiction and you say: our earlier words and concepts do not extend this far. I can not be satisfied with this position and I cannot, accordingly, deduce for myself the right to operate any further with contradictory expressions.

Then Schrödiger contrasted this summary of Bohr's position with his own ideas on the goal of scientific explanation:

What hovers before my eyes is only a single thesis: even if a hundred attempts miscarry, one would not give up the hope of reaching the goal of a representation of the true properties of space-time events through—I do not say classical pictures—but through representations that are free of logical contradictions. It is extremely probable that this is possible.

Away from the competitive strife of Copenhagen, Schrödinger tried to work out this program in the privacy he found necessary for sustained thought. As mentioned in the letter to Bohr, he focused on the space-time representation of events rather than the configuration-space formulation and the physical interpretation which he accorded it and which Bohr had so severely criticized. The first two articles he wrote on his return from Copenhagen reflect an effort to revitalize de Broglie's idea of material waves in real space, rather than Schrödinger's own configuration-space waves.[93] Many physicists independently, and approximately simultaneously, had developed the relativistic wave equation now known as the Klein-Gordon equation.[94] Though Schrödinger was one of the codiscoverers, he used Gordon's development of relativistic quantum mechanics as the basis for his new interpretation. This reason for this is the fact that Gordon gave a relativistically invariant four-dimensional expression of the continuity equation which supplied the justification for Schrödinger's electromagnetic interpretation of the ψ function.[95] For a short time, Schrödinger seems to have banked rather heavily on the acceptability of this new basis:

The Hamiltonian principle from which the exact relativistic differential equation for de Broglie waves can be obtained, appears to justify completely the hopes which I had set upon an intimate blending together of wave mechanics and classical electrodynamics.[96]

This new interpretative basis, unfortunately, had even more flaws than the older one based on nonrelativistic configuration-space wave functions. It did not yield the established relativistic corrections for hydrogen. Any attempt to extend the equation from one to many particles precipitated the problems considered earlier of a choice between de Broglie wave functions in real space and Schrödinger wave functions in configuration space. In his treatment of the Compton effect Schrödinger blurred this distinction by focusing on a steady-state case where only the density distribution need be considered. A few months later, Dirac gave the same problem a more rigorous quantum mechanical treatment that did not involve such simplified assumptions of a steady state.[97] Schrödinger adapted his approach to classical electromagnetic theory, coupling ψ waves to Maxwell fields. Here again, Dirac provided the necessary corrective, one that Schrödinger soon accepted, by substituting a quantized electromagnetic field for the classical Maxwell field.[98] Finally, the new interpretation was no more able to solve the problem of the dispersion of wave packets than was the older one.

In the background behind these attempts at interpretation and reinterpretation there was still the competition between Heisenberg and Schrödinger. As indicated earlier, Heisenberg initially reacted strongly against wave mechanics considered as a competing system. However, he quickly accepted the proofs developed by Schrödinger and Pauli that the two systems were mathematically equivalent. From this time on, Heisenberg carefully distinguished between Schrödinger's mathematical method, which Heisenberg used to supplement matrix methods, and the physical interpretation Schrödinger attached to this method.[99] In the first paper he wrote utilizing both methods, Heisenberg argued that his own interpretation was preferable to Schrödinger's because it treated the many-body problem as a collection of particles in real space rather than as waves in configuration space. Yet, Heisenberg had to admit, the intuitive content of his own theory was not yet clear. The inclusion of Bose-Einstein statistics led to a special solution of the many-body problem which could be interpreted as phase relations between the partial systems or particles. This seemed to accord better with a wave than a particle interpretation.

From this time on, Heisenberg accepted Schrödinger's method and used it to supplement quantum mechanics or even to replace it in solving problems that were forbiddingly difficult by the methods of matrix mechanics, such as the spectra of atomic systems with two electrons.[100] Yet Heisenberg remained unwilling to accept Schrödinger's interpretation of this formalism. Through the fall and winter of 1926 this unwillingness became the focus of a growing rift between Bohr and Heisenberg. During this period Heisenberg wrote two papers which clearly manifest an attempt to show quantum mechanics superior to wave mechanics as a physical interpretation and to counter Schrödinger's criticism that quantum mechanics remained too unintuitive to be acceptable. The first paper, a qualitative nontechnical survey of the new developments in quantum theory, carried the now familiar distinction between Schrödinger's mathematical formalism, which Heisenberg found acceptable, and the physical interpretation, which presented more formidable problems.[101] Here Heisenberg clearly distinguished between de Broglie waves in real space and Schrödinger waves in configuration space. He found Schrödinger's interpretation of the latter unacceptable on the grounds that the assumption of continuity did not seem in accord with such discontinuous phenomena as blackbody radiation and dispersion. This resolution of competing interpretations, however, was still far from conclusive. Heisenberg was forced to admit that the particle interpretation was ambiguous in that the theory did not allow the specification of a particle's position as a function of time, nor did it easily lend itself to an explanation of the interference effects which seemed analogous to interference effects in light.

Fluctuation phenomena, treated in the second paper, were seen as having special significance for bringing out the intuitive consequences of discontinuity in processes occurring in very limited space-time regions.[102] Heisenberg considered two similar atoms a and b having energy states E_n and E_m. When these are coupled together, the interaction between them can be viewed in two ways: as regularly occurring energy jumps in each atom due to energy exchanges between them or as something analogous to the classical interaction between two coupled oscillators. In the latter interpretation the energy exchange is continuous, while in the former it is discontinuous. The two interpretations lead to the same conclusions for all observable effects, chiefly because of the impossibility of measuring individual interactions in such a way as to settle the continuity-discontinuity issue. Yet, the continuity interpretation cannot be accorded any physical significance except in cases where the de Broglie wavelength of the particles is small compared to the size of the interaction volume. Heisenberg interpreted this as favoring discontinuous energy jumps. But here again, as he clearly realized, his argument was far from conclusive. Some three weeks after Heisenberg finished his paper Jordan submitted a paper on essentially the same problem.[103] In it he proved that both quantum mechanics, with the assumption of energy discontinuities, and wave mechanics, which assumes a continuous energy exchange between the two coupled atoms, agree in all empirically testable consequences.

To render the dialectics of this development intelligible it is necessary to disrupt the chronological order of the published papers. Heisenberg's indeterminacy paper was written prior to the Schrödinger paper we are about to consider. Schrödinger's paper, however, answered the Heisenberg papers we have just summarized as well as the objections Bohr had raised, but made no reference to the indeterminacy paper, which was not yet available at the time Schrödinger completed his article.[104]

After returning from a lecture tour in America, Schrödinger made a final sustained attempt to salvage the wave interpretation. He abandoned his efforts to revise de Broglie real-space waves and returned to his own configuration-space waves.[105] Following Heisenberg and Jordan, he considered two interacting systems: I, with energy levels E_k and $E_{k'}$ and II, with energy levels E_l and $E_{l'}$. A necessary condition for a quantum interaction is that $E_k - E_{k'} = E_l - E_{l'}$. Heisenberg had interpreted this in terms of energy jumps. Schrödinger reworked the problem and obtained equivalent results without introducing any assumptions of discrete energy levels and quantum exchanges of energy. If, he argued, frequency is considered basic rather than energy, then the exchange interaction could be interpreted as a resonance phenomenon like the sympathetic oscillations of coupled pendulums. If the basic postulate of energy jumps plays no role, then, in accord with Occam's razor, it should be discarded:

> According to the fundamental principles of research which are commonly regarded as correct, does not the foregoing compel us to exercise the utmost caution with respect to the quantum postulates, even (I almost feel inclined to say) to distrust them—quite apart from their axiomatic unintelligibility.[106]

By extending these considerations to the interaction of a system, which could be called a thermometer, with an extremely large system having a very dense eigenvalue spectrum, or a heat bath, Schrödinger was able to deduce Planck's formula from his wave mechanics, thus answering Bohr's challenge. In passing he also showed that his interpretation yielded Born's probability account, but accorded it a derivative rather than a primary status. Accordingly, the essential aspects of Schrödinger's interpretation could be salvaged if frequency was accepted as a fundamental physical concept and energy accorded the derivative status yielded by the formula $v_{nm} = (E_n - E_m)/h$. A heroic attempt, but one that soon fell victim to Heisenberg's indeterminacy principle.

The Origin of the Indeterminacy Principle

Heisenberg's indeterminacy principle has emerged from the flux of science in progress to become one of the most philosophically pregnant limiting principles established by modern science. Yet, it is only by resituating this principle in the flux from which it emerged that we can see the decisive role Heisenberg's paper played in settling the struggle of competing interpretations. Heisenberg probably would have preferred to concentrate on solving outstanding problems that the new formalism now rendered tractable. He had completed his treatment of two electron atoms. The symmetry considerations involved led him to explore the relation between nuclear spin and band spectra, a topic that will be considered briefly in the next chapter. As his letters to Pauli indicate, he had already begun to develop his quantum theoretical explanation of ferromagnetism. Any one of these achievements would suffice to make a professional reputation.

Yet, there was no way he could dismiss the problems involved in interpreting quantum theory as problems which had already been settled. In the fall of 1926 and on through the winter he and Bohr had interminable discussions of the problematic aspects of quantum theory, focusing especially on the interpretation of various thought experiments. Their initial disagreements, expressed and developed in the pressure-cooker atmosphere of Bohr's think tank, led to increasing friction. The ultimate basis of this disagreement, as Heisenberg later explained it, was:

> The difficulties in the discussion between Bohr and myself was that I wanted to start entirely from the mathematical scheme of quantum mechanics and use Schrödinger's theory perhaps as a mathemetical tool sometimes, but never enter into Schrödinger's interpretation which I couldn't believe. Bohr, however, wanted to take the interpretation in some way very seriously.[107]

Though Schrödinger had left, his influence lingered on.

Heisenberg found more support for his approach in discussions with Pauli and Dirac. What we wish to consider is the contribution each of these men made to the formation of the indeterminacy principle. Pauli found himself serving as something of a sounding board and critic for both Heisenberg and Schrödinger. While Pauli was more sympathetic to Schrödinger than either Bohr or Heisenberg had been, he never expressed any support for Schrödinger's idea of replacing quantum discontinuities by something continuous.[108] Pauli's letters to Heisenberg were more constructive. On 19 October 1926, Pauli sent Heisenberg a long letter presenting his views on recent developments: Fermi-Dirac statistics, Born's collision theory, the way collisions should be treated in three dimensions. He concluded this part of his letter with the reflection:

> So far the mathematics. The physics of this is still quite unclear to me. The first question is: why then only the p's, and in any eventuality not both the p's and also the q's, both should be described with arbitrary accuracy. It is this old question which enters, if one would give the velocity direction *and* separation of the asymptotic path from the nucleus (at least with a determined accuracy). . . . One can look at the world with p-eyes and one can look at the world with q-eyes, but when one attempts to use both eyes simultaneously, then one will err.[109]

Heisenberg reported back that everyone (Bohr, Dirac, Hund, and others) had read Pauli's stimulating letter. Heisenberg himself was enthusiastic about Pauli's suggestions. He linked the symmetry of the p and q representations to the basic equation $pq - qp = hi$. This he saw as one more blow to the wave interpretation. It makes no sense to speak of a monochromatic wave at a definite point or interval. However, if one makes the line not too sharp and the time interval not too short, then it does make sense. Analogously, it makes no sense to speak of the position of a corpuscle with definite velocity. However, such a way of speaking does make sense if one does not take the position and velocity too exactly.[110]

In the letter just summarized as well as in the later letter of 15 November 1926,[111] Heisenberg suggested that such concepts as 'space', 'time', and 'velocity' have meaning for an individual particle only to the degree that they admit of measurement. Then a new element came to dominate Heisenberg's considerations. Dirac had come to Copenhagen and was developing his transformation theory. In a letter to Pauli of 23 November 1926,[112] Heisenberg enthusiastically reported that this extraordinarily comprehensive generalization really contained all the physics of the quantum theory: the Schrödinger function, Born's collision theory, probabilities, Jordan's matrix transformation. Accordingly, the basic question remaining was the physical significance of the transformation function S.

Dirac's idea, as Heisenberg summarized it, is the following. Let p and q be any canonically conjugate magnitudes and $f(p, q)$ be a function of these. In this formulation the question of physical interpretation is, What does quantum mechanics allow one to say physically about f? His answer is that when q is treated as a c number, then one can draw a graph in which the shape of the curve reflects the functional relationship between f and p. When q is treated as a quantum variable, then one cannot have such a graph, but one can set coordinated boundaries to $p + dp$ and $f + df$ (with $q + dq$ as a special case of f).

A week after this letter was written, Dirac sent to England his basic paper on the physical interpretation of quantum theory. In this paper Dirac was quite explicit about the information quantum theory supplied concerning related position and momentum measurements: "One cannot answer any questions on the quantum theory which refer to numerical values for both the q_{r0} and the p_{r0}" (coordinate and momentum measurements). All that quantum theory can do, Dirac argued, is to give associated ranges of values for canonically conjugate variables: "Questions of this type appear to be the only ones to which the quantum theory can give a definite answer, and they are probably the only ones to which physicists require an answer."[113]

Dirac clearly realized that quantum theory could only give limited information concerning coordinated measurements of canonically conjugate variables. But he did not make these limits precise or formulate them in terms of a principle. Heisenberg felt pressured to do so. His confrontations with Bohr had become so acrimonious that Bohr forced him to break down in tears. The two men were no longer able to work together. At the end of February 1927, Bohr left to go skiing in Norway and to think through his own position. Heisenberg remained in Copenhagen. He claimed that on the very evening of Bohr's departure he related the measurement problems that he had been discussing with Pauli and Dirac to the remark that Einstein had addressed to him some eight months earlier: the theory determines what can be observed.[114] He began to write up his views in the form of a long letter to Pauli, presumably in the hope that Bohr would take Heisenberg's views seriously if Pauli found them acceptable. This letter was effectively the first draft of the indeterminacy principle paper.[115]

Schrödinger had accepted the mathematical validity of the matrix formulation of quantum mechanics. His disagreement concerned the physical interpretation accorded this formalism. Schrödinger found Heisenberg's way of interpreting quantum mechanics to be so unintuitive that it was hardly acceptable as a physical theory. Bohr's objections also centered on the issue of the way in which the physical significance of the new formalism is to be determined. Heisenberg thought that one should concentrate on the mathematics and that physical interpretation is essentially a question of making explicit what is implicit in this mathematics. Bohr's own position will be considered in the next chapter. What is pertinent here is Bohr's insistence that Schrödinger's theory of matter-waves must be accorded a basic role in the overall interpretation. Heisenberg's paper "On the Intuitive Content of Quantum-theoretical Kinematics and Mechanics"[116] was intended, inter alia, to answer this criticism. The paper begins with a brief analysis of the notion of intuitive understanding. We believe that we have an intuitive understanding of a theory when we know in a qualitative way the experimental consequences of this theory for all simple cases and when we are assured that it is free from contradictions. Using this rather operationalist account of intuition, Heisenberg went on to explain the significance quantum theory could accord to such concepts as 'location of an electron', 'path of an electron', 'velocity', and 'energy'. By applying these considerations to such ideal experiments as the gamma-ray microscope, Heisenberg derived the indeterminacy principle in the form $p_1 q_1 \sim h$, where p_1 is the degree of exactness in momentum and q_1 that for position. A more formal approach, based on the Dirac-Jordan transformation theory, showed that the minimal uncertainty for the simple case of a Gaussian wave packet is $p_1 q_1 \sim h/2\pi$.

The Schrödinger wave function, Heisenberg argued, admits of the same interpretation. Such an interpretation, in fact, brings out the deep significance of the linearity of the solution. A superposition of solutions is also a solution. There is no prospect of getting around this difficulty by reliance on nonlinear relativistic equations.[117] Though many of his readers might have missed the significance of this reference to nonlinear relativistic wave equations, Schrödinger could not. The interpretation of ψ waves Schrödinger had defended in two papers previously considered, required nonlinear equations in applications to two-electron atoms.

Against the backdrop of these theoretical considerations, Heisenberg examined Schrödinger's contention that it is possible to consider particles as wave packets. First, he considered free particles and argued that the wave packet that Schrödinger had set up for the one-dimensional simple harmonic oscillator does not spread in the course of time only because the Fourier overtones are multiples of the fundamental frequency, something not true in other cases. A more precise treatment of the wave-packet representation of a particle can be had by applying Dirac's transformation theory to the linear motion of a mass-point. Suppose that at the time t the position is measured with a degree of exactness q and the momentum with a degree of exactness p_1 (where $p_1 q_1 = h/2\pi$). The form of the transformation function from a representation in which q_1 is arbitrarily small to a representation in which the position matrix is diagonal shows that future positions are not determined by the theory with any precision at all. More concretely, one can fix the position of an electron at a given time by hitting it with a gamma ray of very short wavelength. But then the high Compton recoil makes future positions quite unspecifiable. The wave packets that relate to measurable information have the same limitations. Accordingly, they cannot be interpreted as waves in real space. Rather, they should be interpreted as probability packets specifying relation between past measurements and possible future measurements.

Heisenberg examined the further idea that wave packets can represent bound electrons by using the same double process of arguing both from theoretical considerations and from an examination of idealized measurements. Suppose an electron is in a very excited state, e.g., $n = 1,000$. Then the path could, in principle, be observed with fairly long wavelengths of light. While this light would not have the energy to knock the electron out of the atom, it would impart sufficient recoil to displace the electron to some other orbit, e.g., between $n = 950$ and $n = 1,050$. Since the precise orbit proper to an excited state cannot be known with greater accuracy than this, one would have to represent the electron's orbit by a superposition of wave packets proper to these hundred orbits. Again, this is really a probability packet and one that disperses in time, though it remains within the atom.

Schrödinger had interpreted the radiation emitted or absorbed by atoms in terms of the simultaneous excitation of excited states. By a detailed examination of different ideal experiments, Heisenberg showed that this notion has no operational significance. In resonance radiation, for example, atoms bombarded with a frequency exactly corresponding to a difference in energy levels, $\nu_{12} = (E_2 - E_1)/h$, will reradiate coherent radiation. Because of this coherence, the phase can be determined precisely. However, because this is a cooperative phenomenon, there is no way of knowing which atoms are in state 1 and which are in state 2. There are ways in which atoms in different energy states can be distinguished, such as passing the atoms through an inhomogeneous magnetic field. But any field that could separate atoms according to their energy states would also destroy the coherence of the radiation and with it any knowledge of phases. It is accordingly physically meaningless to give any precise description of the behavior of electrons in atoms, regardless of whether the description is couched in wave or particle language.

After showing in detail that Schrödinger's interpretation could not be given any operational significance, Heisenberg turned the charge of being unintuitive back against Schrödinger. Quantum mechanics as interpreted by Heisenberg is quite intuitive in the sense that the indeterminacy principle allows one to think through the experimental consequences of quantum theory in all simple cases. Schrödinger's wave mechanics is certainly more intuitive in the sense of having more manageable mathematical forms. In the question of physical principles, however, the popular intuitiveness of wave mechanics must be considered a deviation from the straight path.[118]

After this devastating criticism of Schrödinger's interpretation, the paper concludes with a surprising postscript indicating that wave and particle pictures can be considered complementary. This postscript has a history of its own. When Bohr returned from Norway and received a copy of Heisenberg's hastily written indeterminacy paper, he treated it as a rough draft. It should, he thought, serve as a basis for detailed discussions and then further rewriting. Both Bohr and Dirac pointed out that Heisenberg's treatment of the gamma-ray microscope did not give a proper interpretation of Compton recoil. Heisenberg ignored all such suggestions and sent the paper, with all its imperfections, to the *Zeitschrift für Physik*. He wished this paper to be published as soon as possible.

The friction between Bohr and Heisenberg was exacerbated by the intervention of Oskar Klein, then Bohr's assistant. He had shown, on the basis of correspondence principle arguments, that the concept of matter-waves supplies as good a guide for the interpretation of quantum theory as does the concept of particles. In a letter to Pauli of 16 May 1927, Heisenberg discussed Bohr's criticism of his paper and his insistence on the need for introducing wave concepts in the treatment of the gamma-ray microscope. Heisenberg summarized the situation:

> between Bohr and me there is still an essential difference of taste concerning the word 'intuitive'. Unfortunately these discussions have led in recent days to rude personal misunderstandings between Bohr-Klein and me, and for this I of course bear a share of the guilt.[119]

With tempers strained to the breaking point, trivial incidents could take on an exaggerated importance. Klein thought Heisenberg had ridiculed his work in a letter to Pauli. Heisenberg had Pauli return the letter (now lost) so he could convince Klein that he had not satirized him. Yet, the situation worsened. In a letter of 31 May, Heisenberg explained to Pauli:

> So I have come to be in a fight for the matrices [and also for the primacy of transformation theory, a topic discussed earlier in the letter] and against the waves. In the ardour of this strug- gle I have often criticized Bohr's objections to my work too sharply and, without realizing or intending it, have in this way personally wounded him. When I now reflect on these discus- sions, I can very well understand that Bohr was angry about them. In this personal opposition, which was brought about through my fault, Klein has entered, and Klein's intervention seems to make the situation much worse.[120]

Heisenberg requested that Pauli write to Klein and attempt to clarify the misunderstanding. Earlier Pauli had declined Bohr's invitation to come to Copenhagen for the summer because he had already accepted an invitation to Göttingen. Now Pauli decided that a visit was needed. During the Pentecost vacation in the first week of June, Pauli went to Copenhagen and mediated the rift between Bohr and Heisenberg. Each accepted his face- saving solution that their disagreement did not really concern the basic concepts, but merely the order of precedence given these concepts. Heisen- berg, who was then correcting the proofs of his indeterminacy principle article, made a peace offering to Bohr by adding to the proofs the note mentioning the complementarity of the wave and particle interpretations and indicating that Bohr would clarify this in a forthcoming paper. This paper, which will be discussed in the next chapter, was written, revised, revised again, and discussed for a full two years before it was finally pub- lished. As with matrix mechanics some two years earlier, the Copenhagen interpretation of quantum mechanics evolved from intense, often acrimo- nious, discussions between Bohr, Heisenberg, and Pauli.

Heisenberg had shown that wave packets could not supply a realistic model of particles. Paul Ehrenfest then showed how the concept of a wave packet could serve as a bridge between classical and wave mechanics. He proved that, regardless of spreading, the center of mass of a wave packet obeys the quantum equivalent of Newton's second law of motion: the expectation value of the time derivative of the momentum is equal to the negative gradient of the potential energy function, which may be called a force.[121]

The Demise of the Wave Interpretation

I have been unable to find any source material documenting Schrödinger's immediate reaction to Heisenberg's paper. Yet, this very lacuna may be revealing. After the paper which we summarized as a last attempt to salvage the wave interpretation, Schrödinger wrote no technical papers for a year and a half.[122] This dearth, following upon his remarkable productivity, would seem to indicate a period of deep discouragement. At the Solvay Conference held in Brussels 24–29 October 1927, Schrödinger simply summarized the essentials of wave mechanics and presented his electromagnetic interpretation with no new refinements. In answer to a question de Broglie raised, Schrödinger discussed the introduction of the wave-packet idea and concluded:

> It has since been found that the naive identification of an electron, travelling over a macroscopic orbit, with a wave-packet encounters difficulties and cannot be accepted as literally true. The principal difficulty is this, that a wave-packet is scattered in all directions when it encounters an obstacle, an atom for example.[123]

He also indicated a willingness to go along with a practical, though theoretically unsatisfactory, resolution of the wave-particle duality: "The compromise proposed on different sides consists in admitting an association of waves and point-electrons; I accept it only as a way of resolving the difficulty."

It was not, as far as I know, until December 1928 that he first publicly discussed the indeterminacy principle. Then he indicated that he accepted the indeterminacy principle as valid, but felt that it still allowed a choice of strategies.

> One may believe either (1) that matter has *really* a wave structure. Then the uncertainty principle is an immediate consequence. Or (2) one may think that the uncertainty principle is the more fundamental. The wave theory then is simply an auxiliary construction for the convenience of grasping and representing the principle.[124]

It seems that Schrödinger initially followed the first option. In March 1928, he delivered a series of lectures in London on wave mechanics, which were subsequently published in book form.[125] Though these lectures contain little physics beyond the Q1–Q4 series, which they summarize, they present a slightly more cautious interpretation of the wave model: "The statement that what really happens is correctly described by describing a wave-motion does not necessarily mean exactly the same thing as: what really exists is the wave motion."[126]

In this book Schrödinger defends his interpretation of $e\psi\bar{\psi}$ as the spatial distribution of charge. The defense rests on two arguments that are already familiar. This interpretation allows an intelligible explanation of the emission and absorption of radiation. Here, as earlier, Schrödinger emphasized the predicted, but not yet observed, extra dispersion frequency $(E_k - E_l)/h$. This, Schrödinger argued, should be interpreted as the interaction of two frequencies which are simultaneously excited in the same atom. Second, he felt that an interpretation in which frequency rather than energy is basic removes the unintelligible discontinuities involved in the acceptance of quantum jumps. Thus, his book concludes with the rhetorical question, "Is it quite certain that the conception of energy, indispensable as it is in macroscopic phenomena, has any other meaning in micro-mechanical phenomena than the number of vibrations in h seconds?"[127] For Schrödinger, as for Heisenberg, the attempt to defend an interpretation drove him into a form of conceptual analysis.

Within the next few months after these lectures, advances in physics undercut both of the arguments Schrödinger relied on. The difference between the intensity predicted by the Born and Schrödinger interpretations was put to an experimental test by Enrique Gaviola.[128] Gaviola, an Argentinian physicist who had studied in Berlin and was familiar with the disputes concerning Schrödinger's interpretation, had come to the United States and was working with R. W. Wood. They found that the addition of a few millimeters of nitrogen or water vapor to mercury vapor enhanced the population of the $2\,^3P_1$ and $2\,^3P_0$ levels by a factor of a few hundred because of collisions which leave the mercury atoms in a metastable level. According to the Born interpretation, this should have no effect on transitions to these lines, for the number of transitions depends exclusively on the number of atoms in excited states above these levels. According to the Schrödinger interpretation, however, the intensity of lines to these states should be significantly increased on the grounds that the intensity depends on the simultaneous excitation of both the lower and upper levels. From the experiments he had performed, Gaviola concluded that enhancement of the population of the lower-energy levels produced no change in the ratio of the relative intensities. This conclusion, in turn, could be interpreted as a conclusive refutation of the Schrödinger interpretation. Wessels, who has thoroughly researched Gaviola's writings in English, German, and Spanish, argues (correctly, I believe) that this by itself constitutes a soft rather than a hard refutation.[129] Gaviola's calculations presupposed Born's method of interpreting populations in energy levels. It might have been possible to reinterpret the whole experiment on Schrödinger's basis. Schrödinger apparently never attempted to do so.

Finally, the predicted extra dispersion frequency was detected experimentally by C. V. Raman in India[130] and by Landsberg and Mandelstamm in Russia.[131] The nature of this radiation is worth considering. When a parallel beam of monochromatic light goes through a gas, a liquid, or a transparent solid, a small fraction of the light is scattered in all directions. If the spectrogram is prepared so that the scattered lines identical in frequency with the light source are strongly overexposed, some weak additional lines are found which do not appear in the spectrum of the light source. For a given scattering substance, the displacement of these lines from the line corresponding to the incident radiation is independent of the wavelength of the radiation used. These lines have exactly the form predicted by equation (7.20)

The formula works perfectly; Schrödinger's interpretation of it does not. In a gas or a liquid, some of the extra lines can only be explained as induced rotations or vibrations of the molecules struck by the incident radiation.[132] The idea of a molecule having different rotational or vibrational states simultaneously really has no plausibility at all. The total replacement of rotations and vibrations by frequencies which Schrödinger's interpretation would suggest is one that most physicists would find highly counterintuitive.

Dirac's systematization of quantum mechanics which made both wave and matrix mechanics special representations of his more general theory will be treated in a later volume. Two aspects of his work, however, should be mentioned here. He, and Jordan independently, developed a general transformation theory.[133] In this more general theory the eigenfunctions of the Schrödinger wave equation appear as transformation functions, enabling one to transform Heisenberg's position matrices into a scheme in which the Hamiltonian is diagonal. Second, Dirac developed a relativistic wave equation which, unlike the Klein-Gordon equation, included spin and yielded the proper relativistic corrections for the hydrogen atom.[134] This was an advance that Schrödinger accepted and eventually expanded upon. Though the negative energy states this relativistic wave equation yielded were not correctly interpreted until 1932, the equation itself requires quantum jumps between separated energy levels.

Years later, Schrödinger summarized the basic reasons that led him to abandon his original wave interpretation.[135] Particles cannot be pictured as wave packets because wave packets disperse. The electronic interpretation of the ψ function does not yield a self-consistent interpretation of the hydrogen atom. Planck's radiation law and some nonequilibrium processes such as the photoelectric effect require discrete energy levels and quantum transitions between them. Though, as he admitted, the Copenhagen interpretation has an admirable self-consistency, he could never accept it as an adequate representation of atomic processes. In Schrödinger's view, quantum jumps could no more be a real part of nature than Ptolemaic epicycles. To the end he remained a loner, outside the mainstream of quantum developments, severely critical of the theory he had done so much to shape.

8. The Consolidation of Bohr's Position

The Copenhagen interpretation of quantum mechanics and the subsequent debates between Bohr and Einstein are widely recognized as having a deep philosophical significance. The summaries, analyses, partisan defenses, and rational reconstructions that attempted to dissect, clarify, enlarge, and propagate the positions involved bear ample witness to the import of the problems treated. In spite of such extensive and continuing scrutiny, the philosophical issues involved in the originating debates have, I believe, rarely been adequately interpreted by either physicists or philosophers. Before proceeding to a closer analysis of the issues, I would like to indicate how such systematic misunderstanding might have occurred.

What emerged from Copenhagen in 1926–27 was not a new physics. That had already been accomplished. It was, rather, a set of prescriptions for speaking about the reality revealed by this physics in a consistent, meaningful, and hopefully correct way. This was at least as much epistemology as it was physics. The two pillars of the original Copenhagen interpretation were Heisenberg's indeterminacy principle and the address on complementarity which Bohr first gave in Como, Italy. These pillars eventually supported two related, but rather different, structures: the orthodox interpretation of quantum mechanics and Bohr's epistemology. The orthodox interpretation, summarized in and disseminated through Pauli's 1933 *Handbuch* article,[1] included the indeterminacy principle; the idea that photons, electrons, and other particles exhibit both wave and particle properties; the probabilistic interpretation of the wave function; the correspondence between eigenvalues, derived from the theory, and measured values; the complementary relationship between the Schrödinger and Heisenberg representations; and the guiding principle that quantum mechanics merges with classical mechanics in the limit of large quantum numbers. For the younger generation, those beginning quantum physics after this period, these principles presented pedagogical rather than philosophical problems. They had to be assimilated to master quantum mechanics.

Bohr's epistemological development is more complex. He began his career attempting to give realistic descriptions of atomic structures and elementary processes. By 1924 he had concluded that the old quantum theory could not be interpreted as supplying realistic pictures of submicro-scopic structures or events. The purely formal reinterpretation embodied in the B-K-S paper led to conclusions concerning individual events that were experimentally falsified. This failure induced Bohr to focus on the question of how concepts function in scientific explanations to make physical descrip-tions possible.

It took Bohr about ten years, after the Copenhagen discussions with Schrödinger and Heisenberg, to develop an answer to this question that he found satisfactory. This development can be roughly divided into three phases. The first involves the original presentation of Bohr's position on the interpretation of quantum theory and the ensuing modifications in the formulation of this position. Though these were relatively minor, they serve to bring out the aspects Bohr found unsatisfactory. The second phase is Bohr's work, especially after 1929, to extend the foundational concepts of quantum theory in a way that would allow for an epistemologically self-consistent treatment of quantum field theory, relativistic quantum mechan-ics, and nuclear physics. These efforts induced modifications in Bohr's position on the nature of concepts and on the way in which concepts, which normally function in one conceptual framework, may be extended beyond that framework. The third phase concerns the Bohr-Einstein debate on the completeness of quantum mechanics and the final formulation of Bohr's position. After about 1937 Bohr focused his interests successively on nuclear physics, the atomic bomb, and rebuilding and extending his Institute for Theoretical Physics, and on functioning as an elder statesman for the scien-tific community. Though he frequently gave popular lectures on the inter-pretation of physics, he did not introduce any further modifications in his position.

The first two phases will be treated in this chapter. The 1935 debate with Einstein will be postponed until after we have developed enough back-ground to clarify Einstein's position on the interpretation of science, the task of the next chapter. In the final chapter we will consider the developed position of the two protagonists in a more systematic, less historical fashion. Both men encountered problems in communicating their positions. Each was concerned with a coherent account of scientific knowledge as a whole. Each had developed his position through a long dialectical process of achieving a certain degree of coherence, making revolutionary break-throughs that shattered this coherence, struggling to achieve a new coher-ence, and then reflecting on the significance of such a process of ad-vancement. Neither was able to abstract his developed position from the particular interpretative problems that shaped the development without seeming to trivialize the reasons supporting the final position. For this reason neither was ever accepted as a major philosopher of science.

A detailed examination of the scientific work shaping and supporting each position would seem to serve a double function. It would make the final position of each man more intelligible. It would also supply two exceptional case studies in scientific development. The continuing conflict between the two men forced each to give a detailed justification of the way he interpreted scientific knowledge. In each case, accordingly, assumptions about the nature of scientific explanation continually interract with scientific practice in a way that leads to a dialectical modification of both interpretative theory and actual practice.

Before considering the development of Bohr's thought, one particular interpretative difficulty should be considered. It is helpful to separate the original philosophical issues involved in the interpretation of quantum theory from later philosophizing about these issues. Much of this later philosophizing, however, was done by the original participants, most notably Heisenberg. In an effort to achieve such a separation I will include some discussion of *antecedent* philosophical influences, philosophical positions that may have played a role in shaping some of the positions considered. However, I will prescind from any *subsequent* attempts to relate the philosophical issues involved in the interpretation of quantum mechanics to other philosophical positions, such as Heisenberg's use of terms derived from Aristotle and Kant. Such problems will be treated more systematically in the projected sequel to the present volume.

Bohr and the Philosophers

Max Jammer,[2] Gerald Holton,[3] and Lewis Feuer[4] have stressed the influence of certain philosophical sources on Bohr during his student years. The somewhat Hegelian novel *Adventures of a Danish Student*, written in the early nineteenth century by the Danish philosopher Poul Martin Møller, fascinated Bohr in his youth and later served as his favorite tool for instilling the young with a proper understanding of complementarity.[5] Harald Høffding was both a family friend and Bohr's philosophy professor. It was through him that Bohr came to know both Kierkegaard, who strongly influenced Høffding, and William James, a friend who invited Høffding to lecture at Harvard. It seems, however, that Bohr did not read much of James's own philosophical writings until the 1930s. Even then he probably only read some fragments and never made a systematic study of James's thought.

Søren Kierkegaard clearly seems to have been the philosopher who most stimulated Bohr during his student days. J. Rud Nielsen recounts that he once asked Bohr about Kierkegaard, and Bohr replied, "He made a powerful impression on me when I wrote my dissertation in a parsonage in Funen, and I read his works night and day."[6] In letters written from the parsonage Niels encouraged his brother Harald to read Kierkegaard's *Stages on Life's Road.*[7]

Yet, when Thomas Kuhn was trying to determine the sources that might have influenced Bohr's philosophical interpretation of physics, he asked, "Did you carry on your interest in these problems by reading books of philosophy?" Bohr replied, "No, not at all." Einstein, Heisenberg, Schrö-dinger, and others who wrestled with the problem of interpreting quantum mechanics frequently allude to philosophical sources: the pre-Socratic philosophers, Plato, Aristotle, Kant, or Spinoza (but never contemporary philosophers of science). Bohr never does. I believe that a consideration of the sources in question helps to resolve the apparent conflict between Bohr's youthful enthusiasm for philosophical reading and his later insistence that works by philosophers had no influence on his interpretation of physics.

None of the philosophers cited, Kierkegaard, Møller, Høffding, and William James, can be considered a philosopher of science or an interpreter of physics. Even James, the only scientist in the group, was concerned with psychology rather than any of the natural sciences. A common concern these philosophers shared was one later popularized by the existentialists as crisis philosophy. How does one bridge the gap between life as planned, a rational procedure, and life as lived, an irrational process? How can one carry on when experiencing a breakdown? Does one abandon one's youthful theological certainties, as Høffding did; experience a crisis of neurotic depression, as James did; or search for the true self that is the central concern of Møller's Licentiate? Kierkegaard's answer was that when one stage of life, such as the aesthetic, leads to despair, then one must through his own free choice make a transition to another stage, the ethical. When this stage in turn generates its own antithesis, then one must once again choose to make a further transition and, in Kierkegaard's view, opt for a religious stage.

Bohr's early philosophical interests centered on the problems of language and communication, a topic that always interested him, and on the attempt to explain life scientifically, a problem which Niels's father, a professor of physiology, often discussed in Niels's presence. As students, Niels and Harald helped form and regularly contributed to the Ekliptika Circle, a twelve-member philosophical discussion group. This circle of young students discussed Høffding's lectures, epistemological problems, and the questions concerning the meaning of life that perennially stimulate young students. There is no indication I know of that Bohr in his early days thought of physics as being philosophically problematic. When he read Kierkegaard, it was not to seek an interpretation of physics, but as a relief from the drudgery of writing a dissertation in physics. Even when, in the twenties, Bohr did begin to think of physics as presenting problems that related to his philosophical interest in language and communication, there was no way in which he—or anyone else—could ever draw a philosophy of physics from the sources cited. What he did seem to get, especially in his later writings, was a parallel. Bohr faced an intellectual crisis, a breakdown in the system he had done so much to develop. Eventually he resolved his crisis, not simply by getting a new physics to replace the old one, but by getting a better understanding of the nature of the explanations physics supplies. As we will see in chapter 10, he eventually interpreted the development of physics in terms of a dialectical process. A conceptual framework, adequate to a certain level of experience, breaks down when confronted with radically new types of experience. This antithetical opposition helps to generate a new framework which must eventually be interpretable as a rational generalization of the preceding framework, a new synthesis. There were other parallels, such as complementarity in physics and James's way of interrelating the role of introspection and the thrust toward objectivity in psychology. Yet, these were parallels and illustrations, not sources. The real source for Bohr's interpretation of physics should be sought in Bohr's own struggle to interpret the work he and his contemporaries had done.

Development of the Copenhagen Interpretation

The key event precipitating Bohr's decisive contribution to the Copenhagen interpretation was the failure of the B-K-S paper. His reexamination of its assumptions supplied him with the conceptual framework and basic principles he used to interpret the significance of the new developments, matrix and wave mechanics. As noted at the end of chapter 5 both the Bothe-Geiger and Compton-Simon experiments were interpreted as supporting Einstein's light-quantum hypothesis. Bohr had to accept and accommodate this idea. Yet, he was totally unwilling to follow the lead that de Broglie seemed to have initiated and accept 'wave-particle' as a category term describing some new class of entities.

The previously cited paper, which Bohr delivered at the Scandinavian Mathematical Congress in August 1925, was a general survey intended for people who were not proficient in quantum theory.[8] Nevertheless, it is of interest in its presentation of a distinctively new feature in Bohr's thought, his attempt to accommodate the light-quantum hypothesis within his interpretative framework. He then saw this accommodation as a question of readjusting the relation between formal principles and classical pictures. The doctrine of pictures, stemming from Hertz, Boltzmann, and Planck, may be briefly summarized. Pictures of physical reality are an essential part of the explanations physics supplies. On this point there was a general agreement. There was much less agreement on, or even discussion about, such further questions as whether these pictures should be interpreted as iconic models, functional models, phenomenal representations, conceptual representations, or something else. Since Bohr had not yet faced this issue, we can postpone a discussion of it.[9]

The most foundational pictures classical physics supplies are the picture of particles moving along spatiotemporal trajectories, the picture basic to classical mechanics; and the picture of waves propagating through a medium, the picture basic to electromagnetic theory. The success of the correspondence principle approach shows that there is a profound analogy between classical and quantum physics. Classical pictures accordingly must find some applicability in quantum theory. However, such pictures do not supply a basis for a consistent representation of the atom, its structure, and its activities. Discontinuous quantum transitions, multielectron atoms, the anomalous Zeeman effect; the nonmechanical strain; and the electron spin that replaced it—none of these could be adequately represented by classical pictures or mechanical models. If classical pictures were to have any applicability, they must be given some interpretation different from the standard one of modeling or representing physical reality. For this Bohr returned to his correspondence principle. This, he then felt, should guide both the use and the interpretation of classical pictures in quantum contexts.

Heisenberg's new theory, and especially the interpretation that was given to it subsequent to its development, carried the renunciation of mechanical models to its ultimate conclusion. Bohr accepted it immediately. Yet, he also concluded that Heisenberg's new theory must be considered incomplete. It treated only those quantities which depend on stationary states and transitions between stationary states. Even for the cases it could treat, the calculations were excessively difficult. Later, when Schrödinger's papers appeared, Bohr drew a similar conclusion. The mathematical formalism was helpful; the physical interpretation Schrödinger accorded it was not.

Thus, Bohr's basic interpretative framework was formed prior to his assimilation of the developments considered in the last two chapters. He thought that classical pictures and concepts—the two are not yet clearly distinguished—had played an indispensable role in establishing quantum theory and must still play a role in interpreting it. They must, however, be considered symbolic representations whose proper use is controlled by the correspondence principle. When pictures are so interpreted, there is no difficulty in principle in representing one and the same thing by different pictures for different purposes. Bohr was also clear, prior to these technical breakthroughs, that the use of space-time representations in quantum contexts precluded the simultaneous use of causal laws. However, he had not yet introduced the term 'complementarity.'

Bohr made no technical contribution to the developments in physics considered in the last two chapters. Yet, after matrix mechanics and wave mechanics were developed and shown to be mathematically intertranslatable, he played the decisive role in interpreting the significance of these new breakthroughs. In presenting this, there is some residual uncertainty in tracing the exact stages in the formation of Bohr's position because of his idiosyncratic process of development. His papers were usually dictated to young assistants at the institute, discussed, corrected, and rewritten until every word seemed appropriate and every argument coherent. His original paper on interpreting quantum mechanics, first delivered in the summer of 1927, was not submitted for publication until it had undergone some two years of criticism by Pauli (always Bohr's favorite critic), Heisenberg, Klein, and others and many writings and rewritings. What I am presenting here accordingly is a plausible historical reconstruction that attempts to recapture the dialectical flavor of the original development. It is not likely that the stages in the development of Bohr's position followed the linear ordering we will use for the purpose of exposition.

In the discussions with Schrödinger previously considered, Bohr rejected not only Schrödinger's particular wave interpretation, but also the very idea of according any pictures or models the representational role Schrödinger wished to give them. Bohr's discussions with Heisenberg began with a broader base of shared agreement. Bohr, as noted earlier, had interpreted Heisenberg's method as an extension on his own correspondence principle. Nor was there any disagreement of the mathematical formalism of quantum mechanics. Heisenberg accepted the transformation theory of Dirac and Jordan which unified wave and matrix mechanics. Bohr had no objections to the mathematical formalism. He did object, however, to considering the mathematical formalism an adequate basis for interpreting the physical significance of quantum theory. This was the point where he and Heisenberg diverged.

This divergence is most easily seen by beginning with Heisenberg's initial position, a position more or less shared by most of the physicists then working in quantum theory. The successful new mathematical formalism must supply the basis for any physical interpretation. Heisenberg had originally rejected Born's probabilistic interpretation because it was based on Schrödinger's wave mechanics and because it did not follow as an automatic consequence of the matrix formulation. It did so follow in the Dirac-Jordan transformation theory, where the square of the elements of the transformation matrix, linking the wave mechanical and quantum mechanical formulations, was interpreted as a probability. This Heisenberg would accept. With this there existed a unified mathematical formalism together with a consistent and apparently adequate way of relating the consequences derived from it to all the pertinent experiments. This should suffice. Any further interpretative elements were excess baggage, at best, or, at worst, metaphysical impositions.

Bohr disagreed. Any physical theory depends on the observation and description of events in space and time. The new mathematical formalism together with its probabilistic interpretation had a rather fuzzy relationship to this observational basis. It could give the probability of an electron being at a particular point at a particular time. Yet, it could not give the electron's spatiotemporal trajectory. The Wilson cloud chamber photographs, which were just being produced, literally pictured such trajectories. One must either conclude that quantum theory is incomplete or present some new account of the relationship between quantum theory and its observational basis.

Before examining this development, the novelty of the problem should be noted. Just a few years earlier the outstanding problems concerned experimental data, such as the anomalous Zeeman effect, dispersion, or the doublet riddle, which did not fit the theory. This was no longer the problem. The new mathematical formalism together with its probabilistic interpretation seemed adequate to handle all such previously recalcitrant data. The problem remaining was one of conceptual consistency and of showing how the new theory related to its observational basis. Since this was a consistency problem, the emphasis shifted from actual to idealized thought experiments. These give consistency arguments a physical embodiment. Thus, the ideal gamma-ray microscope and the interpretation to be accorded the Wilson cloud chamber photographs seemed to have played a pivotal role in the Bohr-Heisenberg discussions.

How does the new quantum theory account for the trajectory of an electron? As noted in the preceding chapter, the answer Heisenberg presented in his indeterminacy paper was stimulated by his recollection of Einstein's remark that the theory determines what can be observed. In accord with this principle the concept 'position of an electron' can have significance only to the degree that the position admits of measurement. This approach led to Heisenberg's analysis of the gamma-ray microscope and the indeterminacy in momentum consequent upon any measurement of position. By an extension of this operational approach 'path of an electron' was reduced to the measurement of a series of points corresponding to the position of an electron at a given time and was denied any significance beyond this. The water droplets in the cloud chamber served as position markers. Their spread is larger than the indeterminacy relation's minimal spread in the transverse component of momentum.

Bohr did not believe that the meaning of such terms as 'position' and 'trajectory' could be determined by such operational definitions. These terms already have a determined meaning prior to and independent of any particular atomic experiments. The basic meaning of these terms is determined by ordinary language usage and by the extension of this usage proper to classical physics. The critical problem requiring analysis accordingly is how such already meaningful classical concepts function in quantum contexts. On this point Bohr was not influenced by what philosophers then had to say about language and meaning, but by his own earlier work on the correspondence principle. The technique that worked in extending classical formulas to quantum contexts, where the presuppositions underlying their classical usage did not obtain, should also work for extending classical concepts to quantum domains, where the presuppositions implicit in their normal usage do not obtain.

From his earlier reflections Bohr effectively had two conclusions that now served as boundaries for the problem of concept extension. On the one hand, descriptive accounts of the behavior of an electron could not be interpreted as literally true in a picturing sense. On the other hand, descriptive accounts play an indispensable role in setting up theories and reporting and interpreting data. The notion of elliptical orbits had functioned too well in the Bohr-Sommerfeld theory to be completely wrong. The cloud chamber photographs gave trajectories of particles a visual embodiment. Such pictures must be accorded some sort of validity. Earlier, Bohr had sought the mean between the two extremes of acceptance of pictures as literally true and complete rejection by the expedient of subordinating descriptive accounts to formal principles. This subordination led to the conclusion that conservation laws apply to individual events only in a statistical way, a conclusion that was now clearly unacceptable. A more subtly nuanced mean between these extremes was needed. This came from the necessity of including complications previously excluded.

Bohr's earlier theorizing had relied on the representations of electrons as particles and light as waves. Though he clearly recognized these as classical idealizations, they were the conceptual tools he relied on. After the experimental refutation of the B-K-S paper he was forced to come to grips with the light-quantum hypothesis, which he had steadfastly rejected, and somehow accommodate the idea that light has both wave and particle properties. The success of the de Broglie–Schrödinger wave mechanics convinced him that the wave conception must have some applicability to electrons as well, a point that Heisenberg was not yet willing to accept.

The protracted discussions with Heisenberg forced Bohr to focus these rather abstract epistemological considerations on an analysis of particular atomic experiments. Where Heisenberg attempted to define 'position of an electron' in terms of the operations through which it is measured, Bohr attempted to analyze the same situation in terms of the way 'position of an electron' could function as part of a coherent system of concepts adequate to give a consistent account of both actual experiments and idealized thought experiments. The peculiar problem he then encountered was that, while a consistent account could be devised for any particular experiment, no overall account could be devised which was adequate for all experiments. Heisenberg later recalled the tenor of these discussions:

> I remember discussions with Bohr which went through many hours until very late at night and ended almost in despair; and when at the end of the discussion I went alone for a walk in the neighboring park I repeated to myself again and again the question: Can nature possibly be as absurd as it seem to us in these atomic experiments?[10]

Bohr and Heisenberg so disagreed over the way the meanings of basic concepts are determined and over the partial acceptability Bohr accorded Schrödinger's physical interpretation that they were no longer able to work together. While Bohr went skiing in Norway, Heisenberg wrote the first draft of his indeterminacy paper in the form of a letter to Pauli. Pauli approved. He then mediated the rift between the two men by the somewhat face-saving device of convincing them that their disagreement concerned the order of precedence given the basic concepts rather than the concepts themselves.[11] From this time on, Heisenberg was willing to use both the wave and particle pictures.[12] Only much later, in his quasi-philosophical writings, did he also swing around to an acceptance of Bohr's views on language and meaning.

Bohr profited from the reconciliation through his acceptance of Heisenberg's indeterminacy relation. It supplied the final piece in the puzzle he was assembling. He did not, however, accept Heisenberg's interpretation of this relation. For Heisenberg, physical consequences are derived from and determined by the mathematical formalism. For Bohr, an adequate physical account is prior to and grounds the mathematical formalism. In this interpretative framework the chief function of the indeterminacy relation is to show the limits which must be observed in using such classical concepts as 'position' and 'momentum' in quantum contexts.

Bohr's Initial Presentation of Complementarity

In September 1927, the International Congress of Physics met in Como, Italy, to commemorate the hundredth anniversary of Alessandro Volta's death. There, to an audience that included the leading European physicists except Einstein, Ehrenfest and Schrödinger, Bohr delivered a paper, "The Quantum Postulate and the Recent Development of Atomic Theory," presenting his interpretation of the new developments. In her biography of Bohr, Ruth Moore describes this as a sensational performance which stirred up the congress as the mistral sometimes disturbs the ordinarily calm waters of Como.[13] Max Born, an active participant at the congress, recalled the paper as rather dull.[14] Bohr, a notoriously bad lecturer, was summarizing developments that men like Born were already quite familiar with and then interpreting the significance of these developments in an epistemological terminology that most in the audience found unfamiliar and confusing. Bohr delivered essentially the same lecture at the Solvay Conference held in Brussels a month later and, after further polishing and refining, published it in *Nature* in 1928.[15]

The paper has a deceptive simplicity, giving a qualitative summary with very little mathematics and almost no technical details. Yet, the paper manifests the same spirit as the 1913 trilogy and the 1921 paper on the periodic system. It is a major attempt to achieve a rationally consistent synthesis of elements drawn from radically diverse sources while respecting the differences in the types of elements synthesized. We will attempt to abstract the distinctive epistemological features used to interpret the breakthrough which Bohr summarizes.

Complementarity, a notion Bohr first introduced in this lecture, gradually took on an extended significance in the course of Bohr's development. In the Como lecture it referred primarily to the complementary relationship obtaining between the application of causal principles and spatiotemporal descriptions. Though Bohr does not define 'causality', he clearly intends the Laplacean sense: the future state of a system is predetermined by its present state and the forces acting on it. Causality so conceived is closely linked to the laws of momentum and energy conservation.

As Forman has shown,[16] the sustained attack on causality in Germany in the early twenties was essentially a manifestation of the recurring complaint that a deterministic science is opposed to true humanism. This debate, whatever its merits, brought a renewed consciousness of the role of causality in scientific explanation and of the peculiar problems encountered in reconciling the traditional view of causality with quantum theory. Bohr's earlier attempt at a reconciliation, the B-K-S paper, proved too much of a renunciation. Bohr and Heisenberg undoubtedly discussed this in late 1926. Heisenberg's indeterminacy paper presented a novel solution to the apparent conflict between causal determinism and quantum indeterminacy.[17] Consider the sharp formulation of the causal law: if the past is known exactly, then it is possible to calculate the future. The opponents of causal determinism generally accepted the formulation, but denied the consequent. Heisenberg suggested that it is the antecedent rather than the consequent that is false. The past or present state of a system can never be specified precisely.

Bohr's new interpretation was similar to Heisenberg's inasmuch as it used the indeterminacy principle to restrict the applicability, rather than redetermine the meaning, of the causal principle. However, Bohr thought, the real basis of this principle as it functioned in classical physics was the classical idealizations of the motions of free particles and of radiation in free space. Space-time descriptions, on the other hand, require an interaction between the system observed and the agency of measurement. The ideal of an isolated system cannot be realized. It cannot even be approached in an asymptotic fashion, for quantum interactions embody the quantum postulate that action is in units of h.

Though Bohr did not explicitly relate complementarity to a theory of knowledge, he did relate it to two aspects of knowing he considered crucial. The complementary relation between causal principles and spatiotemporal descriptions supplies a basis for interrelating definitions of crucial concepts and the possibility of observation. When Bohr speaks of 'definition' in this essay, he does not mean *stipulating* the meaning of some term or some concept. He refers, rather, to rules governing the usage, in quantum contexts, of a term whose classical meaning is already determined.

Both matrix mechanics and wave mechanics are accepted as having an established validity, as is the Dirac-Jordan transformation theory unifying them. All are seen as based on symbolic extensions of classical concepts, especially on the classical abstractions of radiation in free space and the motion of free material particles. Bohr's concern accordingly is to use the general idea of the complementarity of spatiotemporal descriptions and causal principles to bring out the way matrix and wave mechanics relate to the complementary epistemological procedures of observing and defining. Matrix mechanics, interpreted as centering around Bohr's own correspondence principle, is easily accommodated in Bohr's new perspective. It explicitly renounces any description of the motions of particles. Its symbols refer directly to the individual processes demanded by the quantum postulate. Because of this renunciation, however, the question of observation is moot.

Wave mechanics cannot be considered as supplying a visualization in terms of space-time pictures. Though Bohr summarized the reasons Heisenberg had given in his indeterminacy paper, he attached a somewhat different significance to them. The Schrödinger formulation supplies an adequate representation of the stationary states of a system, thereby allowing an unambiguous definition of a system's energy. This, in terms of the principle of complementarity, entails a renunciation of any space-time description. When the Schrödinger formulation uses 'space', 'time', and 'momentum', it does so in a purely formal, rather than a descriptive, way. This symbolic significance is strikingly illustrated by symbolic transcriptions through which these terms enter the theory, e.g., $p_x = (ih/2\pi)\partial/\partial x$.

In the light of this clarification Bohr then examined three concepts which present special epistemological problems: 'stationary state', 'interaction', and 'reality'. The first two concepts are interconnected by the polar relation obtaining between the possibility of definition and the possibility of observation. The concept of a stationary state involves a precise specification of energy. This is possible only when a system is undisturbed. Thus, the definition of a stationary state involves the renunciation of a space-time description. An interaction supplies the only possibility for the observation of the behavior of an electron. Since such an interaction involves a change of state over a very short period of time, the energy cannot be specified precisely. This is a consequence of the indeterminacy principle. This observation supplied Bohr with a solution of the shortcoming that undercut the B-K-S paper. Since the interaction between a fast particle and an atom takes a very short time, there is a large indeterminacy in energy. The principle of energy conservation and also the principle of momentum conservation are not applicable to individual interactions with any greater precision.

Bohr's treatment of 'reality' was not yet related to any clarification of what it means to predicate 'reality' in ordinary language discourse. At this stage he simply presupposed the meaningfulness of the term and tried to determine when its predication in quantum contexts was justified. His statement supplies a convenient summary of the principal leitmotiv of this pivotal paper:

> Summarizing, it might be said that the concepts of stationary states and individual transition processes within their proper field of application possess just as much or as little "reality" as the very idea of individual particles. In both cases we are concerned with a demand of causality complementary to the space-time description, the adequate application of which is limited only by the restricted possibilities of definition and of observation.[18]

The paper just summarized might seem to complete the reinterpretation of scientific explanation that Bohr had begun some seven years earlier. Yet, Bohr continued to work intensively on this problem for another ten years. To adopt Bohr's own terminology, two complementary reasons may be assigned for this continued concentration on epistemological problems. First, Bohr thought that his idea of complementarity constituted a significant philosophical breakthrough, one which might contribute to a clarification of murky conceptual muddles in such other fields as biology and psychology. This aspect of Bohr's thought will not be pursued here, though it will be treated somewhat obliquely in chapter 10.

The complementary reason for Bohr's continued concern was the realization that his original formulation was inadequate and not altogether self-consistent. The aspects Bohr found unsatisfactory can best be determined by noting the differences and changes of stress in successive formulations. This attempt, however, presents some peculiar hermeneutical problems. Bohr never explained which aspects of this doctrine were changed or the reasons for the modifications he introduced. Rather, he kept going over and over the same material in his writings, in his lectures, and especially in almost interminable discussions with young scientists at his institute. In doing so, he kept introducing slight modifications of doctrine or terminology which were tested by the reactions of Pauli, Klein, Rosenfeld, and a succession of young assistants. A transformation of this spiraling dialectical development into a linear progression introduces steps whose historical ordering is somewhat flimsy. However, some slight historical oversimplification seems tolerable, provided it contributes to a clarification of Bohr's development.

Bohr's development during the ten years following the Como paper can be divided into two unequal stretches. From 1927 through 1929 he focused on a clarification of the interpretative perspective that he generally labeled 'complementarity'. From 1929 on his interest shifted to the question of how quantum concepts and methods can be extended to the new domains of physics that were just emerging. This attempt modified his view on the nature of concepts and on what is involved in explaining their functioning. This will be considered subsequently.

The shortcomings and subsequent modifications in Bohr's epistemological position can best be seen by focusing on two principles which eventually came to play an organizing role: complementarity and the epistemological irreducibility of atomic experiments. In the Como lecture 'complementarity' referred primarily to the relationship obtaining between space-time descriptions and the application of causal principles. The terminology in which this doctrine was originally couched was shaped by the discussions with Heisenberg and the later debates with Einstein. In both cases the emphasis was on atomic experiments and the type of information they supply.

This terminology had some unfortunate implications which Bohr only gradually became aware of. First, it implicitly presupposed a distinction between atomic systems as they exist objectively and the same systems as known by scientists. Yet, Bohr did not clarify the sense in which one can meaningfully speak about such systems apart from the conditions of their knowability. Second, it presupposed that a system is in a determined state prior and subsequent to measurement, but that we can only come to know about such states by disturbing them. By a fairly straightforward extension of Bohr's perspective, however, there is a complementary relation between objectivity and knowability. The objective state a system has prior or subsequent to measurement is unknowable. A further implication of the distinction between states as they exist and states as they are known by us is that there should be unlimited accuracy in retrodicting past states. No matter how this might be interpreted, it did not seem compatible with the indeterminacy principle.

Bohr was conscious of difficulties in his concept of 'complementarity'. In one of his 1929 essays he substituted 'reciprocity' as a new general term to replace 'complementarity'.[19] This, however, caused even more confusion, because Bohr was attempting to use this term in a sense different from its already established meaning. He returned to 'complementarity', but now attached primacy to a somewhat different application. The reasons for the change in emphasis seem to have been something like the following. Classical concepts are indispensable in any descriptive account, especially in reporting the experimental results on which atomic physics is based. The meaningfulness of the classical concepts used depends on the consistency of the framework in which they function. The introduction of the quantum of action shatters the consistency of this framework. From the point of view of the presuppositions of classical physics the discontinuity introduced by the quantum of action appears as an irrational element. Such an inclusion of irrationalism in a conceptual framework is an obstacle to the unambiguous use of the concepts that function within that framework.[20]

Before, Bohr had focused on concepts and presupposed conceptual frameworks. With the realization that these also presented epistemological problems he shifted his emphasis to concepts which depend on a particular limited framework rather than on a whole general framework, as 'space', 'time', and 'causality' do. Now the primary application of the term was to the complementarity between the wave and particle representations.[21] This new attempt at precision, however, exacted a price. The term 'complementarity' cannot have quite the same meaning in the new case as it had in the old, especially when one attends to conceptual frameworks as well as to individual concepts. The two usages represent different ways of extending classical physics. In quantum physics one uses either one or the other, but not both simultaneously. The wave and particle concepts, on the other hand, represent mutually exclusive conceptual networks embedded within the general framework of classical physics. One rests on the idealization of free particles, the other on the idealization of radiation in free space. Yet, in the new quantum physics, one can use the particle picture to represent radiation or the wave picture to represent what were thought of as material particles.

Two aspects of this shift should be noted. First, Bohr was obviously intrigued by it. He thought that this usage gave 'complementarity' a much wider significance that had a bearing on problems in the interpretation of biology and psychology. Concepts that are mutually exclusive in ordinary usage such as 'alive' or 'dead', 'mechanical' or 'vital', 'determined' or 'free' may have complementary uses in accounting for the properties of living beings or psychological states.[22] The second aspect requiring comment is Bohr's gradual shift from a consideration of concepts to a consideration of the conceptual frameworks in which they function. Only much later did he achieve any significant clarification on what he meant by 'conceptual framework'. This clarification, however, was never made explicitly. It was implicitly manifested by the systematic way in which Bohr came to use 'conceptual framework' in his later writings. I will accordingly postpone any further discussion of this until chapter 10, where I will attempt to present Bohr's final position in a systematic rather than a historical fashion.

Bohr's second basic principle, one more prominent in his later than in his earlier accounts, may conveniently be labeled 'the epistemological irreducibility of atomic experiments'. This also presented terminological difficulties and consistency problems. Bohr's original formulation was:

> Now, the quantum postulate implies that any observation of atomic phenomena will involve an interaction with the agency of observation not to be neglected. Accordingly, an independent reality in the ordinary physical sense can neither be ascribed to the phenomena nor to the agencies of observation.[23]

The basic significance of this principle is clear enough from its standard application. One performing a double-slit experiment cannot ask which slit the electron went through. The only physically meaningful way to ask and answer such a question is to close one slit and then determine whether the electron goes through the other slit. This is not a conceptual division of the two-slit experiment; it is a new one-slit experiment.[24]

The basic idea is clear, but the terminology is more problematic. It took some ten years for Bohr to get the rather subtle problem implicit in this terminology into a clear enough perspective that he could see it and attempt to resolve it. The problem was then seen as an implicit inconsistency between the way Bohr was using terms and the presuppositions governing their normal usage. The types of experiments that supply information about the states of atomic systems can be roughly divided into two groups. The first involves hitting a target with a projectile and then examining the behavior of the resulting fragments. The second involves an examination of the radiation that is emitted, absorbed, or modified in atomic processes. In describing these, one uses either a particle language, based on the idealization of particles in free space, or a wave language, based on the idealization of waves propagating in space.

The key assumption in both cases is not free space; it is determinism. Particles and waves travel in ways determined by conservation laws. In keeping with these laws and the conceptual framework in which they function, anything modifying the motion must do so in a deterministic way. Suppose one wishes to describe a projectile deflected by an atom in such a way as to change the state of the atom. This was the problem Bohr had treated in great detail in his earlier analysis of scattering experiments. As one asymptotically approaches the limit in which classical physics is valid, this interaction can be described in a purely deterministic way. In a quantum context, away from this classical limit, the atom's change of state cannot be determined from the theory. The irreducibility principle seems to cope with this difficulty. If each atomic experiment is an epistemologically irreducible unit, then the change of state can only be determined in a particular case by observation.

This seems satisfactory until one considers the presuppositions underlying such an account. Either the atom's change of state is objectively determined or it is not. If it is objectively determined and yet cannot be deduced from the theory, then the theory must be considered incomplete. This was a conclusion Bohr would not accept. He had insisted that his doctrine of complementarity allowed for as complete a description as was consistently possible. The opposite alternative is to say that the atom's change of state is not objectively determined. When Bohr faced up to this difficulty in 1929, he temporarily adopted some terminology Dirac had introduced and spoke about the need to reckon with a free choice on the part of nature about the state a system enters.[25]

This anthropomorphic terminology was soon dropped. To the best of my knowledge it never recurs in Bohr's later writings. Dropping it, however, did not make the problem of objective determinism disappear. It is possible to discern on the basis of his later formulations the features of this early treatment that Bohr was finding unsatisfactory. Bohr insisted, early, late, and often, that it lies in the nature of physical observation that all experience must ultimately be expressed in terms of classical concepts, neglecting the quantum of action. Earlier, Bohr had accepted the standard empiricist position that the core meaning of the key terms linking what we say to what is, is determined by sensation. If key terms could have meaning in isolation, then there should be no difficulty in selecting the terms necessary to give a descriptive account of some experiment and treating this as an irreducible unit. This, however, led to the dilemma previously considered. If atomic events are objectively determined, then atomic theory is incomplete. To accept the alternative and hold that atomic events are not objectively determined would seem to be opting for irrationalism or some equally desperate alternative.

Here again, Bohr gradually groped his way out of the dilemma by switching his focus from individual concepts to conceptual frameworks. The meanings of the key terms involved in descriptive accounts of atomic experiments, 'particle' and 'wave', are radically underdetermined by sensation. They are idealizations and, as such, depend on the consistency of the framework in which they function. Similarly, 'determinism' and 'objectivity' have meanings which depend on the same conceptual framework. These, in fact, serve as presuppositions which cannot be denied without impairing the consistency on which the meaning of the observational terms depend.

Such considerations seem to have induced Bohr to move beyond standard empiricist theories of meaning. His first significant step beyond this was to relate descriptions, considered as a group, to a manifold of experience:

> We meet here in a new light the old truth that in our description of nature the purpose is not to disclose the real essence of phenomena but only to track down, as far as possible, relations between the manifold aspects of our experience.[26]

He was clearly moving in the direction later given a succinct summary by Quine when Quine introduced a similar shift: "Our statements about external reality face the tribunal of sense experience not individually but as a corporate body."[27] In this perspective the problem of objective determinism gets refocused. When the refocusing was complete some five or six years later, Bohr had effectively transformed the original question of objective determinism into quite a different question. How are the presuppositions of objectivity and determinism implicit in the classical way of describing reality extended to and limited in quantum contexts? At this intermediate stage Bohr was well aware that the problem concerned the nature of concepts: "However, we can no more hope to attain a clear understanding in physics without facing the difficulties arising in the shaping of concepts and in the use of the medium of expression than we can in other fields of human inquiry."[28] What he did not yet have in clear focus was his later concern with the presuppositions implicit in normal usage. Before that happened, some radically different problems induced him to spend a considerable amount of time examining ways in which some particular concepts may be extended to and used in new domains. This is the topic we will consider next.

A Cluster of Problems

As director of a leading research institute and as the presiding figure over the atomic physics community Bohr was inevitably involved in any new breakthrough in physics. This involvement became intensely personal when new developments either required an extension of quantum methods to new domains or seemed incompatible with Bohr's way of interpreting physics. Around 1930 there were three such new problem areas: quantum field theory, relativistic quantum mechanics, and nuclear physics. We will consider just enough background to understand the difficulties Bohr was treating.

Quantum field theory had been partially anticipated by Jordan in the three-man paper considered earlier.[29] Here it was shown that the new matrix formalism could reproduce Einstein's formula for the fluctuations in a blackbody radiation cavity (eqs. [4.2] and [7.8]). This, however, was not really quantum field theory, but merely the quantization of free radiation. Dirac was the first to introduce and develop a quantum field theory that covered the emission and absorption of radiation.[30] Because of the complexity of the problems involved, Dirac introduced a nonrelativistic approximation. He considered an atom interacting with a radiation field. By treating the whole system as enclosed in a finite box, he could decompose the radiation into discrete Fourier components. Through an adaption of the method of action-angle variable he was able to set up a quantum Hamiltonian including terms for the atom, the radiation field, and the interaction between them. He expressed the Hamiltonian corresponding to the Schrödinger equation as

$$F = \sum_{r,s} b_r^* H_{rs} b_s. \tag{8.1}$$

Then a further contact transformation is introduced:

$$b_r = N_r^{1/2} \exp(-i\theta_r/\hbar),$$
$$b_r^* = N_r^{1/2} \exp(i\theta_r/\hbar), \tag{8.2}$$

where $\hbar$ is $h/2\pi$. A mathematical analysis of these new operators led to the interpretation of b_r, b_r^*, and N_r as emission, absorption, and occupation numbers, respectively. The application of this method to a Bose-Einstein gas led to the Einstein coefficients previously considered.

This article was written while Dirac was at Copenhagen and was communicated to the Royal Society by Bohr himself. Its interpretative significance was made explicit in a subsequent article extending the new formalism to cover scattering and dispersion: "One finds then that the Hamiltonian for the interaction of the field with an atom is of the same form as that for the interaction of an assembly of light-quanta with the atom. There is thus a complete formal reconciliation between the wave and light-quanta points of view."[31] The light-quantum hypothesis, which Bohr had long rejected and then treated in an unsatisfactory way through Slater's virtual radiation field, was now included in the formalism of Dirac's quantized fields. This way of handling the hypothesis is proved to be definitive.

Dirac's original treatment had some undesirable features. It was essentially nonrelativistic, relying on the Schrödinger equation and on a clean separation between space and time variables, which allowed Dirac to ignore the Coulomb field. Atoms and electrons were treated as existing in their own right, while photons were treated as manifestations of underlying fields. in this formulation the Coulomb interaction between the nucleus and the electron is absorbed into the electron's Hamiltonian rather than into the radiation field.

The further development of quantum field theory led to formidable formalisms and even more formidable consistency problems. These will be considered only to the extent necessary to situate the conceptual and methodological problems Bohr treated. Two essentially independent paths were developed to quantize the full electromagnetic field, rather than just the radiation component, and to do this in a relativistically invariant way. Heisenberg and Pauli in their first joint paper began with an action principle which allowed them to define field variables in a relativistically invariant way, as derivatives of the Lagrangian for the field.[32] They quantized both the full Maxwell field for radiation and the Dirac field for electrons, i.e., the field derived by quantizing the wave function in the Dirac wave equation. Both fields presented difficulties, difficulties associated with gauge invariance for the electromagnetic field and difficulties associated with negative energy states for the Dirac field. A simpler, independently developed, relativistically invariant generalization of Dirac's quantum theory of radiation was given by E. Fermi.[33]

Both versions generated a fundamental difficulty which was essentially a quantized version of a difficulty previously encountered in the classical theory of the electron.[34] If an electron is pictured as a sphere with a charge e confined to a volume of radius r, then classical theory does not explain how this sphere stays together under the mutual repulsion the parts of this distributed charge exert on each other. If one attempts to avoid this by picturing the electron as a point charge, then this electrical charge has an infinite self-energy. J. R. Oppenheimer showed that the new quantum field theory encountered a similar difficulty when the interaction of an electron with a radiation field is treated in terms of the emission and absorption of virtual particles.[35]

The second new area, relativistic quantum mechanics, also stemmed from the work of Dirac. This not only forced a modification of the concept 'particle', but also reflected a radically different way of doing physics from that advocated by Bohr. Bohr insisted on the interpretative primacy of the basic physical concepts and treated the mathematical formalism as a supplementary tool. In an interview with Thomas Kuhn, Dirac explained how he developed his relativistic wave equation:

> It came just from playing with the equations rather than trying to introduce the right physical ideas. A great deal of my work is just playing with equations and seeing what they give. Second quantization I know came from playing with equations. I don't suppose that applies so much to other physicists; I think it's a pecularity of myself that I like to play about with equations, just looking for beautiful mathematical relations, which maybe don't have any physical significance at all. Sometimes they do.[36]

Since the equations presented the problems of physical interpretation, it will be helpful to see the equations before considering the problems. The Klein-Gordon equation, previously considered, is obtained by taking the relativistic expression for the relationship between energy and momentum

$$E^2 = p^2c^2 + m^2c^4 \qquad (8.3)$$

and using the standard operator substitutions

$$p = -i\hbar\nabla, \; E = i\hbar\partial/\partial t \qquad (8.4)$$

to get the equation

$$-\hbar^2\partial^2\psi/\partial t^2 = -\hbar^2c^2\nabla^2\psi + m^2c^2\psi. \qquad (8.5)$$

This Dirac found unsatisfactory. Since it uses E^2, it is nonlinear in E and also second order in t. Linearity was a requirement of Dirac's transformation theory. Equation (8.5) also presented such problematic features as negative energy states and negative probabilities.

By playing with equations, Dirac discovered a novel way of factoring equation (8.3) to get[37]

$$(p_0 + \alpha_1 p_1 + \alpha_2 p_2 + \alpha_3 p_3 + \beta)\,\psi = 0. \qquad (8.6)$$

He then determined the form (later shown to be one of several possible representations) and the α and β matrices by the requirements that they commute with all the p's (or are not functions of space and time), and that the square of equation (8.6) reproduce equation (8.3). Then the standard operator substitutions, (8.4), led to the Dirac wave equation:

$$(i\hbar\partial/\partial t + (\hbar c\alpha\cdot\nabla + \beta mc^2))\psi = 0. \qquad (8.7)$$

Here α stands for a set of three 4×4 matrices which are generalizations of the Pauli spin matrices, and β is also a 4×4 matrix.

We can consider solutions of equation (8.7) for two basic cases, the free particle and the hydrogen atom. In the free particle case equation (8.7) admits of four equivalent linearly independent solutions, each having the general form

$$\psi(r.t) = \begin{pmatrix} u_1 \\ u_2 \\ u_3 \\ u_4 \end{pmatrix} e^{i(k \cdot r - \omega t)} \qquad (8.8)$$

where $k = p/\hbar$ and $\omega = E/\hbar$. There are two essentially equivalent positive energy solutions. One has the components

$$u = cp_z/(E_+ + mc^2), \; u_2 = c(p_x - ip_y)/(E_+ + mc^2), \; u_3 = 1, u_4 = 0. \quad (8.9)$$

The corresponding negative energy solution is

$$u_1 = 1, \; u_2 = 0, \; u_3 = cp_z/(E - mc^2), \; u_4 = c(p_x + ip_y)/(E - mc^2). \quad (8.10)$$

Following Bohr's correspondence principle this should yield the results proper to nonrelativistic quantum mechanics in the limit in which $v \ll c$. This requirement leads to the interpretation of $E_+ = mc^2$ and $E = -mc^2$. For the positive energy solution, the two small components, those divided by E, become vanishingly small in the nonrelativistic limit. The remaining components, with values 1 and 0, are interpreted as corresponding to a spin-up particle. The equivalent second positive energy solution has components 0 and 1 and corresponds to a spin-down particle. Thus, all the results that had been explained by spin are automatically given by this new formalism without any reliance on the highly suspect model of the electron as a spinning ball.

In applying this new equation to the hydrogen atom, Dirac assumed the approximate form of the Hamiltonian

$$H = C\alpha \cdot p = mc^2\beta - e^2/r \qquad (8.11)$$

and then solved the Schrödinger equation $H\psi = E\psi$. The solution to this equation included all the spin and relativistic contributions to the energy states of the hydrogen atom that had previously been developed as individual correction terms. The Dirac wave equation quickly emerged as, and still remains, one of the most strikingly and, surprisingly, successful equations in the history of physics.

Such success notwithstanding, there were still serious difficulties. First, there was the peculiar feature of negative energy states, the same feature that Dirac had found unacceptable in the Klein-Gordon equation. Dirac's original suggestion for handling this difficulty was to ignore the negative energy solutions as lacking physical significance: "Since half the solutions must be rejected as referring to the charge $+e$ on the electron, the correct number will be left to account for the duplexity phenomena."[38]

Such a simple excision of unwanted solutions proved unacceptable. One reason for this was the paradox discovered by O. Klein, then serving as Bohr's assistant.[39] If an electron is subjected to an electrostatic potential which increases by more than the electron's rest-mass energy mc^2 in a distance of less than the electron's Compton wavelength h/mc, then there is a definite probability that the electron will jump the gap between the positive and negative energy solutions and land in a negative energy state. A related difficulty was discovered by Schrödinger and dubbed 'Zitterbewegung'.[40] If one considers a wave packet of free-particle solutions to the Dirac equation, then the expression for the current includes terms for a rapid oscillation between positive and negative energy solutions. Both difficulties have a bearing on the traditional concept of a particle as a localized entity. A free electron, represented by a plane-wave, does not involve any negative energy states. A plane-wave, however, does not have any definite location. Any attempt, whether conceptually or physically, to localize an electron in a wave packet involves transitions to negative energy states. The concept of the electron as a sharply localized particle seemed to be encountering some sort of a natural limit.

Before 1932, nuclear physics presented two outstanding and apparently related problems, electron confinement and nuclear statistics. It was generally assumed that the nucleus is composed of protons, electrons, and perhaps alpha particles. The fact that the nucleus emits electrons in beta-decay seemed to be convincing proof of the presence of electrons in the nucleus. Yet, the confinement of electrons within a region as small as the nucleus implied that these electrons have such high energy and momentum that relativistic quantum mechanics is required. A relativistic treatment of an electron confined within such a small region inevitably introduces the possibility of transitions to the puzzling negative energy states. Furthermore, it was difficult to understand what forces could confine such energetic electrons in such a small volume.

The development of nuclear statistics emerged from Heisenberg's treatment of the helium atom and from the clarification of the distinction between Fermi-Dirac and Bose-Einstein statistics.[41] Heisenberg showed, in the case of a two-particle system, that the overall wave function must be either symmetrical or antisymmetrical and that no transitions occurred from one type to the other. Bose-Einstein systems have symmetrical wave functions while Fermi-Dirac systems have antisymmetrical wave functions. This overall wave function is essentially a product of the component wave functions for the electronic, spin, rotational, and vibrational components.

For a molecule made up of two identical atoms, a rotation of 180°

changes the rotational wave function by $(-1)^l$, where l is the rotational quantum number. Such a rotation is equivalent to an interchange of the two nuclei followed by a rearrangement of electrons. If each of the two nuclei has N Fermi-Dirac particles (protons or electrons), then this interchange changes the wave function by $(-1)^N$. There is accordingly a strong correlation between the nuclear statistics, determined by the number of particles in the nucleus, and the rotational quantum number. Since the latter is manifested through a pattern of alternating intensities in the molecular band spectra, the former can be determined experimentally.

Nuclear statistics, coupled to the assumption that a nucleus is composed only of electrons and protons, led to very clear predictions for individual nuclei. By the rule of $(-1)^N$ a nucleus having an odd number of particles should obey Fermi-Dirac statistics, while a nucleus with N even should obey Bose-Einstein statistics. Yet, as Heitler and Herzberg showed,[42] nitrogen did not fit this rule. To explain its mass and chemical properties it was assumed that normal nitrogen has fourteen protons and seven electrons, giving N odd. Yet, spectral studies indicated that nitrogen obeyed Bose-Einstein, rather than Fermi-Dirac, statistics.

Beta-decay involves the emission of an electron from the nucleus and a consequent increase of one in a nucleus's atomic number. It also involved failures in the conservation laws for energy and momentum as well as nuclear spin. The emitted electron has an intrinsic angular momentum of $\hbar/2$. Yet, the only possible changes in the nuclear angular momentum are integral multiples of $\hbar(\Delta l = nh, n = 1, 2, 3, \dots)$. In an open letter to Geiger and Meitner, who were attending a physics conference in Tubingen in December 1930, Pauli (who skipped the conference because it conflicted with a dance) introduced the idea of a highly penetrating, neutral particle of vanishingly small mass, which is also emitted in beta-decay. This particle, which Fermi later dubbed the 'neutrino', would serve to save the conservation laws.[43]

Exploring the Concept 'Particle'

During the period when these problems flourished, Bohr remained the presiding figure in the community of atomic scientists. Each spring his institute sponsored an informal meeting where the leading physicists discussed the outstanding problems and thrashed through the various theories and hypotheses offered as potential solutions. Bohr's own approach to these problems was an extension of his previous methods pushed to an idiosyncratic extreme. He left formal theories and mathematical calculations to the younger members of the scientific community. His concern was with the foundational concepts, concepts like 'particle', 'wave', 'position', 'momentum', 'energy', 'spin', and 'magnetic moment' which are used to report experimental results and describe physical reality. He also took an interest in concepts like 'observe', 'report', 'describe', and 'measure', concepts needed to clarify the activities of the physicist. It must, he insisted, be possible to fit these concepts together in a way that is self-consistent and adequate to the data. Otto Frisch, who worked in Bohr's institute during the early thirties, summed up Bohr's attitude: "So it was always: every inconsistency was an enemy who had to be at once attacked, and against whom Bohr turned the full strength of his powerful mind."[44]

As a matter of convenience we will divide the development of Bohr's thought into two stages: pre-1932, when he was analyzing the limitations in the applicability of the concepts on which quantum mechanics is based, and post-1932, when new developments induced changes in his methods. Bohr's initial reaction to the problems we have been considering was presented in a long letter he wrote Dirac on 24 November 1929, after it was clear that the Dirac relativistic wave-equation represented a fundamental breakthrough, but one whose implications remained obscure.

> Recently I have been very interested in these problems and have thought that the difficulties in relativistic quantum mechanics might perhaps be connected with the apparently fundamental difficulties as regards conservation of energy in β-ray disentegration and the interior of stars. My view was that the difficulties in your theory might be said to reveal a contrast between the claims of conservation of energy and momentum on one side and the conservation of the individual particles on the other side. The possibilities of fulfilling both these claims in the usual correspondence treatment would thus depend on the possibility of neglecting the problem of the constitution of the electron in non relativistic classical mechanics. It appeared to me that the finite size ascribed to the electron on classical electrodynamics might be a hint as to the limit for the possibility of reconciling the claims mentioned. Only in regions where electronic dimensions do not come into play, the classical concept should present a reliable foundament for the correspondence treatment.[45]

Dirac replied immediately: "My own view of the queston is that I should prefer to keep rigorous conservation of energy at all costs and would rather abandon even a concept of matter consisting of separate atoms and electrons than the conservation of energy."[46] As a possible solution to the problems considered Dirac sketched the first account of his sea of negative energy hypothesis. All the negative energy states are presumed to be occupied by particles. The Pauli exclusion principle is presumed to apply to these negative energy particles forcing some to occupy less stable states. An electron knocked out of such a state would leave a hole in the negative energy sea. An experimenter would see this hole in the negative energy sea as if it were a particle with positive energy. Dirac concluded, "These holes I believe to be protons."

No hypothesis of Dirac's, no matter how wildly implausible it seemed, could be dismissed or treated lightly. Bohr wrote back that he and Klein had given much thought to Dirac's suggestion, but could not see how Dirac could overcome the difficulties involved in postulating an infinite electrical density in space. Characteristically, Bohr focused on the problem of extending classical concepts rather than on Dirac's concern with finding a coherent physical interpretation for the puzzling features of the new equation:

> In the difficulties of your old theory I still feel inclined to see a limit of the fundamental concepts on which atomic theory hitherto rests rather than a problem of interpreting the experimental evidence in a proper way by means of these concepts. Indeed, according to my view the fatal transitions from positive to negative energy should not be regarded as an indication of what may happen under certain conditions, but rather as a limitation in the applicability of the energy concept.[47]

Bohr was extremely cautious about postulating new particles to explain puzzling phenomena. Neither Pauli's neutrino nor Dirac's sea of negative energy won his initial acceptance. Bohr focused instead on the classical concepts quantum mechanics appropriated. He thought that, by determining the limits of their validity, he could also determine the limits of applicability of quantum mechanics. These limits, in Bohr's view, were determined by the correspondence principle, not by the mathematical formalism. In his letter to Dirac he sought to express this limit in terms of a complementary relationship between the individuality of the electron as a particle and the laws of energy and momentum conservation.

The concept of the electron used in quantum mechanics is one adapted from Lorentz's reformulation of Maxwell's electrodynamics. Even in this classical theory 'electron' has a limited validity signified by the classical radius of the electron, $e^2/mc^2 \approx 2.8 \times 10^{-13}$ cm. Though this radius had originally been associated with the idea that the mass of an electron is of electromagnetic origin, it still seemed to supply a convenient lower limit below which the concept of the electron as a localized particle had no empirical or theoretical significance. Similarly, something of a rough upper limit was supplied by the electron's Compton wavelength $\lambda_C = h/mc \approx 2.4 \times 10^{-10}$ cm. For this and any larger magnitudes, it was certainly meaningful to use the concept of an electron as a localized particle.

Somewhere between these two limits of 2.4×10^{-10} cm and 2.8×10^{-13} cm the applicability of the concept of the electron as a localized particle broke down. The radius of a nucleus of atomic number A is approximately $R = A^{\frac{1}{3}} r_0$, where $r_0 = 1.4 \times 10^{-13}$ cm. Even the very massive uranium nucleus with $A = 238$ only has a radius of about 8.7×10^{-13} cm. Such considerations led to Bohr's tentative hypothesis that electrons do not exist as individual particles within the nucleus, though their charge is conserved. This would seem to obviate the difficulties concerning the infinite self-energy of a point electron and also the problem of beta-decay in that "in this sense we may regard the expulsion of a β-ray from the nucleus as the birth of an electron as a dynamical individual."[48]

This phase of Bohr's development culminated in an address entitled "Atomic Stability and Conservation Laws" which he gave to a nuclear physics convention meeting in Rome in October 1931.[49] The basic interpretative idea of this talk was the limitation of the concept of the electron as an individual particle. This classical concept retains its validity when applied to atomic electrons. The Pauli exclusion principle presupposes particle individuality. However, the electron is also attributed a spin. Not only is this incompatible with classical ideas; it is also unmeasurable.[50] No measurement of a free electron can distinguish between magnetic forces and the effect of an intrinsic magnetic moment. This consideration seems to support the conclusion that the concept of an electron as an individual localizable particle has no legitimate use beyond the limit of the classical electron radius. Since such mechanical concepts as 'energy' depend on this idealization of the electron as a particle, the principle of energy conservation cannot be presumed to be valid beyond these limits.[51]

The limits of the concept of the electron as a localized particle should also set the limits of the applicability of quantum mechanics. Here Bohr relied on the smallness of two ratios: $\alpha = 1/137$, the fine-structure constant, and $\beta = 1/1837$, the ratio between the masses of the electron and the proton. The first ratio guides the applicability of nonrelativistic rather than relativistic quantum mechanics. The second supplies a guideline for the applicability of the concept 'particle'. Using these guidelines Bohr concluded that quantum field theory lies within the limits of applicability of quantum concepts. Such considerations may be extended, within the context of quantum field theory, from 'electron' to 'photon'. The concept 'photon' so interpreted exhibits a complementary relationship to the concept 'field'.

When one adds relativistic to quantum considerations, the foundational concepts begin to lose the possibility of unambiguous applicability. Thus, the Klein paradox requires extremely strong electrical fields over very small distances. Such fields must ultimately be due to the presence of charged particles. Yet, if Bohr's considerations are correct, it is impossible to localize enough particles in a small enough region to produce such a field. The paradox is thus dissolved, rather than resolved. Bohr accorded the Dirac relativistic wave-equation itself a more ambiguous status. It must have an approximate validity, for it yields electron spin and the right relativistic corrections. Yet, it still seems to allow transitions to negative energy states. The basic ambiguity here, Bohr believed, stemmed from the fact that the correspondence principle supplied no guidance to one attempting to understand such transitions.

The application of quantum mechanics to electrons within the nucleus was not at all ambiguous. It was an outright failure. This, Bohr insisted, is because physicists have abandoned the ground

> for the unambiguous application of classical mechanical ideas
> and of any formalism which, like present quantum mechanics,
> is essentially based on such ideas. . . . In this situation we are
> led to consider the capture or the expulsion of an electron by
> a nucleus plainly as an extinction or a creation, respectively, of
> the electron as a mechanical entity. We cannot therefore be
> surprised if these processes should be found not to obey such
> principles as the conservation laws of energy and momentum,
> the formulation of which is essentially based on the idea of
> material particles.[52]

This approach not only accounted for the failure of the conservation laws in beta-decay, but also seemed to account for the apparent anomalies in nuclear statistics. Even within the nucleus the proton, because of its much larger mass than the electron, should have a kinetic energy much less than its rest-mass energy. Unlike the electron, it can still be treated as a mechanical particle. Nuclear statistics accordingly should be determined exclusively by the number of protons in the nucleus. This assumption accounts for the otherwise surprising facts that ^{6}Li and ^{14}N nuclei obey Bose-Einstein statistics.

This correspondence principle approach to quantum field theory and nuclear physics was very similar to Bohr's work some ten years earlier in trying to develop a consistent extension of the old Bohr-Sommerfeld theory by using the correspondence principle as his basic tool. In both cases Bohr was able to discern some underlying conceptual and methodological coherence where others saw only inconsistencies, paradoxes, approximations, and rather arbitrary rules for handling them. Bohr's essentially conservative approach, based on the extension of classical concepts, had some sharp limitations. It did not take theories as explanatory units very seriously. It had difficulties accommodating theoretical breakthroughs, like Dirac's wave-equation, which were achieved by mathematical rather than physicalistic reasoning. It encouraged deep suspicion concerning entities postulated on purely theoretical grounds, such as Pauli's neutrino and Dirac's holes in a sea of negative energy. It also inevitably lacked precision. The year 1932 witnessed a discovery of the neutron and of the positron, quickly identified with Dirac's antielectron.[53] In 1933 Fermi developed his theory of beta-decay, a theory which made essential use both of the neutrino hypothesis and of the principles of energy and momentum conservation as applied to nuclear processes.[54]

Bohr's initial reaction to these developments focused, not surprisingly, on the correspondence principle as a basis of interpretation. However, his emphasis shifted from general interpretative frameworks to problems that could be handled with more precision. We will consider his work in quantum field theory and nuclear physics before discussing the influence this work had in his evolving view on the nature of concepts.

Quantized Fields and Nuclear Models

Early in 1931, L. Landau and R. Peierls, both working at Bohr's Institute, wrote a paper questioning the logical consistency of quantum field theory.[55] The key idea behind the criticism was that the measurement of a field component requires a determination of the momentum the field produces in a charged test body. Quantum field theory assigns values for every point in the field. Such an assignment conceptually entails the use of point-size test particles. This introduces a series of difficulties. The radiation reaction of such a point-size particle would limit the accuracy of any field measurement. A further ambiguity arises from the peculiar zero-point fluctuations.

Bohr's emphatically negative reaction to this criticism was captured by Gamow, also present at the institute, in a pen drawing showing Landau tightly bound and gagged while Bohr stood before him with an upraised finger and said, "Bitte, bitte, Landau, muss ich nur ein Wort sagen." Quantum field theory had, in Bohr's opinion, supplied the only acceptable interpretation of the light-quantum hypothesis. It could not be dropped without warping the conceptual consistency of the interpretative framework Bohr was weaving. To counter the Landau-Peierls criticism Bohr secured the assistance of Léon Rosenfeld, whose first task was to teach Bohr the mathematics of quantum field theory. Though Bohr was quite proficient in mathematics, he never wanted to use any more mathematics than was necessary to extend and make more precise the physical concepts involved in a theory. These, for Bohr, were always the heart of the matter.

The joint paper that finally emerged from their protracted cooperative effort had a very peculiar status.[56] It contains a very detailed mathematical analysis of the measurability of the field quantities used in quantum field theory. Yet, it does not relate to actual measurements, nor even to any measurements that are possible in the sense of being technologically feasible. Nowhere in the paper is there as much as a hint about the way in which the idealized measurements discussed might relate to laboratory practice. Before discussing the contents of this paper, it might be helpful to consider what these authors were really trying to do.

The purpose of the article is to establish the consistency between the presuppositions of quantum field theory and the type of information supplied by any experiment. Whatever the result of an experiment is, it must be something that the experimenter can observe, record, and communicate to others. From this it follows, a point Bohr never ceased to insist on, that the results of any measurement must always be expressed in classical concepts, neglecting the quantum of action. Any attempt to use such concepts as a springboard for a leap into the quantum domain is essentially a correspondence principle argument.

This almost trite observation effectively served to undercut the presuppositions of the Landau-Peierls paper. There are different ways in which classical concepts can be extended to quantum domains. The only universal constants that enter quantum field theory are c, the velocity of light, and h, Planck's constant. These do not suffice to determine any specific space-time dimensions, such as the classical radius of the electron. Therefore, the question of the consistency of quantum field theory is independent of any question concerning the size of the electron or any other unit of charge. To assume electrons or point particles as test charges is to confuse two different extensions of classical concepts to quantum domains.

In place of the assumption of point-size test particles, one should consider the assumption implicit in any process of measuring field values. Such measurements are inevitably averaged values over extensive regions. The issue Bohr and Rosenfeld treated, accordingly, is the simplest extension of classical concepts that covers this operative presupposition. Such an extension is achieved by assuming, as a test particle, a body whose linear dimensions are sufficiently large compared to atomic dimensions so that the charge density can be considered approximately constant over the whole body. To preserve this constancy in changing fields, one must also assume that the test particle is a rigid body.

What this means is assuming a *classical* test particle coupled to a field described by quantum field theory. This is a classical particle in the sense that its atomic structure need not be considered. That any test particle would have an atomic structure is not denied. But this is irrelevant in the context of a correspondence principle argument. Since a classical test particle is assumed, one may also consistently assume models appropriate to classical particles. Bohr and Rosenfeld assume that the test particle can be attached to a rigid frame, released to measure momentum, and attached to other test particles by means of springs.

With these simplifying assumptions it is possible to introduce the constraints proper to quantum field theory in a tractable manner. There are two principal constraints. The first, from the indeterminacy principle, governs any measurement of the position and momentum of the test particle. The second is from special relativity: field effects are propagated at a finite velocity. Further particular constraints may be added when consistency requires them. Thus, field values are measured by the momentum imparted to a test particle. If one wishes another measurement of the same field, one must restore the test particle. One must also compensate for the field produced by the test particle itself. Any attempt to describe such measurements in accord with these constraints leads to limitations in information about simultaneous measurements or in measurements of spatially separated fields. Following these general methods the authors treat the measurability of quantum fields by beginning with the simplest case, charge-free fields, and then adding on complexities. In each case they show that the assumptions of quantum field theory are compatible with the possibilities of measurement, or, more technically, with the commutation relations assumed for quantized field variables.

This argument led to an extension of the idea of complementarity. The relatively large classical test particles used to measure field values smooth out the effect of the quantum fluctuations which, theory predicts, should be present even in a vacuum. The more precisely the field values are measured, the less informative is the measurement concerning the density of fluctuations of photons. As Bohr and Rosenfeld explain it, "The fluctuations in question are intimately related to the impossibility, which is characteristic of the quantum theory of fields, of visualizing the concept of light quanta in terms of classical concepts."[57] It may seem strange or confused to speak about visualizing a concept. Yet, the basic idea is clear and recurs elsewhere in Bohr's writings. Visualizable models are valid only within the limits within which classical concepts may be used to give descriptive accounts.

This paper did not solve or even treat difficulties concerned with infinite self-energy. This, Bohr insisted, is an effect which cannot be treated by correspondence arguments. The paper contributed little, if anything, to the technical development of quantum field theory. Yet, it played an important role in the formulation of Bohr's final position. He had long been concerned with presenting a consistent account of how classical concepts function in quantum descriptions and in the measurements based on them. In this paper he had finally presented his rather qualitative conceptual analysis with quantitative mathematical precision. He had done this for the most obscure

and controverted extension of quantum theory. The attempt seemed to
work in every detail. The doctrine of complementary descriptions, he concluded, must be considered as a permanent part of physics, not simply a
temporary expedient.

The Solvay Conference, held in October 1933, was dominated by consideration of the newly discovered neutron and positron.[58] In his contribution Bohr was characteristically conservative about the way these new
discoveries should be interpreted. In the discussion following the Joliots'
paper he insisted on the importance of attempting to establish as many
conclusions as possible concerning positive electrons without having recourse to Dirac's hole theory.[59] In his paper on the theory of positron Dirac
introduced the idea of polarization of the vacuum.[60] In a letter written to
Bohr just before the meeting Dirac discussed the experimental consequences of this idea. The effective charge of an electron, the value effectively measured in all low-energy experiments, should be slightly less than the
true charge by a factor of 136/137. Scattering experiments involving kinetic
energies of the order of the rest-mass energy should manifest some modification of the observed charge.[61]

Bohr commented on this suggestion in the discussion following Dirac's
paper and expressed doubts about the possibility of any experimental confirmation of hole theory. The reason offered was that such experimentation
could only be understood in terms of successive approximations from a
classical basis. Bohr's own remarks were published under the heading "On
the Method of Correspondence in the Theory of the Electron."[62] Here, as in
the earlier Rome talk, the correspondence principle is seen as the basic tool
both for extending the classical concept of the electron and for determining
the limits of its validity in the problematic areas of relativistic quantum
mechanics, quantum field theory, and nuclear physics. In relativistic quantum mechanics Bohr interpreted the marvelous confirmation of the Dirac
theory accomplished through the discovery of the positron as a removal of
obstacles previously seen as limiting the correspondence principle's applicability. In suggesting this unique interpretation of the new breakthrough, he
did not really explain how the correspondence principle fitted into this
context, except to mark that Dirac's theory of the electron is valid when
applied to dimensions large compared to the classical electron radius.

In spite of the lack of details, the general tenor of Bohr's argument is
clear. The classical concept of an electron constitutes a direct application of
classical physics to a system of material point charges. Its valid application
depends on having sizes significantly larger than the classical radius of the
electron, forces small enough that the concept of the mass of the electron has
an unequivocal usage, and radiation reactions small enough to be ignored in
the first approximation. Then the three problem areas should be seen as

ones in which the classical theory of the electron and the correspondence principle ground the first approximation. Then one adds to this first approximation the correction terms necessary to compensate for the idealizations introduced in neglecting effects due to the size of the electron, to the relativistic increase in mass, and to radiation reaction.

The bulk of Bohr's remarks summarized his work with Rosenfeld. Though Bohr accepted the existence of neutrons, he then insisted that this would not necessitate any change in his conclusion that nuclear physics, where the classical concept of the electron clearly breaks down, is beyond the scope of the correspondence principle and accordingly also beyond the scope of present quantum mechanics. Similarly, Pauli's concept of the neutrino (which Bohr called a 'neutrion') also was suspect because it goes beyond any possibility of a classical description.

In spite of Bohr's reservations, others at the same conference were beginning to consider the nucleus to be within the domain of applicability of quantum mechanics. In place of Gamow's liquid drop model of a nucleus containing protons, neutrons, and alpha particles, Heisenberg suggested that the nucleus is composed of exclusively of protons and neutrons.[63] Building on a suggestion of Bohr, Heisenberg also argued that the mass defect, the difference between the sum of the masses of the particles contained in the nucleus and the mass of the nucleus, would be manifested as a new type of strong nonelectromagnetic force. Here, he also introduced the idea of exchange forces. In spite of these very significant advances, both Heisenberg's paper and the subsequent discussions manifested an underlying uncertainty on an issue that was partly physical and partly conceptual. Should the neutron be understood as an elementary particle or as a novel composite of a proton and an electron? Any answer obviously involves delimiting the applicability of 'particle'.

The discussion on Heisenberg's paper concluded with Fermi's comment that he preferred to follow Heisenberg and Majorana in considering the nucleus as composed exclusively of protons and neutrons.[64] Immediately after this meeting Fermi turned to the problem of developing a theory of beta-decay. He accepted Pauli's neutrino hypothesis. He also found that the creation and annihilation operators he had used in quantum field theory were apt tools for expressing the emission or absorption of these particles from the nucleus. His original note explaining his theory was rejected by *Nature* as being too speculative. A revised version was accepted by *Ricerca Scientifica* and published just two months after the Solvay Conference.[65] Fermi's definitive quantitative treatment was soon published in both a German and an Italian version.[66]

The key assumption grounding Fermi's mathematical treatment is that neither electrons nor neutrinos exist as particles within the nucleus. They are created in the process of beta-decay much like the way a photon is created in the transition of an excited atom to a lower state. The proton and neutron are both accepted as elementary rather than composite particles. These ideas were quickly accepted and effectively governed the way the term 'particle' was used in nuclear physics.

The year 1935 witnessed the decisive debate between Einstein and Bohr, a topic which will be considered in chapter 9. To conclude this survey of the way in which Bohr's changing concept of a particle and of the applicability of the correspondence principle as a guide for the use of such concepts effectively modified his concept of what concepts themselves are, we will consider some work Bohr did in nuclear physics, including work that was not published till after this debate. Bohr soon accepted the idea that a nucleus is composed of only protons and neutrons. Since the kinetic energy of these particles within the nucleus should be in the nonrelativistic range and localization does not present an acute problem, the classical concept of a particle should have an approximate validity for protons and neutrons within the nucleus. So too should the principle of energy conservation. Thus, Bohr was deeply disturbed when an experiment, one which was subsequently shown to be incorrect, seemed to support the idea basic to the earlier B-K-S paper that energy is only conserved statistically.[67] Such considerations led Bohr to modify his approach to nuclear physics. It should not be considered a subdivision of relativistic quantum mechanics. If the classical concept of a particle can play a role in a descriptive account of nuclear behavior, then models of the nucleus should also have some valid application.

The catalyst that precipitated Bohr's new model of the nucleus was Fermi's work, bombarding different elements with neutrons and observing the resulting reactions. This was fairly straightforward until a fortuitous modification of the experimental situation by Fermi himself led to a surprising anomaly. If the bombarding neutrons are slowed down by passage through a block of paraffin, then nuclear activity is greatly increased.[68] How could the collision cross section of a nucleus, manifested in this slow neutron capture cross section be some ten to one hundred times larger than its geometrical cross section? The answer to this question occurred to Bohr while he was listening to a paper on Bethe's theory of the probability of neutron capture. Bohr kept interrupting the speaker. Then he suddenly stopped and turned so pale that people thought that he had become seriously ill. A few seconds later he stood up, smiled, and said, "Now I understand it."[69]

What Bohr understood he soon expressed as the compound nucleus.[70] People had been thinking of the incident neutron as colliding with the nucleus, considered as a unit, or with individual nucleons, considered bound in a potential well. Bohr assumed that a slowly moving incident neutron is absorbed by the nucleus, forming a compound state. After a few random collisions the neutron's excess kinetic energy is shared by many particles, putting the nucleus in an excited state. A low-energy excited state is relatively long-lived (on a nuclear time scale), and has sharply defined and clearly separated energy states. It can decay by any one of a number of competing processes.

This collective model of the nucleus, together with the associated Breit-Wigner formula for energy levels, dominated early research on nuclear structure. When it yielded primacy to a new model, it was to a model also developed by Niels Bohr, the liquid drop model. Though Bohr suggested this model in his early work on the collective model,[71] he did not develop it till later. In a paper published the day World War II erupted in Europe Bohr and Wheeler used this model to explain the mechanism of nuclear fission.[72] In developing both models, Bohr was following the guiding principle that visualizable models are valid within the limits in which classical concepts may be used to give descriptive accounts. The crucial issue with regard to the concepts involved was to know the limits within which 'particle' may validly be used and to know how to use it within those limits.

9. Einstein Disagrees

Einstein concluded his autobiographical notes, which he referred to as "something like my own obituary," with the poignant statement, "This exposition has fulfilled its purpose if it shows the reader how the efforts of a life hang together and why they have led to expectations of a definite form."[1] One of his most repeatedly asserted expectations was that quantum theory would eventually be replaced by a theory capable of providing a complete descriptive account of atomic interactions. This expectation manifested all the signs of a belief syndrome. When the true believer sees the professed reasons for his belief refuted, he does not abandon his belief; he seeks new reasons. From the early 1920s until his death in 1955, Einstein repeatedly and strongly criticized quantum theory in general and Bohr's interpretation in particular. Bohr invariably replied with a demonstration that neither Einstein's technical objections nor the various thought experiments he proposed undermined the orthodox interpretation of quantum mechanics. Yet, Einstein never capitulated. His belief in the type of explanation that should be sought remained steadfast unto the end.

This clash of the titans can be viewed from various perspectives. Sociologically, there was something of a struggle for dominance between two men, each seeking to lead the world physics community. There were profound psychological differences in these two men's makeups. Bohr's thought had an extremely linguistic orientation. He developed his ideas through an almost incessant dialogue. When stymied, he regularly almost ritually repeated key words over and over. Einstein was quite nonverbal in his thinking. He was extraordinarily slow in learning to speak, weak in his mastery of foreign languages, even of the English he used for the last twenty-two years of his life. To Hademard's inquiry concerning the role that language had played in his discoveries, Einstein answered:

> The words of the language, as they are written or spoken, do not seem to play any role in my mechanism of thought. The psychical entities which seem to serve as elements in thought are certain signs and more or less clear images which can be 'voluntarily' reproduced and combined. . . . Conventional words or other signs have to be sought for laboriously only in

a secondary stage, when the mentioned associative play is sufficiently established and can be reproduced at will.[2]

As Holton has noted in his psychobiographical study of Einstein, Einstein's
intellectual strength and uniqueness were rooted in his visual imagination.[3]
Though Einstein occasionally had collaborators and later, in his Princeton
period, had young assistants, at heart he always remained the solitary
thinker, seeking the "secrets of the Old One."[4] Bohr shunned solitude.
Even when he stopped working to see wild West films, one of his pastimes,
he usually wanted his young assistant to accompany him—and explain the
plot to him.

Sociological and psychological studies can be quite helpful in understanding these men, the moves they made, the motives behind the moves,
even perhaps their unconscious motivation. Such studies, however, have no
bearing on the *validity* of the positions presented. And this is our basic
concern. Quantum theory in its developed form did not fulfill the expectations that either Bohr or Einstein originally held concerning the nature of
scientific explanation. Confronted with this gap, Bohr gradually modified
his position on the nature and function of scientific explanation until he had
an account that fitted the scientific theory he accepted as true.

The march of Einstein's thought was guided by a different drummer.
From the efforts of a life spent in science Einstein had built up expectations
of a definite form concerning the properties a theory must have for its
explanations to be accepted as correct and complete. These expectations
were set by his work in kinetic theory, atomic physics, special relativity,
quantum theory, and above all the theory of gravitational and other fields:
"At this point [in his appraisal of the prospects for quantum physics] it is the
experiences with the theory of gravitation which determine my expectations."[5]

The thesis of this chapter is that the ultimate basis of Einstein's disagreement is a position on the nature of scientific knowledge profoundly different
from Bohr's. Einstein's position was shaped primarily by his own sustained
research program, by reflection on his successes and failures and the reasons
for each. He was also influenced by various philosophical currents—he was
certainly much more widely read in philosophy than Bohr. Yet, his
approach to these sources was inevitably eclectic, picking elements that
might help to clarify the interpretative problems he wrestled with. Following
Einstein, I will treat the technical details of his criticisms of quantum
mechanics as the implementation rather than the source of his disagreement.

The development of this thesis, however, presents a peculiar problem.
Einstein's practice of science was unparalleled. His occasional reflections on
the nature of scientific knowledge were, as will be argued, radically inade-

quate even to explain the work he himself had done. This in itself is neither surprising nor disturbing. The practicing scientist is generally no more a philosopher of science than the practicing artist is an art critic. The intellectual orientations are different and, to some extent, mutually interfering. Yet, in the present instance the overlapping of these rather disjointed roles blurs the significance of Einstein's position. He rejected the Copenhagen interpretation because it did not fit his developed expectations of what an ultimate scientific theory should be and do. Whenever he attempted to articulate the reasons for his rejection, he inevitably fell back on his own epistemological analysis of scientific knowledge. Here Einstein was at his weakest. Few, even among his strongest supporters, have found his theory of knowledge to be either coherent or convincing.

I accept this evaluation of Einstein's epistemology. But I do not believe this gets to the heart of the matter. Einstein's expectations on the form an adequate atomic physics should have were shaped by a lifetime of labor and an unprecedented string of successes. He found it extremely difficult to suspend his commitments, step back, and then ask external questions about the nature, functioning, and implicit presuppositions of the work he was doing. One who judges Einstein's rejection of standard quantum theory simply in terms of the reasons Einstein himself offers for this rejection gets only a superficial and rather misleading view. For this reason this chapter will focus on Einstein's practice of science and attempt to show how it led to certain developed expectations concerning the nature of scientific explanations that could be accepted as adequate. We will try to see his epistemological position against the background of this development. Finally, we will consider, as the culmination of this aspect of Einstein's development, the paper he wrote in conjunction with Podolsky and Rosen. In tracing these developments, we will try, as much as possible, to present them from Einstein's perspective.

Formative Years

By the mid-1920s atomism and quantum theory were inextricably intertwined. It was not so in the early part of the century, when quantum theory was more closely related to thermodynamics than to atomism. Since this earlier era conditioned Einstein's initial orientation, it is helpful to separate the two fields. Earlier, in chapter 4, we considered Einstein's technical contributions to these fields. In returning to them, we wish now to focus on the presuppositions underlying his work and on the interpretative framework manifested and modified in the course of his development.

In his early period Einstein was influenced by both Mach and Boltzmann. For one trying to do foundational philosophy there is a radical conflict between Mach's *epistemological* reductionism and Boltzmann's *ontological* reductionism. If knowledge can be reduced to sensations and constructs based on sensations, then insensible atoms must be accorded a derivative rather than a foundational status. However, for one more concerned with intellectual stimulation and suggestive ideas than with a resolution of hoary metaphysical debates, this clash may be rather blithely ignored. In common with other young German-speaking physicists, Einstein was influenced by both Boltzmann and Mach and apparently felt no compulsion to harmonize the underlying discord.

Einstein's earliest papers, discussed in chapter 4, manifest a strong commitment to proving the existence and determining the properties of real atoms. His original approach, manifested in his first two papers, was one of focusing on observable macroscopic phenomena, such as capillarity or the electrical potential of dissolved salts, and then attempting to explain some otherwise puzzling aspects in terms of atoms responsible for these phenomenal manifestations. This approach did not by itself lead to any hard results concerning atomic properties.

The alternative approach, manifested in his redevelopment of Boltzmann's kinetic theory, proved more successful. Instead of trying to infer the properties of atoms, Einstein relied on very general principles such as energy conservation, determinism, and continuity. Much of this duplicated work Gibbs had already done. Yet, the two men thrusted in rather different directions. Gibbs's statistical mechanics kept assumptions about atomic behavior to the absolute indispensable minimum.[6] After Einstein had set up a similar framework, he searched for problems where assumptions about the behavior of atoms would lead via statistical reasoning to testable macroscopic manifestations. This was the general method underlying his treatment of Brownian motion and the specific heat of solids.

Mach's contribution to Einstein's intellectual development will be considered later. For present purposes we need only note that Einstein was influenced both by Mach's epistemology and by his critique of the foundations of Newtonian mechanics, especially the ideas of absolute space and time. As Einstein later summarized it:

> It was Ernst Mach who, in his *History of Mechanics*, shook
> this dogmatic faith [in classical mechanics as foundational for
> all of physics]; this book exerted a profound influence upon me
> in this regard while I was still a student. I see Mach's greatness
> in his incorruptible skepticism and independence; in my youn-
> ger years, however, Mach's epistemological position also in-
> fluenced me very greatly, a position which today appears to me
> to be essentially untenable.[7]

Einstein was influenced by Boltzmann and Mach, but was never dominated by either man's thought. He soon emerged as an independent thinker with his own distinctive method. The year 1905 was Einstein's *annus mirabilis*, the year he produced his papers on special relativity, on the light-quantum hypothesis, and on Brownian motion. The details pertinent to the present study have already been discussed. What we wish to consider now is what this monumental breakthrough and its aftermath manifests of the young Einstein's ideals of scientific explanation.

Though Einstein was writing about three different problem areas, he saw them as interrelated on a deep foundational level. Physics should supply an account that is self-consistent and increasingly more adequate to physical reality. The physics Einstein had learned was incapable of doing this. Hence Einstein's concern with new foundations interrelating these otherwise separated problem areas. It will be helpful to get a general overview of the particular inadequacies and inconsistencies that bothered Einstein around 1905 before considering his subsequent attempts to rectify these shortcomings.

Thermodynamics was and, Einstein was convinced, would remain secure. As he later summarized his early position:

> A theory is the more impressive the greater the simplicity of
> its premises is, the more different kinds of things it relates,
> and the more extended is its area of applicability. Therefore,
> the deep impression which classical thermodynamics made on
> me. It is the only physical theory of universal content concern-
> ing which I am convinced that, within the framework of the
> applicability of its basic concepts, it will never be overthrown
> (for the special attention of those who are skeptics on
> principle).[8]

Thermodynamics was secure, but it was so because of its generality. It did not rest on any assumptions concerning the composition of matter. It supplies the prototype of what Einstein was later to call a principle theory. To describe the structure of matter and radiation one had to turn to electromagnetic theory and kinetic theory—which was still applied classical mechanics. In the special theory of relativity Einstein had interrelated these two in a new formal way, in terms of constraints that these and any theory of physical reality had to meet. Though special relativity led to the concept of 'mass-energy', it was also a principle theory, an evaluation Einstein made as early as 1907.[9] Special relativity provided laws about the laws of physics. It did not of itself provide any information about the structure of matter and radiation.

Thus, matter and radiation remained essentially disjoint topics. Electromagnetic theory described radiation in space. It did not include the sources of radiation. Though Einstein had more respect for Lorentz than for anyone else he ever met, he found Lorentz's attempt to include sources inadequate. Theories of matter, on the other hand, had never explained the emission and absorption of radiation.

Then quantum theory appeared. Einstein was the first to sense its potential for a unified treatment of matter and radiation. As noted earlier, his original approach to quantum theory was essentially a qualitative one, working out the implications of his light-quantum hypothesis. He was thinking of radiation as a kind of gas of light quanta. As such it should be amenable to the general statistical methods he had developed. Soon Einstein accepted Planck's radiation formula, though not his derivation of it, as basic and found that this also supported something of a dualistic theory of light.

Our present concern is not with the technical details, but with the underlying thrust of this development. Einstein was attempting to develop a unified self-consistent account of physical reality. For this purpose he needed concepts and methods different from those already functioning in physics. In particular he needed, as he then appraised the situation, a quantum electrodynamics, or a more adequate quantum theory of radiation. Though Einstein did not succeed in this endeavor, he did publish enough to show the direction of his striving. In the first of two papers, both written in 1909, which treat this problem, Einstein showed how Planck's law applied to radiation fluctuation leads to both a particle-type term and a wave-type term, as previously shown in discussing equation (4.33).[10] This development led him on to some further ideas on how the quantum of action, proper to radiation, might relate to the charge of the electron, presumably basic to matter. The key points were that h has the same dimensions as e^2/c and that e does not fit into electromagnetic theory in any natural way. This theory deals with continuous rather than discrete quantities. Einstein suggested that the same modification that yields the charge of the electron will also contain, as one of its consequences, a quantum theory of radiation. Though this was only offered as a suggestion rather than a developed position, it clearly revealed a desire to develop a fundamental quantum theory in a way that would allow similar treatments of the structure of matter and of radiation.

This suggestion was extended in an address given to a convention of scientists and physicians meeting at Salzburg in September of 1909.[11] In electricity one thinks of an electron as a singularity surrounded by an electrostatic field. Similarly, one might think of a light quantum as a singular concentration of energy surrounded by a field of force. Fields and singularities had come to replace wave-particle duality in Einstein's speculative probing for new foundations.

These simple suggestions were just the tip of the iceberg. Concealed from public view, then as now, were the intensive efforts Einstein put in to develop a quantum field theory which would explain something about the structure of matter and the interaction of matter and radiation. Around 1910–11, he abandoned these efforts, because there seemed so little hope of success. Instead of attempting to construct a unified theory of both matter and radiation, Einstein returned to the more pedestrian, but also more promising, task of attempting to infer something about the behavior of atoms from macroscopic phenomena whose explanation required detailed assumptions concerning the properties and activities of atoms.

His first new paper of this type dealt with an obscure and highly complicated issue, the opalescence of homogeneous fluids and fluid mixtures near the critical state.[12] Variations in the density of a fluid induce changes in the dielectric constant. This, in turn, changes the way light is scattered by the fluid. Near the critical point of condensation of a homogeneous fluid, or for a mixture of two fluids, minute density fluctuations can have a significant effect on the intensity of the rays of opalescence. To determine the density fluctuations near the critical state Einstein first had to show how statistical fluctuations, a reversible phenomena, could explain irreversible effects in the macroscopic parameters thermodynamics treats. Next he had to apply Maxwell's equations to an electromagnetic field whose dielectrical constant is rapidly fluctuating in both space and time. Then he had to determine the ratios between the intensities of the incident and scattered light. This finally could explain why light scattered through a mixture of nitrogen and oxygen makes the sky look blue.

The calculations are highly complicated. Yet, the underlying assumptions are quite straightforward. The Boltzmann equation, which Einstein writes in the form $S = (R/N) \log W + $ const, seems to presuppose a complete theory (perhaps a complete molecular-mechanical theory) of a completely described system. How does the Boltzmann principle work without a complete molecular-mechanical theory of elementary particles? Einstein's answer to this has two features that are significant in bringing out the presuppositions implicit in his arguments. First, provided one presupposes that the ultimate processes are really reversible, one does not need a detailed theory of elementary interactions. For an isolated system the principle of energy conservation suffices to characterize the phenomenological state of a system. Second, Einstein introduced the idea of probability he had developed earlier. The probability that a system is in a particular state is set equal to the fraction of time the system spends in that state over a sufficiently long period of time.

Einstein's treatment of observable phenomena as manifestations of the underlying atomic events causally responsible for these phenomena presupposed objectively determined elementary processes. His treatment of macroscopic states presupposed the absolute validity of the law of energy conservation for an isolated system. These two assumptions of objective causal determinism and energy conservation ground the introduction of probabilities. Probability is treated as a derivative rather than a foundational concept. It would be dispensable if one had a complete theory of atomic behavior.

Another macroscopic phenomenon for which Einstein sought a depth explanation was the surface tension of a fluid. An empirical law fitting this phenomenon had already been established by Roland Eötvös. To fit a molecular account to this established law Einstein assumed that molecules interact only with their near neighbors.[13] In subsequent papers Einstein introduced similar molecular interactions to adapt his earlier treatment of specific heat to low-temperature behavior. This, however, was soon superseded by the revisions of Einstein's specific heat theory developed by Debye, and by Born and von Karman.

In a 1919 article, written for the (London) *Times*, Einstein distinguished between constructive theories and principle theories.[14] Constructive theories attempt to build up a picture of the more complex phenomena out of the materials of a relatively simple formal scheme from which they start out. Principle theories employ the analytic rather than the synthetic method. The basic elements are discovered principles. These supply the basis for mathematical criteria which the separate processes or the theoretical representations of them must satisfy. These principle theories are not as complete, adaptable, or explanatory as constructive theories, but have greater logical perfection. Thus, thermodynamics uses the principle of the impossibility of perpetual motion machines of the first and second kinds to impose general constraints. In this section we are treating primarily Einstein's work on constructive theories. His work on principle theories will be treated separately when we discuss the theory of relativity.

In his work on—or rather toward the development of—a constructive theory of matter Einstein was manifesting a kind of Aristotelian methodology. He would begin by accepting observable phenomena and empirical laws systematizing this data. Then he sought a depth account to explain the phenomena in terms of atomic behavior causally responsible for it. This entailed accepting the real existence of atoms and molecules, making simple plausible assumptions about atomic behavior and the nature of interatomic forces. Then, through a reliance on kinetic theory, electromagnetic theory, technical competence, and incredible ingenuity, Einstein would work out the consequences of his assumptions in a way that would hopefully give a causal account of, or at least a strong correlation with, the phenomena that were to be explained. In doing this Einstein invariably presupposed that the ultimate interactions are governed by objective laws of universal validity. These laws might not yet be discovered. Nevertheless, he presupposed that there were such laws and that they could, when properly interpreted, supply a causal account of observable phenomena. Thus, a year before writing the *Times* article, Einstein had declared, "The supreme task of the physicist is to arrive at those universal elementary laws from which the cosmos can be built by pure deduction."[15]

Einstein in his early years seemed to have an internal seismograph enabling him to detect the tremors produced by faults in a theory. Few faults registered quite as high on his scale as Planck's derivation of the radiation law. Even after Einstein's attempt to develop a more fundamental quantum theory of radiation was abandoned as hopeless, he returned again and again to the problem of deriving Planck's law. Here the term 'derive' is somewhat misleading. Planck had derived it, but his derivation rested on classical principles incompatible with the physical significance Einstein attached to the law. Accordingly, what Einstein was trying to do was to determine what physical assumptions implied or were implied by the radiation law. This, in turn, should clarify and unify the assumptions that were used in constructive theories of matter or the interaction of matter and radiation.

In 1910 Einstein, with the assistance of L. Hopf, wrote two papers examining the question of whether classical statistics could be modified to yield Planck's law. In the first paper they noted that a straightforward following of the assumptions proper to radiation theory and mechanics led to results in contradiction with experience, viz., the Rayleigh-Jeans law.[16] Accordingly, they considered the possibility that the guilt for this failure lay with the standard statistical assumption of the independence of atomic events. Here again, Einstein used his interpretation of probability as the average time a system spends in a state. The independence assumption took the form of assuming that mean values of the Fourier coefficients in the expansion of an oscillator's electrical or magnetic field are independent. Then their calculations showed that this escape route led nowhere. One is led back, they concluded, much more to the physical problems than to the pure mathematics.[17]

In a companion paper they tried to work out another derivation of the radiation law without relying on any physical assumptions that seemed open to doubt.[18] The most dubious assumption at that time was still the law of the equipartition of energy. Its application to vibrational or rotational states of molecules seemed to involve inconsistencies. Its application to electronic motion did not seem to be in accord with the laws for the specific heat of solids. Einstein and Hopf accordingly presupposed the validity of the equipartition law only for translational motion, an application whose validity was beyond doubt. Even with this limitation, their derivation still led to the Rayleigh-Jeans law. Their article ended with the statement, "we must accordingly conclude, that only a more principled and deepergoing change in the fundamental intuitions can lead to the experimental knowledge of a better formulated radiation law."[19] Einstein was clearly willing to make changes, even radical ones, in received opinions. What he was most reluctant to do was to accept as basic a law whose physical significance he did not understand.

Even while developing general relativity, Einstein kept returning to the problem of what physical assumptions entail, or are entailed by, Planck's law. In 1913, he and O. Stern wrote a paper arguing for the existence of some molecular agitation at the absolute zero point.[20] At the conclusion of their article they point out a surprising implication of this assumption that there is some thermal agitation at absolute zero. If one assumes that the zero-point energy is $h\nu$ rather than the $h\nu/2$ Planck had introduced in his second radiation theory, and also uses the Einstein-Hopf treatment of an oscillator, one obtains Planck's radiation law without introducing any assumptions concerning discontinuity.[21]

Einstein returned to this problem again in 1914, deriving Planck's law on the assumption that the molecules in a chemically homogeneous gas are related to oscillators which can only have energies of the form $E = nh\nu$.[22] This was, in effect, a coupling of Planck's quantized oscillators to rather classical considerations of molecules. Bohr's theory of spectra, Einstein soon realized, supplied a much better basis for a consideration of molecular states. Using the general features of Bohr's theory, especially transitions between discrete energy states, he attempted to show how Planck's law can be obtained by a consideration of the interaction of matter and radiation instead of a reliance on the type of electromagnetic-mechanical considerations Planck had used.[23] This was the initial version of the paper considered earlier on the quantum theory of radiation.[24] In the initial paper Einstein introduced probability coefficients for the emission and absorption of radiation. However, he did not yet introduce the momentum exchange considerations that were used in the later paper to support his light-quantum hypothesis.

Einstein did succeed in deriving Planck's equation from considerations involving the general features of the Bohr theory. Yet, he was still not satisfied, for even this derivation did not indicate how matter and radiation interact. Earlier we discussed this paper from the perspective of its role in Bohr's development. Here we wish to bring out two aspects of the paper which played rather conflicting roles in Einstein's development. The first, the attribution of momentum to light quanta, completed the particle interpretation of light and strengthened Einstein's belief in the validity of the light-quantum hypothesis. The second feature, one which haunted Einstein for the rest of his career, was the introduction of chance factors in quantum theory. Einstein's earlier probabilistic arguments, like Boltzmann's before him, had rested on deterministic foundations, whether actual or presumed. Einstein's new probability coefficients were geared to spontaneous as well as induced transitions. Spontaneous transitions would seem to constitute a pure chance factor. Indeed, this is the way they are generally interpreted. This introduction was not altogether novel. Both radioactive decay in nuclei and Planck's second quantum theory, which involved continuous absorption but discontinuous emission of radiation, could be interpreted as introducing chance factors. At this time, however, it seemed much more reasonable to hold that the apparent chance factors would eventually be explained by some to-be-discovered causal mechanism.

This long and rather frustrating series of attempts manifests the underlying thrust of Einstein's approach to constructive theories. He had long since come to accept Planck's law as valid on empirical grounds. It was the only radiation law that fitted all the data. It yielded the other laws in the limits in which they were valid. Planck's law admirably fitted all the criteria Mach had set for the acceptance of a law of physics. Yet, even before he began to reflect on the divergence between Mach's criteria and his own practice, Einstein's search for a depth explanation of Planck's law reflected ideas on the nature of scientific laws quite different from those given by Mach, Duhem, Pearson, and the positivist tradition. If Planck's law is true, it is so because it reflects something about the structure of matter and radiation and their interaction. Einstein never relinquished his determination to learn more about this elusive something.

The Theory of Relativity

The theory of relativity has provoked almost as many philosophical problems as the quantum theory. None of them will be treated here. We will simply consider the way in which the development of relativity, both the special and general theories, modified Einstein's views on the relation between scientific knowledge and reality. Einstein's interpretation of the philosophical significance of the special theory of relativity varied somewhat in the many summary accounts he gave. Yet, two aspects distinguished this from the other types of theories we have been considering. First, Einstein took the theory of relativity to be an example of a principle, rather than a constructive, theory. Second, even while his summary accounts of special relativity gradually came to reflect a different philosophical perspective, these accounts invariably stressed the primacy of epistemological considerations. Only relative motion could be known through experience. Nothing more could be said to exist. It seems that such epistemological considerations were not exactly Einstein's own starting point. He claimed that this was his reasoning at the age of sixteen, after reading Mach: "If I pursue a beam of light with the velocity c, I should observe such a beam as a spatially oscillatory electromagnetic field at rest. However, there seems to be no such thing, whether on the basis of experience or according to Maxwell's laws."[25] When Einstein transformed this initiating intuition into a developed theory, any talk about the spatial and temporal separation of two events was reduced to discussions of measurements with rods and clocks. There is no doubt but that, in this area especially, the young Einstein thought of himself as a tough-minded empiricist. It is for this reason that a discussion of Einstein's epistemology has been deferred until now.

Fortunately, it is possible to know the philosophical sources that influenced Einstein and to trace something of their changing effect on his intellectual development. In Bern he and some friends, especially Maurice Solovine and Conrad Habicht, formed the "Olympia Academy," essentially a dinner-discussion group of a few ambitious and rather impecunious young intellectuals.[26] They read Karl Pearson's *Grammar of Science*. Pearson was like Mach in discounting metaphysical statements and in interpreting science as an economical resumé of actual and possible experiences. Though not as thorough or systematic as Mach, he went beyond him to some extent in the importance he attached to theoretical constructs.

The young searchers went on to study Mach's *Mechanics*, a work Einstein was already familiar with; Mill's *Logic*; Hume's *Treatise on Human Nature*; Spinoza's *Ethics*; selections from the works of Helmholtz, Ampère, Riemann, Avenarious, Clifford, and Dedekind; and Poincaré's *Science and Hypothesis*.[27] It is difficult to know how much influence these sources had on Einstein's development. The group's study was anything but systematic. At their frugal meal someone would read a passage and then this would serve as a point of departure for an extended discussion. In most cases they probably only had one copy of each book and passed it around. When Einstein referred back to this period, he generally cited Hume, Mach, and Poincaré as the sources that influenced him most strongly.[28]

Around 1919 Einstein wrote a paper, "Fundamental Ideas and Methods of Relativity Theory, Presented in Their Development." Written, apparently, for his own benefit, it has not yet been published except in excerpts. Here he explained the origin of his work on general relativity:

> When, in the year 1907, I was working on a summary essay concerning the special theory of relativity for the *Yearbook of Radioactivity and Electronics*, I tried to modify Newton's theory of gravitation in such a way that it would fit into the theory. Attempts in this direction showed the possibility of carrying out this enterprise, but they did not satisfy me because they had to be supported by hypotheses without physical basis. At that point there came to me the happiest thought of my life, in the following form.

Just as in the case where an electrical field is produced by electromagnetic induction, the gravitational field similarly has only a relative existence. *Thus, for an observer in free fall from the roof of a house there exists, during his fall no gravitational field*—at least not in his immediate vicinity. . . .

The extraordinarily curious empirical law that all bodies in the same gravitational field fall with the same acceleration immediately took on, through this consideration, a deep physical meaning. For if there is even one thing which falls differently in a gravitational field than do the others, the observer would discern by means of it that he is falling in it. But if such a thing does not exist—as experience has confirmed with great precision—the observer lacks any objective ground to consider himself as falling in a gravitational field. Rather, he has the right to consider his state as that of rest, and his surroundings (with respect to gravitation) as field-free.

The fact, known from experience, that acceleration in free fall is independent of the material is therefore a mighty argument that the postulate of relativity is to be extended to coordinate systems that are moving non-uniformly relative to one another.[29]

With this text as a point of departure, it seems possible to reconstruct the reasoning Einstein followed. In accord with the principles of special relativity gravitation had to be thought of as a field rather than as action at a distance. The Poisson equation $\nabla^2\phi = -4\pi\rho$, where ϕ is the gravitational potential and ρ the density of the gravitational source, can be included in the framework of special relativity by using the four-dimensional equivalent of ∇^2 and a time-varying density

$$\Box^2\phi = -4\pi\rho(x, y, z, t). \tag{9.1}$$

An immediate difficulty with equation (9.1) is that the left side is relativistically invariant while the right side, in general, is not. It becomes invariant only when one is using the rest-mass density ρ_0. Then the attempt to include a mass equivalent of kinetic energy $m = E_{kin}/c^2$ immediately generates a difficulty. In the mechanics stemming from Galileo and Newton, the vertical acceleration due to gravity is independent of the horizontal component of velocity. This could not be so in a relativistic approach because gravitational attraction depends on mass and mass varies with velocity.

At this juncture Einstein was faced with a choice. He could have interpreted the equivalence of gravitational and inertial mass as an approximation valid within the same limits as classical mechanics. Or he could have retained the strict equivalence of gravitational and inertial mass and sought a modification of the Newtonian laws of gravitation. He immediately and decisively opted for the latter course and intuitively thought of the idea of general covariance as a guide in the search for the modification needed. He did so, as he later recalled, without knowing at that time of the high-precision experiments in which R. Eötvös had established the equivalence of gravitational and inertial mass.

Even at this initial stage Einstein already had some of the key principles that eventually went into the formulation of general relativity: the equivalence of gravitational and inertial mass, and the stipulation that it should be possible to formulate the laws of physics in such a way that they are covariant under general (not just inertial) transformations. He was also aware of a consequence of these ideas, the conclusion that the path of a light ray should be bent in a gravitational field. Before considering further developments, we will say something about the epistemological implications of the principles that served as his point of departure.

Epistemologically, Einstein still considered himself to be in the tradition of Hume and Mach. All knowledge is founded on sense experience. There is no process of logical inference leading from sense experiences to the general concepts and laws used to systematize these experiences. Nevertheless, a confrontation with experience is the only scientific way of determining the scope of a concept's applicability or a law's validity. This much Einstein still accepted.

The way in which Einstein actually treated laws, however, was already quite at variance with the manner in which Mach interpreted their significance. According to his principle of economy, laws should be interpreted, not as expressions of some objective order in nature, but as concise summaries of actual and possible experiences. One of his clearest expressions of this doctrine was given in a popular lecture:

> The grandest principles of physics, resolved into their elements, differ in no wise from the descriptive principles of the natural historian. . . . Nature exists only once. Our schematic mental imitation alone produces like events. Only in the mind, therefore, does the mutual dependence of certain features exist.[30]

By this norm Planck's law of radiation and the principle of the equivalence of gravitational and inertial mass should have been considered perfectly acceptable. Each fitted all the known data with the highest precision science was capable of achieving. Yet, Einstein was clearly not functioning along such positivist lines. He accepted Planck's law as valid, but never rested content with a natural historian's descriptive account. He kept trying to find a depth account of why it obtained. He accepted the equivalence of gravitational and inertial mass as empirically established. Yet, he now spoke of it as the "extraordinarily curious empirical law" and was "in the highest degree amazed at its existence."[31] This is very strong language from a man who is quite sparing in his use of superlatives. In expressing such reactions, Einstein, at this stage of his development, seemed to be reflecting his own orientation toward physics more than any as yet consciously articulated change in his own empiricist epistemology.

Between 1907 and 1911, Einstein worked chiefly on the problems in atomic physics and the foundations of quantum theory that we have already considered. In 1911, he returned to the problem of general relativity and wrote his first paper on the topic.[32] In this paper, Einstein relied on two basic principles. The first was the physical equivalence of two reference systems: a stationary system K with a gravitational field of acceleration γ, and a gravity-free field K' moving toward K with acceleration γ. The second principle was the conservation of mass-energy. From these he concluded that light emitted in the K' system with energy E would have a measurable energy in K of $E(1 + \gamma h/c^2)$, or $E(1 + \phi/c^2)$, where h is the distance separating the two systems and ϕ is the gravitational potential. On the basis of these principles he predicted that light just grazing the sun would suffer a deflection of 0.83 seconds of arc.

Einstein did not as yet have a *theory* of general relativity. He merely had some rough first-order corrections to Newtonian gravitational theory. Partially because of such perceived limitations but also because of criticisms by M. Abraham and competition from G. Nordström, Einstein began intensive work on the problem of gravitational theory.[33] In 1912, Einstein returned to Zurich after a brief teaching stint at Prague. On 29 October 1912, he wrote to Arnold Sommerfeld: "I am presently concerning myself exclusively with the problem of gravitation and now hope that with the help of a local friendly mathematician all the difficulties will be mastered."[34] The friendly mathematician was Marcel Grossmann, whose meticulous notes had enabled Einstein, when a fellow student at the ETH (the Swiss Federal Institute of Technology), to prepare for examinations. When Einstein was searching for a mathematics adequate to express the physical principles he was relying on, Grossmann introduced him to the absolute differential calculus of Ricci and Levi-Civita.

Einstein and Grossmann wrote three joint papers in which Grossmann supplied the mathematics and Einstein the physical interpretation. The basic idea is that a geodesic in space is described by

$$ds^2 = g_{\mu\nu}dx^\mu dx^\nu, \tag{9.2}$$

with the Einstein convention of summing over repeated indices. The presence of a gravitational field is then manifested by a departure of the $g_{\mu\nu}$ coefficients from the Euclidean values proper to flat space. This departure, in turn, should be determined by a generalization of the energy-momentum tensor $T_{\mu\nu}$. Since $g_{\mu\nu}$ and $T_{\mu\nu}$ were both second-order tensors, Einstein and Grossmann looked for second-order equations interrelating them. They found that with such second-order equations it was not possible to preserve both the generalized conservation laws for energy and momentum, and the requirement of general covariance. At this juncture Einstein decided to sacrifice the principle of covariance and regard the conservation laws as absolute requirements.

By the time Einstein moved to Berlin, in April 1914, he had mastered the novel mathematics well enough so that he no longer needed Grossmann's assistance. Einstein not only faced the problem of developing an adequate mathematical formulation, but also found that he had to rethink the way in which he had been relating mathematical expressions to physical reality. In the special theory of relativity, coordinates had a clear physical significance. Differences between coordinates correspond to measurable lengths. However, when nonlinear coordinate transformations are allowed, as in equation (9.2), any correspondence between coordinate differences and measurable lengths is not covariant. Furthermore, the tensor calculus seemed to allow an almost indefinite number of possibilities for expressing the physical principles Einstein was treating as basic. Einstein was unable to determine which of these many possible formulations was correct on any physical or observational grounds. He gradually came to look on the *formal* properties of mathematical expressions as decisive clues to whether or not the equations have physical significance.

The form this striving took is best seen by considering the final result of his labors, "The Foundation of the General Theory of Relativity."[35] To have the right relation between the energy tensor and the field metric, Einstein needed tensors of the second rank. He found, however, that to have equations which are covariant with respect to any transformation rather than simply inertial transformations, he needed tensors of the fourth rank, e.g., $B^{\rho}_{\mu\sigma\tau}$. The vanishing of this Riemann tensor is a necessary condition that, by an appropriate choice of the system of reference, the $g_{\mu\nu}$ may be constant. This corresponds to the case in which, with a suitable choice of the system of reference, the special theory of relativity holds good for a finite region of the continuum.[36]

By using the Riemann-Christoffel tensors, $B^{\rho}_{\mu\sigma\tau}$ may be contracted with respect to the indices τ and ρ to obtain the second-order covariant tensor

$$G_{\mu\nu} = B^{\rho}_{\mu\nu\rho}. \tag{9.3}$$

The required equations for a matter-free gravitational field will be satisfied if all $B^{\rho}_{\mu\sigma\tau}$ vanish. This condition, however, is too strong, since it involves transforming away the gravitational field in the neighborhood of a material point. Hence, Einstein suggested that the needed equations be determined by the requirement that the contracted tensor $G_{\mu\nu}$, defined by equation (9.3) vanish. The only real justification for this choice came from the formal mathematical properties of the tensors involved, rather than from any physical considerations. Besides $G_{\mu\nu}$, there is no tensor of the second rank which is formed from the $g_{\mu\nu}$ and its derivatives, contains no derivatives higher than the second order, and is linear in these derivatives.[37] This choice, in turn, led to a set of equations which yielded Newton's law of attraction in the first approximation and the correction to the perihelion of mercury in the second approximation. The introduction of the special theory of relativity now led to the correct result for the deflection of light grazing the sun, 1.7 seconds of arc.

In deriving the final form of these equations, Einstein was, as Mehra has shown,[38] influenced by Hilbert's attempt to construct an axiomatic system, which merged Einstein's earlier formulation of the gravitational equations and G. Mie's field theory of matter. Einstein did not accept Hilbert's unification, but he was impressed by the generally covariant forms that Hilbert was then first to develop. He was also impressed by Hilbert's ideal of a rigorously developed axiomatic system. This stress on such formal aspects of theories as mathematical simplicity and aesthetic appeal as criteria for judging a theory's physical significance came to play an increasingly important role in Einstein's attempts to extend general relativity. The two directions this attempted generalization took will be briefly indicated.

The first direction was cosmology, an attempt to develop a theory of the universe as a whole. Here Einstein's attempts ran into the same difficulty that Newton had discussed in his letters to Bentley, the apparent contradiction between the assumption of an essentially static, spatially finite distribution of matter in the universe and the implication of the theory of gravity—all matter should gravitate toward the center of mass. Einstein's initial suggestion for obviating this difficulty was to modify the gravitational equations by the introduction of a cosmological constant λ. In the non-Euclidean, finite but unbounded model of the universe Einstein proposed, the cosmological constant would correspond to a very weak cylindrically symmetric repulsive force which, at large distances, would so balance gravitational attraction that the universe would remain essentially static. As Einstein admitted when he introduced this constant, "In order to arrive at this consistent view, we admittedly had to introduce an extension of the field equations of gravitation which is not justified by our actual knowledge of gravitation."[39]

In introducing this cosmological constant, Einstein hoped that general relativity would yield a unique solution for the cosmos. Later solutions by de Sitter, Friedmann, Lemaître, and others showed that Einstein's solution is not unique. There can be time-dependent models as well as static models. The type of expanding universe proper to the Friedmann and Lemaître solutions won widespread acceptance after Hubble's discovery of the recession of the galaxies in 1929. The existence of such dynamic solutions which admitted of a reasonable physical interpretation obviated the need for the cosmological constant, which Einstein subsequently dismissed as a logical blemish. Other cosmologists, most notably F. Hoyle, thought that the presence or absence of such a cosmological term should be determined by observation. Einstein steadfastly rejected this, which he considered the worst hypothesis of his own career, on essentially aesthetic grounds: "The introduction of the 'cosmological member' into the equations of gravity, though possible from the point of view of relativity, is to be rejected from the point of view of logical economy."[40]

The second direction of expansion came from the pioneering work of G. Mie and the later attempts of D. Hilbert and H. Weyl to develop a field theory of matter. Though Einstein rejected the particular solutions they proposed, which involved only electrical forces, he responded warmly to the general idea. It bore a strong resemblance to his earlier attempts to develop a quantum electrodynamics. When Einstein returned to this problem after developing the general theory of relativity, he saw a new possibility. If an electron were purely negative electricity confined to a small volume, then the mutual electrostatic repulsion of the distributed 'bits' of negative electricity should tend to blow the electron apart. Perhaps at very short distances gravitational attraction could so balance electrostatic repulsion that there would be a uniquely stable configuration.[41]

Here the implementation of the field equations of general relativity took the same mathematical form as in the cosmological solution:

$$G_{\mu\nu} - \lambda g_{\mu\nu} = -kT_{\mu\nu}, \tag{9.4}$$

where k is a proportionality constant proper to the right side of the amplified Poisson equation (9.1) and, as in the cosmological solutions, λ is a repulsive force balancing gravitational attraction. In this case, however, λ is not characteristic of the fundamental gravitational law itself. It should be determined by the way in which electromagnetic repulsion fits into this framework (in the form of a constant of integration). In this theory the tensor representing the matter-energy density $T_{\mu\nu}$ has two parts, one representing electromagnetic and one gravitational force. This ambitious attempt encountered a technical difficulty. Even when electromagnetic theory supplemented the gravitational field equations, the resulting set was one equation short of the number required for determining the unknowns in the case of a static, spherically symmetric solution. Therefore, the structure of the electron is underdetermined by the theory.[42]

Pauli once commented as follows on the various attempts to fuse electromagnetic and gravitational forces in one theory: "What God has rent asunder, let no man put together." Gravitational forces are so many orders of magnitude weaker than electromagnetic forces that when dealing with ordinary matter (rather than such exotic matter as neutron stars or black holes), gravitational forces are totally negligible compared to electromagnetic forces. A more significant point, at least from the perspective of the problematic proper to the present study, is the implication of Einstein's method of treating the problem. In spite of the mathematical complexity proper to tensor calculus, what is under consideration is a theory of matter in which differential equations are basic. Differential equations presuppose continuity, e.g., continuous fields of force. In this approach, accordingly, continuity is already present as an implicit presupposition of the methodology used. The hope is to explain atomicity as something derivative. In these later theories the electron is not thought of as a singularity in a field, as in the earlier unsuccessful electrodynamics, but as a locus of balancing force fields. This stress on continua as basic characterized Einstein's subsequent theorizing.

Epistemological Reflections

In the summer of 1918 Einstein was incapacitated with a gastric disorder. While recuperating, he read Kant's *Prolegomena* (it seems that he never read the *Metaphysical Foundations*, where Kant also treated ultimate particles as equilibrium regions of competing forces). This was not Einstein's first exposure to Kant. He claimed to have read *The Critique of Pure Reason* when he was thirteen.[43] However, even the most precocious youth of thirteen could not be expected to assimilate the labyrinthine subtlety of Kant's argumentation. When Einstein was in Prague, he was a member of a "Kant-Abende," a group including Max Brod and occasionally Franz Kafka, which met to discuss Kant's philosophy.[44] One cannot but be intrigued at the idea of Einstein and Kafka discussing Kant's views on appearance and reality. Yet, Einstein did not seem to have taken this discussion group seriously. He later wrote to Max Born about Max Brod: "I think he belongs to a small circle there of philosophical and Zionist enthusiasts, which was loosely grouped around the university philosophers, a medieval-like band of unworldly people."[45]

When Einstein studied the *Prolegomena* in 1918, he had already altered the received concepts of 'space' and 'time' and had given non-Euclidean geometry a foundational role in general relativity. His reaction to this reading is indicated in a letter to Born:

> I am reading Kant's *Prolegomena* here among other things, and am beginning to comprehend the enormous suggestive

power that emanated from that fellow, and still does. Once
you concede to him merely the existence of synthetic *a priori*
judgments, you are trapped. I have to water down the '*a
priori*' to 'conventional' so as not to have to contradict him,
but even then the details do not fit. Anyway it is very nice to
read, even if it is not as good as his predecessor Hume's work.
Hume also had a far sounder instinct.[46]

In chapter 2 we tried to show that Kant's doctrine of the role of synthetic
a priori judgments is far more subtle and complex than textbook Kantianism. The proper interpretation of Kant's doctrine is not, however, the point
now at issue. Einstein definitely seems to have taken the classical doctrines
of space and time and the unquestionable physical validity of Euclidean
geometry as prime examples of what Kant meant by the a priori. In support
of this we might cite the biography pseudonymously written by Einstein's
son-in-law. It is generally presumed that the editorial and evaluative opinions stemmed from and were probably checked by Einstein himself. This
book claims:

> Kant's critical philosophy, despite its skeptical character,
> contained one inviolable dogma: the absolute accuracy of
> mathematics and of its allied study, natural science. . . . New-
> ton's system of mechanics and Euclidean geometry were there-
> fore, to Kant, indisputable truths influencing the relation of
> philosophy to exact science.[47]

It must be admitted that Kant's *Prolegomena*, a deliberately oversimplified
and popularized work, lends more support to such views than any other
writings of Kant.[48]

In spite of the judgment that Hume's epistemology is superior to Kant's,
a judgment Einstein never seems to have altered, Einstein's 1918 reading of
Kant seems to have modified the epistemological terminology he used.
Hume's epistemology was based on a distinction between 'sense impressions' and 'ideas'. Mach had made a very similar distinction between
'*Empfindungen*' (sensations) and '*Vorstellungen*' (representations). From
this time on, Einstein relied on a sharp distinction between '*Sinneserlebnis-
sen*' (sensible experiences) and '*Begriffen*' (concepts). His understanding of
concepts is much more Kantian than Machian. Thus, discussing the question
of whether one could visualize a finite, yet unbounded universe, Einstein
said in 1921:

> A geometrical-physical theory as such is incapable of being
> directly pictured, being merely a system of concepts. But these
> concepts serve the purpose of bringing a multiplicity of real or
> imaginary sensory experiences into connection in the mind. To

'visualize' a theory therefore means to bring to mind that abundance of sensible experiences for which the theory supplies the schematic arrangement.[49]

In the next few years, Einstein gave various talks on the significance of relativity in which he invariable stressed its epistemological basis.[50] His basic epistemological doctrine at this stage was clear and unambiguous. Concepts are distinct from the sense experiences. The concepts basic to physics cannot be derived from sense experiences through any process of induction. Nevertheless, these concepts have meaning and physical significance only to the extent that observable facts can be assigned to them in an unambiguous way.[51] The verification of concepts and the theories built upon them ultimately depends exclusively on sense experience.

This was still Einstein's professed epistemology in the early 1920s. It did not really fit his scientific practice. Scientific theories may be verified by sense experiences, but they are not *about* sense experiences. The theories Einstein was then attempting to develop were theories about the universe as a whole and about the true constitution of elementary particles. Even for the general theory of relativity itself Einstein had repeatedly, and only semi-facetiously, claimed that if the experimental tests did not check out, it would have been too bad for the cosmos, or the good Lord; the theory was correct. How do such theories relate to the subject matter which they are theories of?

Einstein probably found, as have many scientists before and since, that extensive concern with such metascientific questions is a distraction from the task at hand, doing rather than discussing science. The two require quite different intellectual perspectives. He commented on such issues chiefly in his popular writings and does not seem to have attempted any systematic development of a theory of scientific explanation. However, in the early 1920s he did spend some time on two special facets of this general question. The first was the relation between geometry and reality, a question precipitated by the use of non-Euclidean geometry in general relativity. His basic doctrine was: "As far as the propositions of mathematics refer to reality, they are not certain; and as far as they are certain, they do not refer to reality."[52] Mathematics, as Einstein then saw the issue, had been clarified by the trend known as axiomatics. One must make a sharp separation between the logical-formal aspect of mathematics and its objective or intuitive content. Mathematics is properly concerned only with the formal logical aspects. In expressing such views, Einstein shows not only a strong influence from David Hilbert and his work on the axiomatization of geometry, but also his affinity for the developing thought of Moritz Schlick. He thought that Schlick was correct in characterizing axioms as implicit definitions of the terms used in the axioms.[53] On epistemological issues Einstein always retained an affinity with the empiricist and positivist traditions.

Geometry, whether Euclidean or non-Euclidean, is best interpreted as an abstract axiomatic system. A physical interpretation of this abstract system can be had by identifying practically rigid bodies with straight lines. Granted such a physical interpretation, the question of whether the universe is Euclidean is to be settled by *empirical* means. If one were to carefully measure, in terms of rigid rod lengths, the distance from a mass-center to a shell surrounding this center and then measure the circumference of the shell with the same rigid rod, one would not find the Euclidean result, $C = 2\pi r$. This proposed test presents serious difficulties, which have been discussed by A. Grünbaum and others.[54] The point to be noted here is that in spite of Einstein's claim that such empirical tests represent the only scientific way to settle the question of whether space is Euclidean, Einstein's own opinion on this issue was actually settled by purely theoretical considerations. There was a growing gap between what Einstein practiced and what he articulated as scientific method.

The second aspect of this developing metascientific quest concerned the interpretation of scientific effort as a search for absolutes. Here the influence of Max Planck, Einstein's colleague since 1914, seems unmistakable. Thus, in a 1918 celebration of Planck's sixtieth birthday, Einstein put Planck in the camp of the select few scientists who find favor with the angel of the Lord because they pursue science for its intrinsic value rather than for power or utility. The intrinsic value of science is both negative, escape from the soul-crushing dreariness of routine life (he uttered at a time when Germany was collapsing), and positive: "Man tries to make for himself in the fashion that suits him best a simplified and intelligible picture of the world of experiences, and thus to overcome it." To achieve precision the scientist must simplify, select, and focus on the appropriate qualities: "supreme purity, clarity, and certainty at the cost of completeness."[55]

The idea that science should aim at supplying a unified intelligible picture of the cosmos was one of the basic themes in Planck's philosophical writings after 1909. Could the simple unifying principles basic to the world-picture also be identified as principles built into the construction of the cosmos? This was a question Planck had explicitly faced in his earlier conflict with Mach: "What do we really mean when we speak of a physical world-picture? Is it merely a convenient but basically arbitrary intellectual concept, or should we take the opposite view, that it reflects material processes quite independent of us?"[56] Planck's basic answer to this question is that history shows that scientists gradually both discovered laws of objective validity and learned how to strip the expression of such laws of anthropomorphic formulations. The objective validity of the first and second laws of thermodynamics is evidenced by the impossibility of constructing a perpetual motion machine of either the first or second kind.

Einstein was not altogether of Planck's persuasion. Though he attributed objective validity to the first and second laws of thermodynamics, he did not interpret these as constructive laws, but as systematizations of the phenomena which constructive laws must explain. Relativity, similarly, was explained as a principle rather than a constructive theory. Einstein accordingly did not think that in developing the general theory of relativity he had discovered objective laws about the content of the universe. Cosmological solutions or elementary particle solutions of the laws of general relativity might contain such information. Such solutions, however, were not determined by the general formalism of relativity theory. In spite of such qualifications, Einstein certainly shared Planck's view that the order of the universe is ultimately to be explained through laws of objective validity rather than as a phenomenal order arising out of some deep-lying chaos of random events.

This submerged issue came to the forefront in the early 1920s with the crisis over causality. The underlying issue here, as Paul Foreman has shown,[57] was a postwar disenchantment with science and with the idea that physical reality could be understood as a deterministic system. Next to Planck, Einstein emerged as one of the strongest defenders of causality.[58] Here again, Einstein's practice was in conflict with his ostensible commitment to Humean ideals of explanation.

Conflict over Quantum Theory

The goal of science is a coherent world view based on general laws which have an objective validity. The confirmation or falsification of scientific laws is ultimately a matter of experiment and observation. Between these two positions, both of which Einstein had explicitly defended, there exists a logical gap. Experiment and observation test particular consequences of hypotheses introduced as laws, not their objective validity in a strong, or ontological, sense. Einstein was thoroughly cognizant of this gap and of the various attempts to breach it without transgressing the limits of empiricist epistemology: the interpretation of laws as economical summaries of actual and possible observations, developed by Mach and Pearson; Poincaré's conventionalism; Duhem's doctrine of an explanatory system as a hypothetical-deductive model, buttressed by auxiliary hypotheses. Einstein was obviously becoming dissatisfied with such epistemological reductionism. Thus, he defended causality as a principle of nature, not merely of man's reasoning about nature. He certainly held that the two laws of thermodynamics have an objective validity.

However, neither these instances nor even the general theory of relativity has a deep ontological significance. Causality is a general a priori requirement. Thermodynamics and relativity are principle theories. Only when one turns to descriptive accounts of atoms and particles and their interactions does one have a constructive theory with a deep ontological significance. The development of quantum mechanics has already been treated in some detail. Rather than repeat it, we will merely attempt to bring out Einstein's reaction to these developments. As indicated earlier, Einstein regarded Bohr's 1913 trilogy as a fundamental breakthrough. He accepted Bohr's idea that atoms and molecules exist only in discrete states with characteristic internal energies and that the emission and absorption of radiation are associated with changes of state. After Sommerfeld's revision of the Bohr theory, Einstein used his extensive knowledge of how invariants are formulated in mechanics to show how the Hamilton-Jacobi equation used to express the Sommerfeld-Epstein quantum conditions could be transformed in a way that allows separation of variables. Such a separation facilitates the solution of problems involving many degrees of freedom.[59]

Einstein clearly took an interest in the development of the Bohr-Sommerfeld theory. Yet, he did not consider it as giving a solution to the foundational problems that perplexed him. His reaction is manifested in his correspondence. Sommerfeld sent Einstein a copy of his paper redeveloping the Bohr theory. Einstein replied, on 8 February 1916, that he was enchanted with the theory of spectral lines.[60] Yet, two years later he expressed his deep frustration with his attempts to achieve a clear understanding of the quantum principle.[61] A year later he wrote to Born: "The quantum theory gives me a feeling very much like yours. One really ought to be ashamed of its success, because it has been obtained in accord with the Jesuit maxim: 'Let not thy left hand know what thy right hand doeth.' "[62]

The difficulty that bothered Einstein was clearly not the limited success of the Bohr-Sommerfeld theory in fitting observational data. Einstein delighted in theories that were natural, simple, logically coherent. In appraising general relativity and its applications to cosmic and particle problems, he had come to think of such properties as criteria of a theory's objective validity. To him the Bohr-Sommerfeld theory, especially in its later stages, must have seemed a thing of rags and patches. Even more basic than these aesthetic criteria was the fact that Einstein interpreted the Bohr-Sommerfeld work as an attempt to develop a theory about the structure of the atom, not as an attempt to come to grips with the fundamental conceptual and consistency problems involved in the quantum postulate itself. After his introduction of statistics into quantum theory and the public debate concerning the validity of the causal principle, Einstein began to think about the possibility of acausal statistical accounts of quantum theory. On 27 January 1920, he wrote to Born:

> That business about causality causes me a lot of trouble, too. Can the quantum absorption and emission of light ever be understood in the sense of the complete causality requirement,

> or would a statistical residue remain? I must admit that there I
> lack the courage of my convictions. But I would be very un-
> happy to renounce *complete* causality.[63]

He suggested an experiment that might resolve the question of how light is emitted, an attempt to determine whether light emitted by moving particles of canal rays (protons) is strictly monochromatic. This, he thought, would indicate whether emission is an instantaneous process. When an experiment by Bothe and Geiger seemed to confirm this, Einstein wrote to Born that this experiment "has been my most impressive scientific experience in years."[64] Slightly later, Einstein admitted that his interpretation of this experiment was a monumental blunder.

After his trip to Japan and Israel, Einstein returned to his intensive efforts to understand the foundations of the quantum postulate by adapting the ideas of general field theory. The technique he was then experimenting with involved overdetermining the laws by having more differential equations than field variables. The excess determination might, he wrote to Besso, explain the reality of quanta without a *sacrificium intellectus*.[65] Slightly later, on 24 May, he wrote that he was "plunged almost without interruption in the problem of quanta."[66]

The B-K-S paper appeared at a time when Einstein was intensively preoccupied with the foundations of quantum theory. It attempted to present an epistemologically consistent account of the interaction between radiation and matter in a way that resolved the wave-particle duality. The details of this paper have been discussed earlier. What we wish to consider now are the reasons for Einstein's rejection and the bearing this has on his later criticism of quantum theory.

Einstein was concerned with the paper, not only because it treated the foundational problems he was wrestling with, but also because the ideas involved were ones that Einstein himself had either introduced or probed for their explanatory value. Einstein more than anyone else had made wave-particle duality a basic problem. It was he who had introduced statistical considerations into quantum theory. He had repeatedly, though unsuccessfully, attempted to explain the quantum postulate by altering the assumptions underlying the statistics used. Even the most novel feature of the paper, Slater's virtual radiation field, was no stranger to Einstein's thought. Bohr himself reported that at their first meeting in 1920, Einstein spoke of ghost waves guiding the particles.[67]

Einstein was familiar with all the ingredients. Yet, he decisively rejected the way in which the B-K-S paper cooked them together. Apart from technical details, there seem to have been three underlying points of disagreement. The first and most conspicuous was the authors' rejection of Einstein's light-quantum hyothesis, a hypothesis that had just won support from Compton's experiments—and from the Nobel Prize.

The second, less obvious, point of disagreement requires more of an explanation. The B-K-S paper used Einstein's statistical reasoning in a way that seemed to be incompatible with the presuppositions grounding this reasoning. Einstein's use of statistics presupposed that events in nature follow determined laws, though not necessarily already known laws. Statistical considerations are introduced either because the underlying laws are not precisely known or because of the difficulties involved in handling large numbers. Such assumptions were not reflected solely in the papers considered earlier. Einstein had repeatedly shown a concern for the proper interpretation of the presuppositions grounding statistical reasoning. At the 1911 Solvay Conference he began his comments on Planck's paper, "It seems a bit shocking to apply the Boltzmann equation as Planck wishes, by introducing a probability W without giving it a physical definition."[68] Einstein, generally the kindest of critics, was uncharacteristically harsh. This appraisal, however, came after the papers previously considered in which Einstein together with Hopf had shown that no reasonable change in the foundations of statistics or in the interpretation of probabilities would lead to Planck's law. In a similar vein, in a 1918 letter to Sommerfeld Einstein insisted that the assignment of equal a priori probabilities to equal regions of phase space is not at all arbitrary. It is a direct consequence of the underlying mechanics.[69]

Einstein's statistical reasoning always presupposed a deterministic foundation underlying and grounding statistical inferences. Since probabilities are defined in terms of statistics, 'probability' must be considered to be a derivative rather than a foundational concept. However, the B-K-S paper accorded 'probability' a foundational rather than a derivative status. In doing this, it did not even explore, much less solve, the foundational problems that troubled Einstein.

The third underlying difference concerned conservation laws. Here the difference in attitude that continued to characterize the two men can, I believe, best be understood in terms of their differing epistemological orientations. In spite of his empiricist stress on the primacy of sensation, Einstein had come to attach increasing importance to theories as ordered bodies of knowledge and to the principles which played an organizing role in such theories. In his protracted efforts to develop general relativity, Einstein had always required that the to-be-developed theory incorporate, in a suitably generalized form, the principles of energy and momentum conservation. Where the requirement of general covariance was seen as a law about the laws of physics, the conservation laws were seen as laws of objective validity.

Einstein's attachment to the conservation laws seems to have been strengthened by two external influences. The first, Max Planck, has already been sufficiently considered. The second influence stemmed from David Hilbert and his school. In 1916, Emmy Noether, who was then working as Hilbert's assistant, established a general equivalence between the invariance of equations of motion under symmetry transformations and conservation of certain qualities that generate the transformation groups.[70] In the simplified terminology now customary, if a system is invariant under time transformation, energy is conserved; under linear displacement, linear momentum is conserved; under rotation, angular momentum is conserved. These conclusions flow from a general consideration of the effect of transformations; they do not depend on any internal details of the systems under consideration. Though the details of atomic systems and their interactions may be unknown, any account of such systems must be in accord with the general principles governing all transformations. To many, this would seem like an abstract and unfamiliar argument. In developing the general theory of relativity, Einstein had spent years wrestling with the proper form of physical laws under general transformations.

Even when Bohr was struggling to put the Bohr-Sommerfeld theory into a deductive form, he was never able to treat it as an isolated theory in anything like Hilbert's method. Einstein had objected to Planck's derivation on the grounds that it employed a classical law, equation (4.9), to derive results incompatible with the presuppositions of classical physics. Bohr's practice was much more extreme than Planck's. Through the correspondence principle he systematized the use of classical principles, derived from mechanics and electrodynamics, in quantum contexts incompatible with the presuppositions governing the standard applications of classical principles. In judging the acceptability of the resulting theories, Bohr attached very little weight to such factors as the naturalness, logical simplicity, or aesthetic appeal of a theory. His emphasis, rather, was on adjusting theories to facts and interpretations to theories to obtain the most coherent and epistemologically consistent descriptive account. In Bohr's eyes this represented something of a consistency argument, though of an unfamiliar sort. Nature is self-consistent. The inconsistencies are due to our way of speaking about nature. When apparent inconsistencies arise, we must be prepared to alter our ways of speaking.

One might highlight the contrast by saying that by 1924 Einstein's approach to scientific explanation was very much theory centered, while Bohr's was observation centered. He and his associates had collected precise, extensive, and highly varied data concerning spectral lines, the effect of weak and strong electrical and magnetic fields on these lines, scattering data, periodic properties of elements, and similar data. All had to be accommodated. The proper procedure in doing this is, Bohr insisted, to begin with a descriptive account that is adequate and self-consistent.

Mathematics is supplementary. If accepted theories or laws do not seem to allow such a descriptive account, then modifications must be introduced until one has an account that is adequate and self-consistent. In making modifications, Bohr attached much more importance to experimental data than to any general principles held on purely theoretical grounds.

To the physics community at large, this first clash of the giants centered on Einstein's defense of and Bohr's rejection of the light-quantum hypothesis. For the two men themselves, the underlying issue was deeper and more subtle. At stake was the question of how theoretical physics is to be interpreted as a branch of human knowledge. In the public forum, Einstein won this first round and won decisively, one more success in an unprecedented string of scientific successes. Though this victory did not settle the underlying differences concerning epistemology and scientific explanation, it strengthened Einstein's belief in the essential correctness of his way of doing and interpreting physics.

Einstein's final positive contribution to quantum theory, the development of Bose-Einstein statistics and of a new ideal gas law, has already been treated in chapter 7. Here we will only add a couple of points whose significance was brought out by Pais's account.[71] First, though Bose did not seem to appreciate the novelty of his assumptions, he introduced, at least implicitly, the idea that the light quantum has two polarization states and that light quanta are not conserved. Second, Einstein's adaption of Bose's method effectively reverses his earlier reasoning. Einstein's 1905 argument had the form

$$\left.\begin{array}{l} \text{Wien's law} \\ \\ \text{gas analogy} \end{array}\right\} \longrightarrow \quad \text{light quantum}$$

Between 1905 and 1924–25, the concept of light quanta had developed to the stage where light quanta were attributed momenta and polarization states. Einstein felt secure enough about this hypothesis that he could effectively follow the line of reasoning

$$\left.\begin{array}{l} \text{Bose statistics} \\ \\ \text{particle–light quantum analogy} \end{array}\right\} \longrightarrow \quad \text{the quantum gas}$$

As noted in chapter 7, Einstein's gas papers had a significant, though indirect, role in the development of wave mechanics. Though Einstein was never in complete accord with either de Broglie or Schrödinger, he definitely favored their approach over matrix mechanics.

The Solvay Conferences

In popular histories of atomic physics, the Solvay Conferences of 1927 and 1929 centered around the conflicts between Bohr and Einstein. This is somewhat misleading. In all the Solvay Conferences he attended, Einstein only gave one paper, the paper on the specific heat of solids at the 1911 conference. At the 1927 conference, Einstein made one intervention after Max Born's paper and later asked one brief question.[72] Bohr did not respond to either his intervention or his question. After they delivered their papers, both de Broglie and Schrödinger encountered critical opposition from Bohr's followers. Einstein never intervened on their behalf. No other major participant in the 1927 Solvay Conference had as small a part in the published proceedings as Einstein.

Yet, Bohr's summary of his epistemological debates with Einstein stresses the importance of these two Solvay Conferences. It is not difficult to reconstruct the reason for this. When Bohr was convinced that he had discovered the truth about some basic issue, he felt a strong need to have others share his conviction. The reminiscences written by his associates stress the intense and sometimes almost interminable discussions that ensued when Bohr attempted to bring Heisenberg, Pauli, Landau, or others to agree with his opinion. By the spring of 1927, Bohr was convinced that he and Heisenberg had achieved a perspective in which the new developments in quantum theory could be given a coherent interpretation. This new perspective, centered on Bohr's doctrine of complementarity, also seemed to offer a satisfactory resolution to the difficulties encountered in the B-K-S paper. Bohr was convinced. But he would not feel satisfied until Einstein shared his conviction.

Where Bohr was effusive, Einstein was elusive. On 13 April 1927, Bohr sent Einstein a copy of Heisenberg's indeterminacy paper together with a four-page letter summarizing Bohr's new way of resolving the wave-particle duality. One of his opening sentences touched the issue that was to become the pivotal point of epistemological disagreement between the two men: "For a long time there has been a realization of how intrinsically the difficulties of the quantum theory are bound up with the concepts, or even more with the words, which are used in our customary descriptions of nature and which all have their origin in the classical theories."[73]

This was the only philosophical letter the two men ever exchanged. Einstein did not reply. After this the only contacts Bohr had with Einstein were the direct contacts at the Solvay meetings and the indirect contacts through Paul Ehrenfest, their mutual confidant. This, I believe, accounts for the importance Bohr attached to his discussions with Einstein at the Solvay Conferences. They presented the only real opportunity Bohr ever had to win Einstein over to his persuasion.

Einstein's involvement in the new quantum theory was indirect, chiefly through the stimulus he gave de Broglie and Schrödinger. His reaction to the new developments can be gleaned from his correspondence. He was very favorably impressed by de Broglie's thesis, for it seemed like an extension of his own earlier theorizing. On Christmas day 1925 he wrote to his old friend Besso that the most recent development is the Heisenberg-Born-Jordan theory. Einstein thought it ingenious (rather than true) and protected from error by its great complexity.[74] On 21 August 1926, he wrote to Sommerfeld that Schrödinger's seems the best way to a deeper formulation of quantum laws. He continued, "The Heisenberg-Dirac theory certainly forces me to wonder, but it does not smell to me like the true one."[75] Four months later, he wrote to Max Born: "Quantum mechanics is certainly imposing. But an inner voice tells me that it is not yet the real thing. The theory says a lot, but does not really bring us any closer to the secret of the 'old one'. I, at any rate, am convinced that *He* is not playing dice."[76]

Einstein's theology will be considered briefly later. The point to be noted here is that Einstein followed and respected the new breakthroughs. However, from the beginning he rejected the claim that this breakthrough represented the definitive solution to the problem of developing a coherent quantum theory. It, especially in its matrix formulation, was too far removed from Einstein's idea of what a theory should be and do. When he first confronted the Copenhagen interpretation in Heisenberg's indeterminacy paper and Bohr's accompanying letter, he undoubtedly thought of it as essentially an extension of the discredited B-K-S paper. It too accorded probabilities a foundational rather than a derivative status. Now, however, there seemed to be a technical justification for the new interpretation through Heisenberg's indeterminacy principle. This principle became the initial target for those who wished to retain a deterministic foundation for physics. Such a reaction is most clearly shown in Lorentz's initial remarks after Bohr's Solvay paper:

> For me an electron is a corpuscle which, at a given instant,
> is at a determined point of space, and, if I had had the idea
> that at a subsequent moment the electron is elsewhere, I
> should imagine its trajectory to be a line in space.[77]

In private discussions at the conference, Einstein expressed a similar view quite strongly.[78] His only public intervention, however, was much more cautious. He did not comment on Bohr's paper, though Bohr presented his position as a revision of the B-K-S paper. However, Einstein did comment on the paper in which Born presented his statistical interpretation of the wave function. Here Einstein was not directly attacking the Copenhagen interpretation. He was, rather, attempting to clarify the presuppositions underlying the statistical reasoning employed.

The problem Einstein raised is often referred to in the general literature as 'reduction of the wave packet'. The situation in question can be represented by two complementary diagrams, figure 9.1, which Einstein used,[79] and figure 9.2, which Bohr used.[80] Figure 9.1 represents electrons hitting a

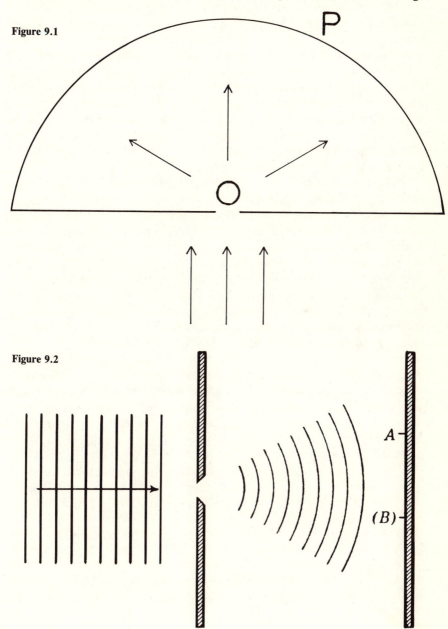

Figure 9.1

Figure 9.2

screen, being diffracted through a narrow hole O and then striking a photographic plate at some point P. Before the electron strikes the plate it is represented by a spreading wave packet. In the collision it is represented as a particle localized at P.

If this reduction of the wave packet is the only point at issue, then Bohr's reply surely meets Einstein's objection. Though Bohr never used the term 'reduction of the wave packet' and did not reply to Einstein's intervention, he had already worked out his own solution to this and similar problems. On such particular problems his early solutions were essentially the same as in his later retrospective account. Even in drawing simple illustrative diagrams, Bohr was always conscious of how the measuring apparatus was to function in a particular experimental situation. A rigid diaphragm supplies a basis for measuring paths, but not momentum transfer. A movable diaphragm absorbs momentum and so supplies a basis for its measurement. In this case, however, there is no fixed framework for position measurements.

In Bohr's illustration of Einstein's objection (figure 9.2), there is a rigid plate with a slit opening. There must be an uncertainty Δp in the component of the electron's momentum parallel to the plate after it has passed through the slit. This is a diffraction effect which requires a wave representation. The incident electron is represented by the cone of a spherical wave train emanating from the slit. If the slit width is accepted as the position uncertainty Δq, then it is easy to show that $\Delta p \Delta q < \hbar$. Because of this uncertainty, one cannot predict the position at which the electron will strike the plate, or, retroactively, explain why it struck at A rather than B. If Einstein's difficulty had simply been one of relating reduction of the wave packet to the indeterminacy principle, then Bohr has surely solved this problem.

While discussing this experiment, Bohr says in passing, "Surely, we all recognized that, in the above example, the situation presents no analogue to the application of statistics in dealing with complicated mechanical systems.[81] This, however, was precisely the point that Einstein was unwilling to accept. His intervention occurred as a comment on Born's paper. Einstein accepted Born's probabilistic interpretation of the Schrödinger wave function as adequate to the experimental data. The question Einstein was probing was what sort of hidden determinism could ground these statistics. In his analysis of the experiment he considered two modes of interpretation: Conception I is essentially an ensemble interpretation. The de Broglie–Schrödinger waves correspond only to clouds of electrons and give no detailed information about individual processes. In Conception II the theory is interpreted as giving, through the wave packet, a complete account of individual processes.

Conception I easily fits the probability interpretation in the sense that $|\psi|^2$ corresponds to the probability of finding some particle within a particular region. This conception, however, neither preserves the conservation laws nor explains such things as the Bothe-Geiger experiments and the Wilson cloud chamber pictures, the very results that refuted the B-K-S paper.

Conception II expresses through $|\psi|^2$ the probability of one definite particle being in a certain region. Though Einstein took this interpretation much more seriously than Conception I, he raised two objections to it. First, the sudden localization of the particle at P seems to presuppose some sort of action at a distance. This objection is really directed at Schrödinger's conception of electrons as wave packets. The second difficulty concerned the use of configuration, rather than real, space and an apparent difficulty this presented for Bose-Einstein statistics. This technical difficulty was soon clarified, and the objection does not recur.

A text precedes explication. Action precedes analysis. Even in 1927, Einstein and Bohr were already clearly manifesting the grounds of their epistemological divergence, though neither had as yet given an adequate explanation of or justification for his distinctive epistemological orientation. Bohr was explaining how we should understand the significance and scope of *our descriptions* of atomic situations. Though he had not yet concentrated on the analysis of language and its functional presuppositions, his emphasis was on how language functions to make such descriptions possible. Einstein was asking in effect what sort of reality could serve as a ground for the appearances which quantum mechanics seems to schematize. His orientation was distinctly ontological. He had not yet given this orientation the clear formulation it received in his "Autobiographical Notes": "Physics is an attempt conceptually to grasp reality as it is thought independently of its being observed. In this sense one speaks of 'physical reality'."[82] He had not yet expressed this goal that clearly. Yet, the goal represented the direction of his striving since his earliest published papers.

Einstein's reaction to quantum theory was extremely puzzling to many of his contemporaries. The key elements in the Copenhagen interpretation, wave-particle dualism, and the introduction of probabilities into the functioning (rather than merely the derivation) of quantum theory had been Einstein's own contributions. The pivotal new principle, Heisenberg's indeterminacy principle, was, Heisenberg insisted, stimulated by Einstein's remark to him that the theory determines what can be observed. Even the role attributed to measurements in determining the significance of position and momentum attributions seemed to be a logical extension of the role the young Einstein attributed to similar measurements in determining the physical significance of simultaneity. When such relations were pointed out to Einstein, he replied that a good joke should not be repeated too often.[83] If Einstein had died in 1925, the *Kopenhagener Geist der Quantentheorie* might well have been called the *Einsteinischer Geist der Quantentheorie*. He was the spiritual father of the new conception.

Yet, Einstein rejected his spiritual offspring, a rejection that was emphatic and sustained for the rest of his life. Why? One may attempt various psychological answers: a conflict for priority between two prime males, an innate conservatism in Einstein's character, or simply the hardening of the categories that often accompanies middle age. Yet, no psychological or sociological explanation of Einstein's attitude can adequately account for his conviction that the Copenhagen interpretation of quantum mechanics is bad physics, which must eventually be replaced.

Of the various answers offered to this question the only one I find convincing is that given by Einstein himself. After discussing the undoubted success of the new quantum theory in his "Autobiographical Notes," he continues:

> However, the question which is really determinative appears to me to be as follows: What can be attempted with some hope of success in view of the present situation in physical theory? At this point it is the experiences with the theory of gravitation which determine my expectations. These equations give, from my point of view, more warrant for the expectation to assert something *precise* than all the other equations of physics.[84]

The basic idea behind the program he set for himself was clear. The various wave-particle duality models de Broglie had proposed had all failed. Something more fundamental and more natural was needed. This should come from a unified field theory. Einstein had twice before attempted and reluctantly abandoned similar efforts. He now returned to his earlier attempts to extend general relativity with a new approach based on the idea of distant parallelism.

No one knows how many unsuccessful attempts Einstein attempted and discarded in the next twenty-five years. It was certainly the most sustained as well as the most unsuccessful effort of his scientific career.[85] In spite of the complexity of the various theories attempted, the basic idea behind the effort is fairly straightforward. A field theory, such as Maxwell's, relies on differential equations which depict a continuous field. Because the equations are linear, the fields involved obey a superposition principle. Such field theories are incomplete in one essential respect. They do not explain the sources of the field. The most distinctive feature of quantum theory is the discontinuity associated with Planck's constant h. If the field equations of general relativity were expanded to include the electromagnetic field as well as the gravitational field, then sources should emerge as quasi singularities, i.e., limited regions of space in which the field strengths or energy density is particularly high. Einstein' hope was that such quasi singularities in field equations would explain both particles and their associated fields in a unified way. If this to-be-found field formalism also yielded such quantum conditions as $E = hv$, $\oint p \, dq = nh$, and $p = h/\lambda$ (or appropriate relativistic generalizations of these), then it would supply the underlying reason why the phenomena systematized by quantum theory are as they appear.

The program encountered difficulties that eventually proved insoluble—or at least insoluble to Einstein and his various associates. A field theory that includes singularities requires nonlinear equations, a difficult and inadequately understood subject. Einstein found no Ariadne's thread leading him from general relativity to a unified field theory in the way that the principle of equivalence and the requirement of general covariance had guided his path from special to general relativity. In the light of later knowledge concerning strong and weak as well as electromagnetic and gravitational fields, and the interpretation of particle interactions in terms of fermions (quarks or leptons) exchanging bosons (gluons, photons, or weak interaction bosons), Einstein's approach looks misguided and implausible. Yet, at the time of its inception, it was neither mysticism nor disguised metaphysics. It was a logical extension of the type of physics Einstein had been practicing since 1901.

Einstein persevered unflaggingly in this attempt to develop an alternative to quantum mechanics. At the same time he continued his spasmodic criticisms of the new reigning view. The next dramatic confrontation between the two men came at the 1930 Solvay Conference. Though neither gave a paper, they had extensive semiprivate discussions. It was here that Einstein presented his famous clock experiment, attempting to show that Heisenberg's indeterminacy principle did not represent an unsurpassable limit. He considered a radiation-filled box containing a clock and attached to a weight scale. If a shutter in the side of the box were open for a sufficiently short time ΔT, one photon would be emitted. This emission would decrease the radiant energy in the box by an amount ΔE. In accord with the relativistic equation $E = mc^2$, one could determine ΔE by a precise weighing of the box before and after the emission of the photon. Thus, ΔE could be determined precisely. Since ΔT is determined by the clockwork mechanism, this could also be determined precisely. This would seem to imply that the product $\Delta E \Delta T$ could be made arbitrarily small, contrary to Heisenberg's indeterminacy principle.

After a hectic, sleepless night, Bohr returned with his answer, showing that if the weighing process took account of Einstein's own general theory of relativity, then the product $\Delta E \Delta T$ yielded the standard indeterminacy relation. Bohr's answer convinced Einstein. It did not, however, convince many subsequent critics who argued that the general theory of relativity is irrelevant and misleading in this context. The electromagnetic and mechanical interactions proper to the thought experiment require only a simple weighing, not the general theory of relativity. As Francis Everitt put it, "The experiment should work even in a universe in which the principle of equivalence does not hold."[86] I believe that Bohr's refutation is valid and will endeavor to present it in a way that answers the type of objection mentioned.

First, a point on which Einstein and Bohr agree completely. One must make a sharp distinction between the system being measured, the radiant energy trapped within the box, and the measuring apparatus. The latter, which includes both the box and the scales, is not treated as a quantum system. When general relativity is introduced, it is only for the measuring system, not for the quantum system measured. Second, the refutation does not use the field equations of general relativity. It simply uses the equivalence principle and uses it only in a first approximation. This is compatible with Newtonian gravitational theory. The only irreducible non-Newtonian element in the treatment of the measuring apparatus is the theory of *special* relativity.

Suppose the box containing the shutter mechanism and the trapped radiation is suspended by a spring connected to some weighing scale. To weigh the box with an accuracy Δm involves fixing the position of the weighing scale needle with an accuracy Δq. This, in turn, relates to the uncertainty in the box's momentum Δp, i.e., the uncertainty in momentum due to the box's oscillations about the zero position of the scale. Since the uncertainty in momentum cannot be greater than the impulse communicated to the box by the mass change Δm, one has

$$\Delta p \approx (h/\Delta q) < Tg\Delta m, \tag{9.5}$$

where T is the time of the balancing procedure and g the acceleration due to gravity.

When the shutter is opened and the photon emitted, the box loses energy and mass and rises to some point where the gravitational field is less. Then it gradually reaches a new equilibrium position through damped oscillations. To calculate the effect of this change in gravitational potential we will use the physics that Einstein presented in his 1911 paper, i.e., when he had the principle of equivalence but not the gravitational equations proper to the general theory of relativity.[87]

Suppose that the clock, originally at S_1, rises a distance l to a point S_2 after emitting a photon. To determine the change in the rate of measuring time we may consider equivalent clocks at S_1 and S_2. We may also assume that S_1 emits light which can be represented by a simple wave of length λ.

Newtonian potential theory gives a gravitational potential difference between these two points of $\Delta\phi = gl$. By the principle of equivalence this difference in gravitational potential is equivalent to an acceleration g. In the time $t = l/c$ it takes the light wave emitted at S_1 to get to S_2, S_2 would have equivalently accelerated to a velocity $v = gl/c$. If S_2 is considered as a system moving with this velocity relative to S_1, then there is a first-order Doppler shift, $\delta\lambda/\lambda = v/c$. Since λ is proportional to T, we have $\delta T/T = \delta\lambda/\lambda = gl/c^2$. If we consider this l the uncertainty in the position measurement and δT the uncertainty in the time measurement, we have

$$\Delta T = gT\Delta q/c^2. \tag{9.6}$$

When this value of Δq is substituted in equation (9.5) and the special relativistic relation $E = mc^2$ is used, the result is $\Delta T\Delta E > h$. All that was needed for this derivation was the special theory of relativity, the basis of Einstein's own argument, and the principle of equivalence, a principle Newtonian physics uses but does not justify. The gravitational field equations were neither needed nor used.

From this time on, Einstein abandoned the attempt to show that it is possible in principle to go beyond the limits set by Heisenberg's indeterminacy principle. In 1931, he wrote a letter to the Nobel Committee nominating Heisenberg and Schrödinger for the Nobel Prize.[88] Einstein's intellectual disagreements were never characterized by personal pettiness. This new attitude is strikingly shown in a collaborative letter to the *Physical Review* Einstein, Tolman, and Podolsky wrote during Einstein's summer in Pasadena.[89] In yet another idealized thought experiment, figure 9.3, they considered a box containing molecules, a switch controlling two shutters, and the trajectories of two simultaneously released molecules, one traveling to O directly and one reflected from a curved surface to reach O at a later time. A simple argument suggests that one could determine the momentum of the SO molecule from the time interval between the opening of the shutter and its impact at O. One could also weigh the box before and after the shutter is opened and then, through a simple energy balance, determine the momentum of the SRO molecule. Since the time of departure, the momentum, and the trajectory are known, this would seem to suggest that one could predict this molecule's exact arrival time at O. Such a prediction, however, would contradict the indeterminacy relation between position and momentum. On this ground the authors conclude that momentum cannot be measured as precisely as the ideal experiment may suggest.

It is not now possible to determine how much of this argument is due to Einstein. This matters little. Einstein would not have attached his name to a paper whose content, especially on this sensitive point, he disagreed with. In keeping with Einstein's practice, the starting point of the argument is an

Figure 9.3

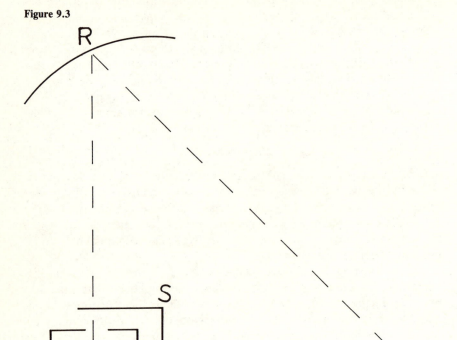

assumption concerning the way molecules behave and only subsequently a consideration of how this behavior can be measured. The paper presents two novel features. First, indeterminacy is now accepted as a principle from which one draws a conclusion rather than as a principle to be challenged. Second, Einstein, Tolman, and Podolsky are using thought experiments to test the implications of quantum mechanics for systems containing two particles, particles which are initially in one system, become separated, yet remain coordinated. This was the precursor of the Einstein-Podolsky-Rosen (EPR) paradox.

By the early 1930s, Einstein had come to accept quantum theory as internally consistent and adequate to the observable phenomena, though not to the underlying reality that should explain these phenomena. He studied Dirac's systematic development of quantum theory and accepted it as the most logically consistent presentation of the theory.[90] He also studied Dirac's relativistic quantum mechanics and saw in it the possibility of a new way of developing his program. With the assistance of the mathematician Walter Mayer, he wrote a series of three articles concerned with the properties of semivectors, a generalization of Dirac's spinors.[91] The general thrust of this work was explained in the Herbert Spencer lecture, which Einstein gave while the work was in progress:

> Dirac found in the spinors field-magnitudes of a new sort, whose simplest equations enable one to a large extent to deduce the properties of the electron. Subsequently, I discovered, in conjunction with my colleague, Dr. Walter Mayer, that these spinors form a special case of a new sort of field, mathematically connected with the four-dimensional system, which we call 'semivectors'. The simplest equations which such semivectors can satisfy furnish a key to the understanding of the existence of two sorts of elementary particles, of different ponderable mass and equal but opposite electrical charge.[92]

After discussing quantum mechanics he concluded:

> I cannot but confess that I attach only a transitory importance to this interpretation. I still believe in the possibility of a model of reality—that is to say of a theory which represents things themselves and not merely the probability of their occurrence.[93]

The best hope, he explained, of grasping the objectively real was through simple concepts and mathematically simple forms relating them, such as the concepts and methods of field theory. More particularly, what he and Mayer attempted was to set up a general mathematical formalism which would allow them to apply to semivector fields the tensor transformations of general relativity supplemented by the idea of distant parallelism. If the structure of Dirac's relativistic wave equation could be deduced from such transformations, then one would have good reason for believing that the mathematical representation of the field corresponded to something more basic. At the end of their third paper they concluded that their formalism plays no role in the structure of the Dirac equation.[94]

The citation from the Spencer lecture mentioned Einstein's hope that his new formalism offered a key to the understanding of two sorts of elementary principles. This he worked out in collaboration with Nathan Rosen, the third contributor to the EPR paradox.[95] They attempted to construct an atomistic theory of matter and electricity making use only of the $g_{\mu\nu}$ of general relativity and the ϕ_μ of Maxwell's theory. The basic plan was to modify the gravitational equations for the case of a spherically symmetric field. Space is represented by two identical sheets with a particle as a bridge connecting them. This approach allowed of two types of particles: a neutral particle, which might have negative mass, though they wished to exclude that; and a charged particle, which the theory seemed to accord a zero mass. They concluded that though their formalism did not yield any quantum phenomena, the possibility of its doing so could not be excluded a priori.

Einstein's primary purpose in this ongoing research program was clearly not one of attacking quantum mechanics as such. This he had come to accept as the most successful theory of modern times. What he could not abandon without, as he saw it, abandoning the very thrust of scientific inquiry was the idea that science seeks to explain physical reality as it exists objectively, or independent of our observations. Science relies on observation and experimentation. But the phenomena observed through our instruments must be explained through the underlying reality objectively responsible for the phenomena. The only way of representing such an underlying physical reality that has any real hope of success is through mathematical forms that are simple, natural, and aesthetically pleasing. This was the intellectual orientation leading to the EPR paradox.

The Einstein-Podolsky-Rosen Paradox

A work of art, once cut from the conceptual chords linking it to its creator, has a life of its own, sometimes a life quite different from anything the creator envisaged. Hamlet, under psychoanalysis, illustrates an Oedipus complex. Nefertiti sells perfume. The Night Watch advertises cigars. And the music of Richard Strauss orchestrates humanity's entrance into the space age. Something similar has happened to the argument introduced by Einstein, Podolsky, and Rosen. Following the now accepted tradition, this will be referred to as the EPR paradox, though 'paradox' is a somewhat misleading term. It is undoubtedly the most analyzed and reinterpreted thought experiment of modern times, one that ranks with Newton's rotating bucket and Maxwell's demon. It has launched hidden variable theories; suggested locality theorems; tested logical reformulations of quantum theory; and inspired various dialectical refutations of the Copenhagen interpretation. What is easily obscured in this transformation is the significance the original paper had as the culmination of a confrontation.

This is the aspect we wish to consider here. Since, as far as I can determine, neither Bohr nor Einstein introduced any further changes in their interpretations of quantum theory after this period, the historical summary will end with this pair of papers.[96] In the final chapter I will try to summarize each man's final position on atomic physics as a crucial testing ground for revised theories of scientific explanation. There will be one further point of emphasis. Technical reformulations of the EPR paradox have tended to draw attention away from the conceptual issues basic to the original papers. Here I will endeavor to reverse this trend and bring out the underlying differences in as clear and nontechnical a way as possible.

The most striking feature of the paper by Einstein, Podolsky, and Rosen is the primacy accorded *ontological* considerations. This may be summarized rather succinctly. Physical reality exists independently of man's knowledge of this reality. The goal of science is to describe reality as it exists objectively. Fulfilling this goal requires the coordinated use of percepts and concepts, of descriptive language, experimental investigation, and mathematical reasoning. Any attempt to explain how these elements function in an interrelated way presents notorious problems. But these problems are epistemological rather than scientific. They may have to be solved to explain how science functions; they need not be to *do* science. The scientist accordingly should not let inadequacies in the accounts supplied by epistemologists—deter him from his proper goal, to describe reality as it exists objectively.

It is only within this interpretative perspective that the real significance of the EPR insistence on completeness can be properly appreciated, a significance quite different from the meaning this term has when one speaks of an axiomatic system as being complete. A scientific description of a physical situation is incomplete if there are elements in the physical situation not included in the description. A simple comparison may bring out the significance of this concept and the difficulties it generates.

A college catalog would be considered incomplete if there were courses taught in the college that were not listed in the catalog. However, it would not be considered incomplete on the grounds that the catalog failed to categorize the courses listed as 'good', 'mediocre', or 'poor'. Though such information would be of interest to students, it is not the sort of information a catalog is expected to supply. Testing completeness in this sense of the term is a very straightforward procedure. One simply compares the courses listed with the courses actually taught.

This simple test involves a definite epistemological presupposition. One can know the courses as listed in the catalog and also have an independent knowledge of the courses actually taught. This hardly needs stressing, since its implementation presents no problem, at least none of an epistemological nature. In the context of atomic physics, however, the implementation of such a norm presents formidable problems. One cannot first look at physical reality and them compare this with what the theory says about it. Rather, using 'theory' in a broad enough sense to include interpretation of experiments, one knows the reality only through the theory. There is no way to sneak a peek at the objectively existing reality and then compare this with what the theory says about it.

Before considering the way in which EPR attempts to get around this difficulty, we should consider why it seems plausible to think of this as a hurdle, rather than as an insurmountable obstacle. Even without a theory-independent peek, one knows that physical reality exists and is causally responsible for the particular manifestations of this reality which an experiment records and a theory systematizes. When, for example, a uranium atom emits an alpha particle, it may be possible to measure or infer the precise energy involved. According to the indeterminacy principle, such a measurement precludes a precise determination of the time of decay. Yet, it would seem quite unreasonable to argue that the decay did not occur at a definite time. If this is granted, then the theory would seem to be incomplete if it cannot specify this time. As with the college catalog, the theory would seem to be incomplete in a correspondence sense. There is an element of reality that is not included in the theory. Further, the missing element seems more analogous to the unlisted course than the the unlisted student evaluations of the courses listed.

The EPR paper begins:

> Any serious consideration of a physical theory must take into account the distinction between the objective reality which is independent of any theory, and the physical concepts with which the theory operates. These concepts are intended to correspond with the objective reality, and by means of these concepts we picture this reality to ourselves.
>
> In attempting to judge the success of a physical theory, we may ask ourselves two questions: (1) "Is the theory correct?" and (2) "Is the description given by the theory complete?" It is only in the case in which positive answers may be given to both of these questions, that the concepts of the theory may be said to be satisfactory.

Bohr, as shown in chapter 8, had gradually come to insist that concepts correspond to words and, though he did not anticipate the later Wittgensteinian slogan, that the meaning of a term is essentially its use in ordinary language and the idealization of that usage proper to classical physics. The EPR idea is quite different. Concepts, at least the crucial physical concepts in question, correspond to aspects of physical reality. If the correspondence is incomplete or misleading, then the concept is inadequate. This correspondence idea, in turn, leads to a necessary condition for the completeness of a physical theory: "Every element of the physical reality must have a counterpart in the physical theory."

It is precisely here that the epistemological difficulty previously noted enters. One cannot sneak a peek at physical reality to compare it with what the theory says. Again using 'theory' in a broad sense, one must use the theory itself to determine whether there are elements in physical reality not represented by the theory. The EPR paper does not attempt to solve the epistemological problems this generates. Nor does it attempt to give a general account of how a theory as a whole corresponds to some domain of reality. It simply presents, as intuitively reasonable, a criterion which supplies a sufficient condition for the correspondence between elements of physical reality and quantities, i.e., terms in the theory with fixed dimensions and variable numerical values: "If, without in any way disturbing a system, we can predict with certainty (i.e., with probability equal to unity) the value of a physical quantity, then there exists an element of physical reality corresponding to this physical quantity." Henceforth, this will be referred to as RC, the reality criterion.

Most discussions of the EPR paradox focus on the particular thought experiment presented. From the perspective of a quantum physicist or a philosopher of science, such an emphasis is eminently reasonable. The thought experiment presents problems that require explanation regardless of whether one shares the interpretative presuppositions of Einstein or Bohr. From the historical perspective in which we are viewing this, however, the presuppositions are crucial. The thought experiment merely serves as one particular illustration of general principles. When Bohr showed how this particular experiment could be explained, Einstein did not abandon the principles behind it. He sought better illustrations for the same principles. For this reason the thought experiment will only be treated briefly.

In the orthodox interpretation of quantum mechanics, the Schrödinger wave function ψ characterizes the state of a system. (Neither paper discusses Hilbert space representations.) The question of completeness accordingly reduces to a question of whether there are physically significant aspects of a system not represented by ψ. The contrasting answers are illustrated by a simple example. Suppose A is an operator corresponding to some observable such that

$$A\psi = a\psi, \tag{9.7}$$

where a (the eigenvalue) corresponds to the numerical value one would get if one measured the observable corresponding to A for a system in state ψ. A could be, for example, the operator $(\hbar/i)\partial/\partial x$ corresponding to the momentum p_x. If ψ is a momentum state of an electron, e.g., $\psi = \exp(ip_0 x/\hbar)$, then the eigenvalue equation becomes $p_x \psi = p_0 \psi$ corresponding to an electron with a momentum p_0. What position does this electron have? Since p and x are noncommutative operators, there is no eigenvalue equation relating this wave function to the operator x and yielding one eigenvalue. Instead, one must compute its probability. The relative probability that a measurement of the x-coordinate will yield a result lying between a and b is

$$P(a,\ b) = \int_a^b \overline{\psi}\ \psi\ dx = b - a.$$

This means that for any interval all values of the x-coordinate are equally probable. What physical significance does the attribution of an x-position have under these circumstances?

This question admits of two possible answers. First, the electron does really have some definite location. Since the wave function (plus the formalism of quantum mechanics) does not represent this location, the wave function must be considered incomplete in the EPR sense of 'incomplete'. The second alternative is that an electron in a definite momentum state has no definite location. This, of course, is the orthodox quantum mechanical answer, though the corresponding question is not usually cast in such ontological terms. In slightly more general terms these two alternatives are *"(1) the quantum mechanical description of reality given by the wave function is not complete;* or (2) *when the operators corresponding to two physical quantities do not commute, the two quantities cannot have simultaneous reality."* Position (2), the orthodox quantum answer, is generally justified by the argument that the information obtainable from the wave function is a complete description of the state of a system because it contains all the information that can be obtained without altering that state. The EPR paper is defending position (1). What it offers is a *reductio ad absurdum* argument directed against position (2). In at least one case, position (2), coupled to RC, leads to a contradiction. If RC is accepted as valid, then position (2) must be judged incorrect. If positions (1) and (2) represent the only alternatives, then this refutation of position (2) serves to establish position (1).

The example chosen to refute position (2) is one of two systems, I and II, which interact from the time $t = 0$ to $t = T$, after which there is no longer any interaction. Contrary to some later hidden variable theorists, the authors are definitely not considering hidden interactions between the separated systems. It is further assumed that the states of the two systems before $t = 0$ are known. If the interaction is also known, then one can use the Schrödinger equation to calculate the state of the combined system, I + II, at any subsequent time including times after T when the two systems are again separated. Let ψ stand for the state of the two systems considered as a unit. This ψ-function for the combined systems does not give definite information concerning the state of either individual system considered separately.

The further technical details concerning reduction of the wave packet should not obscure the basic argumentative point. If it is reasonable to suppose that after separation each of the two systems is in a definite state, then quantum mechanics is incomplete. The wave function for the combined system can admit of different expansions such as

$$\psi(x_1, x_2) = \sum_{n=1}^{\infty} \theta_n(x_2) u_n(x_1), \qquad (9.8)$$

$$\psi(x_1, x_2) = \sum_{s=1}^{\infty} \phi_s(x_2) v_s(x_1). \qquad (9.9)$$

Both expansions are chosen to fit the form of equation (9.7) for system I or particle 1. That is, for this particle, $u_1(x_1), u_2(x_1), \ldots, u_n(x_1)$ are eigenfunctions with eigenvalues $a_1, a_2, \ldots, a_n$ for some quantity A. Similarly, $v_1(x_1)$, $v_2(x_1), \ldots$ are different eigenfunctions with eigenvalues $b_1, b_2, \ldots$ for some other quantity B. In this case $\theta_n(x_2)$ and $\phi_s(x_2)$ are treated as expansion coefficients.

If a measurement of A for particle 1 yields the value a_k, then particle 1 is in the state characterized by $u_k(x_1)$. It follows from the vanishing of all other expansion coefficients that particle 2 is in a state characterized by $\theta_k(x_2)$. The infinite expansion given in equation (9.8) has been reduced through the measurement. Similarly, if a measurement of B yields b_r, then system 1 is in the state $v_r(x_1)$ and system 2 is in the state $\phi_r(x_2)$. The crucial point here is that measurements are performed only on particle 1 (or system I), and then the values proper to particle 2 (or system II) are inferred from equations (9.8) and (9.9).

If A corresponds to the first particle's momentum and B to its position, then A and B are noncommuting operators. The standard interpretation of quantum mechanics holds that if the quantity corresponding to A has a definite physical reality, then the quantity corresponding to B lacks physical reality. This is where the paradox enters. If one measures the position of particle 1, then one can determine the position of particle 2 without in any way disturbing particle 2. By RC the position of particle 2 must be considered real. If, however, one had determined the momentum of particle 1, then one could have determined the momentum of particle 2. By RC the momentum of particle 2 must also be considered real.

This argument leads to a conclusion concerning particle 2 (a special case of system II) which directly contradicts position (2). Objectively, this particle has both a definite position and at the same time a definite momentum. This conclusion does not contradict the indeterminacy principle. Any attempt to measure the position of particle 2 would produce an uncertainty concerning its momentum, and vice versa. However, the conclusion of the argument, if accepted, strongly supports the further conclusion that quantum mechanics is incomplete. A system in a state has physically real aspects not represented by the wave function for that state.

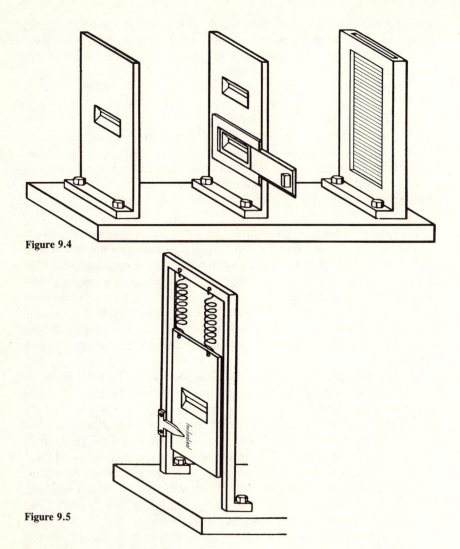

Figure 9.4

Figure 9.5

Bohr's Reply

Bohr's paper, bearing the same title as the EPR paper, is characterized throughout by an epistemological, rather than an ontological, mode of reasoning. Bohr is not trying to determine a relation between things as they exist objectively and our concepts of those things. He is, rather, concerned with the possibility of a completely rational self-consistent description of physical phenomena. The classical presuppositions of objectively existing physical reality and objective causal relations require radical revision when one switches to the epistemological mode. In such a revision the pertinent problem is the degree to which an unambiguous meaning can be attached to the expressions 'physical reality' and 'causality'. This is a context-dependent issue, with experiments supplying the physical context. In this interpretative perspective the idea of a unique criterion of physical reality becomes essentially ambiguous when applied to the atomic domain.

When Bohr uses 'complete', he is not using as a standard, or even as an ideal, any comparison between terms in a theory and some objective reality existing independently of what any theory may say about it. His idea is, rather, that a scientific explanation is complete if it is a rational and consistent utilization of all the available sources of information. Quantum mechanics does this. Therefore, it is complete. Any purported incompleteness inevitably involves an ambiguity in defining the physical quantity allegedly omitted, rather than an ignorance about the value of that quantity.

Bohr illustrates this general doctrine by an examination of selected examples in a semirealistic fashion. The term 'semirealistic', which Bohr stresses in his contribution to the Einstein volume, deserves some comment. None of the disputes between Einstein and Bohr concerned actual experiments. Even the single-slit, double-slit experiment, routinely discussed in introductory texts, had not yet been performed.[97] Bohr's discussion is semirealistic in a special sense best exemplified by his own illustrations reproduced here.[98] In figure 9.4, the diaphragm containing the slits and the plate holders is pictured as a solid rigid object firmly bolted to a common frame. When Bohr wishes to consider the change in the vertical component of a particle's momentum as it passes through a slit, the diaphragm is pictured as suspended from flexible springs, as in figure 9.5.

Neither of these, Bohr's own diagrams, depicts an actual or even an experimentally feasible experimental arrangement. Like the semirealistic discussions of measurements in the Bohr-Rosenfeld paper treated earlier, they serve to illustrate Bohr's *conceptual* analysis of the way in which quantum mechanics makes use of all the possible, mutually compatible sources of experimental information. Any observation of experimental results, Bohr never tired of repeating, must be expressed in classical terms. Hence, his blatantly, almost belligerently, classical measuring apparatuses: rigid bars firmly attached to a common support to measure distances; springs obeying Hooke's law to measure momentum transfer; clocks with massive gears showing to measure time. Our classical concepts are geared to—and best illustrated through—such devices and measurements.

To go beyond such classical concepts one needs the correspondence principle, which enables one to extend classical concepts to quantum domains, and the doctrine of complementarity, which specifies the set of concepts that is applicable for a particular experimental arrangement. Thus, what the figures illustrate is

in the phenomena concerned we are not dealing with an incomplete description characterized by the arbitrary picking out
of different elements of physical reality at the cost of sacrific-
ing other such elements, but with a rational discrimination between essentially different experimental arrangements and procedures which are suited either for an unambiguous use of the
idea of space locations, or for a legitimate application of the
conservation theorems of momentum.[99]

Figure 9.4 illustrates an unambiguous use of the idea of space locations.
In this experimental context 'position' means distance from a fixed position.
The diaphragm and plate, firm bodies rigidly attached to the common
frame, supply a clear basis for such measurements. In Bohr's analysis no
consideration is given either to the technical question of how positions are
actually measured or to the conceptual question of how a particle is located
at a definite position. The question at issue is, What does it mean to assign
positions in an unambiguous way?

Suppose the particle passing through the diaphragm loses momentum.
By the law of momentum conservation the diaphragm should pick up the
lost momentum. Since, however, it is rigidly attached to the common frame,
the apparatus as a whole picks up the lost momentum. This common frame
also absorbs any other momentum lost by the particle elsewhere in its path.
In this situation there is no experimentally meaningful way in which one can
speak about the momentum transferred from the particle to the diaphragm.

Figure 9.5 illustrates the same diaphragm in an experimental setup
where it can measure momentum transfer. In this case it measures the
vertical component of the momentum transfer, the component responsible
for interference effects. Since, after the momentum-transferring collision,
the diaphragm oscillates up and down, it cannot supply an unambiguous
basis for measuring the vertical deflections of the particle.

When Bohr finally considers the particular example given in the EPR
paper, it is treated as one more relatively simple illustration of the same
basic doctrine. As a simple illustration of a two-particle system one can
consider two particles passing through the slits in the middle diaphragm of
figure 9.4. The appropriate coordinates for the combined system are the
differences in position (initially determined by the distance between the two
slits) and the sum of the momenta (with a rigid diaphragm there is no way to
determine how much of the momentum transfer was due to a particular
particle). This can be presented in the more general form of a canonical
transformation. Focusing on this, however, obscures the simple *conceptual*
point Bohr is making.

If, after the particles have passed through the slits, one measures q_1, the
position of particle 1, and knows $q_1 - q_2$ (or has a wave function expressing

the separation of the particles), then one may easily infer q_2. One could do this, in Bohr's semirealistic fashion, by having the photographic plate rigidly fixed and using this as a spatial framework relative to which positions are measured. In this case, the plate, like the diaphragm, would have to be rigidly attached to the common frame. Since any momentum absorbed goes into the apparatus as a whole, there is no experimentally meaningful way in which one could speak about the momentum proper to particle one.

Suppose one did want to measure the momentum proper to particle 1. The semirealistic illustration proper to this would be another adaption of figure 9.4. First, we would have to consider particles 1 and 2 as a two-particle system and consider the momentum transfer between this two-particle system and the diaphragm with two slits. For this purpose the diaphragm would have to be free to move back and forth. It could be connected to the general framework by some sort of horizontal springs that could serve to measure the momentum transferred. A knowledge of the original momenta coupled to a knowledge of this momentum transfer gives us $p_1 + p_2$, the combined momenta of the two-particle system after it has passed through the diaphragm.

Next, we wish to measure the momentum of particle 1 separately. This could be done, as in ballistic tests, by having it hit and be absorbed by a test particle of known mass, whose deflection could be observed (thus, a classical test particle). From a knowledge of p_1 and $p_1 + p_2$, one can easily infer the value p_2 without performing any measurement on particle 2. Now, however, there is no unambiguous way in which one can meaningfully assign a position along the horizontal axis to either particle. It is not simply a case of a failure to perform a measurement that it is possible to perform. Oscillating diaphragms and sliding test particles, the only relevant position markers, cannot supply a spatial framework in which positions can be assigned. Thus, the very notion of 'position of a particle' becomes essentially ambiguous in any experimental arrangement designed to measure the conjugate components of momentum.

On the basis of this analysis, Bohr concludes that the EPR reality criterion is itself essentially ambiguous. It is based on the notion 'without in any way disturbing the system'. However, what is at issue here is not some mechanical disturbance during the process of measurement. At issue, rather, are the very conditions that make measurement possible:

> Since these conditions constitute an inherent element of
> the description of any phenomenon to which the term 'physical
> reality' can be properly attached, we see that the argumenta-
> tion of the mentioned authors does not justify their conclusion
> that quantum-mechanical description is essentially incomplete.
> On the contrary this description, as appears from the preced-

ing discussion, may be characterized as a rational utilization of all possibilities of unambiguous interpretation of measurements, compatible with the finite and uncontrollable interaction between the objects and the measuring instruments in the field of quantum theory.[100]

10. Two Theories of Scientific Knowledge: Bohr and Einstein

Most practicing scientists are not preoccupied with reflections on the nature of scientific knowledge. Such reflective preoccupation, in fact, often constitutes an impediment to more direct scientific work. When a scientist makes public pronouncements about the nature of scientific knowledge, the motive is often something extrinsic to the scientists' own work: the desire to make some scientific breakthrough intelligible to a wider audience; the need to convince public figures that scientific research deserves financial support; the attempt to prove that real science is compatible with humanism, or religion, or environmentalism, or socialism, or democracy, or Eastern mysticism. Philosophers, professionally concerned with an analysis of the nature of scientific knowledge, generally find little that is helpful or enlightening in such pronouncements.

The nontechnical writings of Bohr and Einstein may seem to fit into this general category. Each presents an epistemological position supported by only the sketchiest of philosophical arguments. In their cases, however, there is a difference. Their philosophical pronouncements, especially after 1935, articulate positions on the nature of scientific knowledge which were forged in the fires of scientific revolutions they themselves had engineered, and then shaped and tempered through sustained opposition. Neither position can be adequately understood or even rendered plausible on the basis of the philosophical reasons offered in its support. They are only intelligible against the background of the dialectic of development from which they emerged. Now that this background has been surveyed, the emergent positions can be considered in a more systematic, less historical, manner.

Before beginning, this one point should be stressed. Atomic physics proved to be the crucial test case for the theories of scientific knowledge each man developed. Yet, each tried to make his position intelligible as part of an overall comprehensive theory of knowledge. To detach what either had to say on scientific explanation and atomic physics from the overall framework in which it is embedded is to lose the coherence and sense of perspective which each protagonist thought essential for a proper understanding. When Bohr defends his interpretation of quantum mechanics by insisting that we are suspended in language, one who wishes to understand that position must consider what Bohr had to say on language. When Einstein counters this interpretation by insisting that God does not play dice, it becomes important to see how Einstein's conception of God relates to his theory of scientific explanation. For these reasons our discussion will include topics not generally considered relevant to either atomic physics or scientific explanation.

Neither man was a professional philosopher. Neither developed his theory of knowledge through the sorts of arguments philosophers are accustomed to consider. To make each position intelligible as a philosophical position it is necessary to go beyond what each man said. I will attempt to do this by relating the position each man held to positions and considerations philosophers consider pertinent. I will attempt not to impose my views on either man, but simply to clarify their positions. Since I am going beyond the text in both cases, the only real justification I can offer is whatever coherence and plausibility the ensuing accounts possess.

Bohr as a Conceptual Analyst

By the late 1930s, Bohr's views on the interpretation of quantum theory and on the nature of scientific explanation had jelled into definitive form. In subsequent years, Bohr repeatedly presented his views on the interpretation of quantum theory through popular lectures which took on a kind of standard format. A sketchy history of scientific progress led to a very nontechnical survey of quantum theory and an explanation of the doctrine of complementarity. Then, depending on the audience, Bohr usually drew an analogy between complementarity in quantum physics and a complementary relationship in some other field: subject and object in psychology, simple and advanced cultures in anthropology, science and religion, or the relation that should obtain between the authority of the United Nations and the sovereignty of individual nations. This seemed so to trivialize the interpretative problems involved that philosophers generally paid as little attention to Bohr's epistemology as historians paid to his thumbnail sketches of the development of physics.

Bohr, however, seemed to feel that such lectures supplied an adequate basis for presenting his views, provided philosophers would take them seriously enough to accord them the study he thought they deserved. In attempting to show how the fabric of human knowledge could be woven into a coherent pattern, he gradually built up a consistent, though somewhat idiosyncratic, terminology. Interpreters of Thomas Aquinas adopted the hermeneutic principle: "Thomas semper formalissime loquitur." Something similar holds of Bohr. One who wishes to learn his position from his popular lectures must attend carefully to the precise meaning he attaches to such terms as 'classical', 'objective', 'subjective', 'experience', 'unambiguous communication', 'concept', 'conceptual framework', 'picture', 'observe', and 'experiment'. I fear, however, that one more summary of Bohr's doctrine in Bohr's own terms would do little to close the communications gap between Bohr and the philosophers. For these reasons I will attempt to present his views in a terminology that should be more intelligible to contemporary philosophers. Any non-Bohrian terminology will be clearly indicated as such. However, the doctrine presented will, as much as possible, be Bohr's rather than mine.

Bohr's epistemology was not presented in traditional philosophical terms because it was not worked out as a philosophical exercise. Though he did read some philosophy in his student days, he seems to have read very little in his later years. He preferred to obtain his information through dialogue or through hearing papers on which he could comment. Though he did read technical articles, he did not in his mature years seem to enjoy reading books. In spite of this almost total lack of philosophical scholarship, he intended his epistemology to be a coherent overall account of human knowledge. In the development we have considered Bohr gradually progressed from his early attempts to give a descriptive account of atomic structure to the more general questions of how concepts express meaning and how language relates to both thought and reality. His primary focus, however, was always on the concepts and conceptual structures basic to atomic physics. The broader aspects of human knowledge, to which Bohr related these particular concepts, were radically underdeveloped in his writings. For these reasons we will begin with his epistemology of quantum mechanics and work from this to more general aspects of his epistemology.

In chapter 8 we traced the gradual modification of the two pivotal points of Bohr's doctrine: complementarity and the epistemological irreducibility of atomic experiments. Then we showed how his later attempts to interrelate quantum theory, relativistic quantum mechanics, quantum field theory, and nuclear physics focused on an analysis of the limits of applicability of the foundational concepts these theories shared. The differences in the presuppositions of these theories heightened Bohr's concern with the problem of relating concepts to the frameworks in which they function. It also sharpened his awareness of two points in which they function. First, to analyze concepts Bohr relied, not on any process of introspection, but on an examination of how the term expressing this concept is used in different contexts. He was effectively relying on a doctrine of meaning as use, though he never introduced this terminology. Second, the concepts he found problematic were almost invariably classical concepts extended to new domains. Earlier, Bohr had taken the classical meanings as essentially nonproblematic. His concern was with the way in which the correspondence principle guides the extension of such concepts to new domains. After his consideration of differing conceptual frameworks, he realized that classical concepts also require some sort of analysis. Their meanings are also problematic, though in a different way. He did not judge, however, that giving such explanations was his proper task.

After his reply to the EPR paper, a reply which Bohr interpreted as a complete victory, his epistemological position seemed to take on a more or less definitive form. In the years subsequent to 1937, he focused his attention successively on nuclear physics, the atomic bomb, rebuilding his institute after the war, and the concerns proper to an elder statesman serving as spokesman to the world for the scientific community. Except for some rather embarrassingly staged encounters with Einstein at Princeton and his contribution to the Schilpp volume on Einstein, he seems to have thought of his conflict with Einstein as a definitively settled issue.

In his reply to EPR and even more clearly from 1937 on, Bohr found his definitive formulation of the doctrine of complementarity in terms of 'mutually complementary phenomena'. In his earlier writings, Bohr had used 'phenomena' as a rather loose term for whatever was observable. After 1937, he advocated and followed the practice of using 'phenomena' to refer exclusively to observations obtained under specified circumstances, including the type of conceptual analysis of experiments he presented in his reply to EPR. If scientific information is concerned with phenomena revealed by experiment rather than with some unobservable noumenal reality, then it does not make sense to speak about disturbing phenomena by conducting an experiment.

With this restricted usage the doctrine of complementarity now fits quite well with Bohr's holistic principle of the epistemological irreducibility of atomic experiments, a principle he sometimes referred to as 'an individuality totally foreign to classical physics'. Each set of phenomena revealed through a particular experiment has a wholeness not susceptible of further conceptual subdivision. Any attempt at such conceptual reduction effectively replaces the experiment in question by a different experiment. A different experiment involves different phenomena, not a different view of the same phenomena.

The manner in which this transforms ontological into epistemological issues is highlighted when one considers the ontology implicit in the classical concepts Bohr is discussing. If there's a tree in the meadow, then the meadow may serve as a basis for specifying the tree's location. However, the tree's size and shape may be considered properties which it has regardless of where it is located. Both ordinary language and classical physics treat primary qualities as objectively real properties of bodies. If there's an electron in the bubble chamber, then bubbles in the chamber may serve as a basis for specifying the electron's location. Can the electron also be attributed a size and shape, e.g., a particle of 10^{-13} cm, independent of this localizing process? This is the sort of ontological question that Bohr thought to be misguided and that he avoided by switching from the language of property attributions to talk about phenomena. Instead of asking what properties the electron has, considered as an isolated object, one must specify the experimental circumstances under which the phenomena of localization or size are discussed.

Because of this insistence, Bohr appears to many critics to be a positivist, or even an idealist, rather than a realist.[1] Such categorizations distort the issues Bohr was attempting to elucidate. If one prescinds from dogmatic metaphysics, then any physically meaningful doctrine of realism relies on presuppositions implicit in the use of 'real'. These presuppositions are conditioned by ordinary language usage and classical physics. One cannot simply extend such usage to quantum domains without adverting to the way such presuppositions function and the limits of their valid applicability. This was Bohr's doctrine. Its presentation, however, was radically incomplete. Though he discussed some problems involved in extending ordinary language concepts, he never explained how the ontology implicit in ordinary language functions. This must be made explicit, if one wishes to examine it as a presupposition for extended usage. Our attempt to make this explicit necessarily goes beyond what Bohr said. This will be done only to the degree necessary to fill in gaps in Bohr's doctrine.

Ordinary Language Analysis and Descriptive Metaphysics

To complete this presentation of Bohr's epistemological position it is necessary to discuss issues which Bohr considered an integral part of this position, but which he never developed in a systematic fashion. These issues include the ground of meaning of classical concepts, the relation between language and ontology, the extension of conceptual frameworks, and subject-object duality. These topics are definitely not a part of what is generally accepted as the Copenhagen interpretation. Most quantum physicists would consider them rather irrelevant fringe issues. Nor, it must be admitted, did Bohr ever develop any of these issues in an adequate philosophical fashion. Yet, when these issues are removed from Bohr's overall position, the residue that remains is neither complete nor even coherent as an epistemological position.

The linguistic orientation of Bohr's thought was clearly summarized by A. Peterson, Bohr's final assistant:

> As far as I can see, the doctrine that we are, philosophically speaking, suspended in language, that we depend on our conceptual framework for unambiguous communication in the

way illustrated by mathematics, forms the general basis of
Bohr's philosophy. In his writings he never gave a detailed exposition of this view. Nor did he discuss its relation to other
philosophical conceptions of the philosophical status of
language.[2]

The philosophical problems this perspective posed had the general form
of a transcendental deduction. The concepts used in atomic physics are
already adequate for the unambiguous communication of information. This
is clearly shown by the success of this physics. The philosopher's task
accordingly is not one of *replacing* these concepts by others. It is, rather, one
of explaining how the concepts in use function as they do. This, in turn,
requires some clarification of what it means for a term to have meaning.
Bohr never intended or attempted a general doctrine of meaning. His
problem was the more limited one of clarifying the conditions of the possibility of unambiguous communication. This is a transcendental deduction in
the sense that one goes from a fact, the unambiguous communication of
information concerning atomic systems, to the conditions of possibility of
this fact.[3]

The best that Bohr was able to do in developing a theory of word
meaning was to offer a mathematical analogy. The way a word functions in
different contexts is analogous to the way multivalued functions may be
represented through different sheets on a Riemann surface. This analogy
brings out the idea that, at least for the foundational concepts at issue, it is
misleading to speak of *the* meaning or to explain word meaning by a
correspondence between terms and things. An n-dimensional Riemann
manifold can be considered as an aggregate of its tangent spaces fused into a
whole. Each tangent space is connected with its neighboring tangent space.
In the neighborhood of a point any tangent space is approximately Euclidean. In a similar fashion one may use logical analysis, like Euclidean
geometry, to explain how a term functions in a specified context, e.g., a
particular atomic experiment. But the term also functions in other contexts,
and the logical analysis that works *within* one context does not quite apply as
one goes from one context to another.

This analogy illustrates a basic, though underdeveloped, point in Bohr's
doctrine of meaning. Clarifying the meaning of a word involves showing
how it relates to other words through logical structures somehow implicit in
language. Any attempt to consider a term in isolation from such logical
structures as having a precise meaning is essentially misguided. In a particular context an analysis of a term's meaning focuses on some aspects of the
network of logical relations connecting this term to the body of language. In
another context one may focus on other aspects of this logical network, just
as one can consider different tangent spaces in the neighborhood of a

particular point in a Riemann surface. It follows that the language we use to describe the world as experienced is not susceptible to any sharp all-purpose distinction between content and logical form. Any form-content distinction is invariably context-dependent.

This is a crucial point in interpreting Bohr's doctrine. It is the reason why he thought that no type of logical analysis could clarify the problems of meaning he was wrestling with. Logical analysis presupposes terms with precise univocal meanings. This presupposition is valid only in specialized and somewhat artificial contexts. It is radically misleading when one is facing the problem of extending words whose meanings are specified in one context to a context incompatible with the presuppositions governing the word's normal usage. This was Bohr's problem. He could find no way to analyze it except through analogies, examples, and explanations that involved a mélange of philosophy and psychology. Thus, Heisenberg, who had extensive discussions with Bohr on this problem, summarized Bohr's theory of word meaning:

> Bohr said that when any word is produced, that word raises something into the light of consciousness, and at the same time it raises many other things which are only in a shaded light and are almost completely covered, and all these things enter a consciousness at the same time. Therefore, a word is such a complicated thing that one could not hope to copy it by a mathematical letter, because the mathematical letter can then only mean, let us say, that thing connected with the word which is in the center of consciousness.[4]

These analogies and psychological explanations introduce their own obscurities. To avoid them it is helpful to compare Bohr's development with some roughly parallel developments in the intellectual career of Ludwig Wittgenstein. His *Tractatus*[5] carried the picture theory of meaning to its ultimate and rather bizarre conclusion. Before considering this, it should be noted that the picture theory of meaning is not a reliance on intuitive or imaginative representations. It is, rather, the type of correspondence between language and reality that was thought to ground the applicability of logic and mathematics to physical reality. Hertz gave this picture theory its classical expression in the introduction to his *Principles of Mechanics*:

> We form for ourselves images or symbols of external objects; and the form which we give them is such that the necessary consequents of the images in thought are always the images of the necessary consequents in nature of the things pictured. In order that this requirement may be satisfied, there must be a certain conformity between nature and our thought.[6]

As noted earlier, this interpretation of science as embodying a unified coherent picture of physical reality was stressed, with somewhat varying emphases, by Helmholtz, Boltzmann, and Planck, and served as the intellectual background for Bohr's early views on science as supplying descriptive accounts. Wittgenstein leaves no doubt that this was the tradition of picturing he refers to:

> 4.04. In a proposition there must be exactly as many distinguishable parts as in the situation that it represents.
> The two must possess the same logical (mathematical) multiplicity. (Compare Hertz's *Mechanics* on dynamical models.)

For our purposes we need not consider Wittgenstein's early position on the foundations of language and logic, but simply two polar aspects of this picturing relation. On a global basis it meant that the world as a whole and language as a whole shared a logical form, but a form that could not be expressed in language. On an individual basis it meant a correspondence between propositions, considered as names in relation, and states of affairs, considered as objects in relation. Thus, Wittgenstein had: The world is the totality of facts (1.1). A fact is the existence of a state of affairs (2), which is a combination of objects (2.01). A proposition shows the logical form of reality (4.121). An elementary proposition consists of names in immediate combination (4.221). It depicts a state of affairs when the combination of names represents the combination of objects. This is logically the same sort of picturing relation between language and the world that holds for a gramaphone record, the musical idea, the written notes, and the sound waves (4.014).

When Wittgenstein began to modify his views in the late 1920s, he did not abandon this picture theory of meaning. However, he began to realize that some basic propositions describing physical reality, such as color ascriptions, could only be understood as part of an interrelated system of propositions. Instead of language and the world sharing *one* common logical form, it became necessary to speak of subsystems containing distinctive logical structures. Thus, in a conversation with F. Waismann in December 1929, Wittgenstein said:

> I once wrote "A proposition is like a ruler laid against reality. Only the outermost graduating marks touch the object to be measured." I would now rather say: "a *system* of propositions is laid against reality like a ruler."[7]

From a series of such modifications gradually emerged the new idea of a *Philosophische Grammatik*, an analysis of the depth grammar of language. A simple example illustrates Wittgenstein's concerns in the early 1930s. In its surface grammar the sentence 'The chair is brown' parallels 'The chair is heavy'. However, the implicit criteria governing the use of these predicates allow 'The surface of the chair is brown', but not 'The surface of the chair is heavy'. The implicit but operative criteria governing the application of both physicalistic and mentalistic predicates are, in their turn, governed by norms of coherence more subtle and sophisticated than those proper to a logical reconstruction of language. In the *Philosophical Investigations* Wittgenstein concentrated on an analysis of such mentalistic terms as 'understand', 'read', 'know', and the terms related to expressions of pain. We must turn elsewhere for an analysis of physicalistic terms. But in the *Investigations* Wittgenstein also developed a doctrine of meaning as use that has a special appropriateness for Bohr's analysis.

For both men the abandonment of a picture theory of meaning led to a realization that the meaning of a term is its use in language. This, in turn, raised the question, What is the minimal unit in which terms can function in a meaningful way? Wittgenstein's celebrated answer is his doctrine of language games. For our purposes, we may distinguish between *pure* language games, where the meaning of a term depends only on its relation to other terms, and *impure* language games, which include the physical and social environment. It is through the latter that we learn basic terms for shapes, colors, observational categories, actions, etc. Impure language games accordingly are basic in setting meaning and pure language games are parasitic upon them. Wittgenstein worked out this doctrine for ordinary language, while Bohr was primarily concerned with scientific discourse. If we postpone consideration of the problems involved in going from the conceptual framework of ordinary language to that of science, we may rephrase one of Bohr's basic doctrines in Wittgensteinese. The principle that each atomic experiment must be considered an epistemologically irreducible unit may be reformulated as: in any situation, experimental or conceptual, in which the quantum of action is basic the minimum framework which has epistemological significance is an impure language game constituted by an observed system linked to a measuring apparatus whose recordings can be expressed in classical concepts. This minimal quantum language game differs from the minimal Wittgenstein language game, e.g., the worker calling "Slab" to his assistant, in that Bohr's language does not serve to establish meaning. Bohr is concerned with classical concepts, terms whose meanings are already determined through their usage in ordinary language and classical physics. Bohr's language games supply a minimal quantum context in which such terms can function meaningfully.

Before considering particular examples of applied language games, some comment is needed on the relation between language games and formal systems. Formal logic, whether standard or nonstandard, involves rules whose validity is independent of the content to which they are applied. According to the received tradition in the philosophy of science, any body of scientific knowledge can be rationally reconstructed along the lines of a formal axiomatic system. If this is believed, then it makes sense to begin reconstruction with a formal logical structure on which an interpretation is imposed. Thus, semantics, which treats meaning and truth, presupposes syntax. The syntactical structures are independent of this imposed semantics.

In conceptual entailment, semantics is basic and logical structures are derivative. Thus, if something is categorized as an X, one may infer that it has property Y or a disposition toward activity Z, if the concept of being an X entails having property Y or disposition Z. A network of conceptual entailments supplies an inference mechanism. In this case, however, the inference is *material* rather than *formal*. That is, inferences depend on meanings, not on rules whose validity is independent of meanings. In the *Philosophical Investigations*, a language game is presented as the minimal unit in which a cluster of interrelated conceptual entailments can function. This language-game approach to meaning constitutes a radical break with Wittgenstein's earlier view. Now he is explicitly repudiating the idea that there is one logical form which all meaningful use of language presupposes. In a somewhat similar fashion, Bohr insists that the ultimate basis for interpreting quantum mechanics is the type of informal semantics he developed rather than the mathematical formalism. A fortiori, no logical reconstruction of this formalism could supply a proper basis for interpretation. Though Bohr never discussed the correspondence between the changes in his and Wittgenstein's positions, Heisenberg did. In a letter to Heisenberg I suggested a relation between Wittgenstein's later position and the Copenhagen interpretation. He replied:

> I was also glad to see from your letter that you prefer the later work of Wittgenstein as compared with the *Tractatus*. This was a point of strong dissent between Bertrand Russell and myself. Russell loved the *Tractatus* and told me that he could not understand what Wittgenstein meant in his later work. I told him that I felt the *Tractatus* was wrong or trivial in all essential points, but I had a high esteem of the late Wittgenstein.[8]

Bohr's analysis focused on what he called 'classical concepts', or ordinary language concepts which have been given an idealized form and a foundational role in classical physics. 'Particle' and 'wave' are prime examples of such classical concepts. To call anything a particle implies that its internal structure, if such there be, is irrelevant in the context of discussion. A particle has a location and moves in a space-time trajectory. When this trajectory encounters an obstacle, such as a target, the particle can strike the target, penetrate it, recoil from it, or be deflected by it. 'Travel in a trajectory', 'pass through a slit', 'hit a target', 'impinge', 'recoil', 'be deflected', 'strike a photographic plate at a point'—these and related terms from a cluster of conceptual entailments which center around 'particle'.

The term 'wave' is at the nub of a different conceptual cluster. A wave does not have a precise location or a space-time trajectory. It propagates through vibrations in a medium. It does not collide with a target, recoil from it, or hit a plate at a point. Waves are reflected, refracted, or absorbed. They may be diffracted. Two waves may interfere with each other. Waves are characterized by such things as frequency, wavelength, amplitude, and phase.

An atomic experiment, including both the system studied and the measuring apparatus used to study it, constitutes the minimal language game proper to discourse in atomic physics. Depending on how the experiment is arranged and the type of information that is sought, either the 'particle' conceptual cluster or the 'wave' conceptual cluster supplies an appropriate basis for describing what happens and reporting the results. Two aspects of these quantum language games deserve stressing. First, the network of conceptual entailments hinges in each case on *classical* concepts. It is here that the impure aspect of the language game is basic. The experimental apparatus, at least the aspects of it which supply the basis for recording results, must be something that can be described in strictly classical terms. With such an apparatus and a network of entailments based on classical concepts, one has a conceptually consistent impure language game. Secondly, though one of these clusters of concepts is applicable to any atomic experiment, neither is applicable to every atomic experiment. The choice of which cluster is appropriate is determined by the way the experiment is arranged and the type of information that is sought.

These are minimal language games in the sense that they are the smallest units of discourse in which such concepts can meaningfully function. The significance of this can be seen by contrast with what is sometimes spoken of as a more realistic approach. In this view, the electron already has a position, a size, a shape, a velocity proper to some framework, or similar properties, regardless of whether these are measured. This is defended as realism in the sense that its denial would seem to constitute a species of idealism, to make the properties of the electron dependent on the thought of the observer. Realism and idealism so conceived are ontological positions. Thus, in this brand of realism one is speaking of the electron as a kind of object and asking what properties this object has.

Bohr is not an idealist. To think of his doctrine in such ontological terms is to misconstruc what he is doing. He is insisting on the primacy of epistemological analysis over ontological doctrines. Such terms as 'position', 'size', 'shape', 'velocity', and even 'reality' already have determined meanings. Any questions about the location of an electron presupposes the applicability of 'location' to 'electron'. Bohr examines the grounds and limits of validity of this presupposition. He concludes that it cannot be meaningfully used in isolation, but only in the context of a possible experiment, or quantum language game. This he tried to convey in his later years through the technical use he made of the term 'phenomenon'.

One phrase which Bohr used repeatedly and emphasized consistently was that quantum mechanics represents a 'rational generalization of classical physics'. It is likely that no other claim of Bohr's occasions as much confusion as this one, for he also insists that quantum physics represents the historically decisive break with classical physics. To understand what he meant we must also consider one other expression Bohr used repeatedly, especially in his later years, but never adequately explained. It was the 'conditions of the possibility of unambiguous communication'. In his typical chiaroscuro fashion, he never really explained what these conditions are or how they function. Instead, he assumed, as something too obvious to require explanation, that ordinary language and classical physics certainly meet these conditions. People do use language to communicate both ordinary information and the results of experiments in such a way that their hearers or readers understand them without ambiguity. What is actual must be possible. What is possible must fulfill the conditions of its possibility. The only problem Bohr focused on accordingly was one of explaining how quantum physics can also fulfill these same conditions. It is this fulfillment that makes quantum physics a rational generalization of classical physics.

To understand what is going on in this argument it is first necessary to illuminate some of the shadowy recesses in Bohr's sketchy picture, in particular to explain the way in which ordinary language and classical physics fulfill the condition of the possibility of unambiguous communication. Fortunately, in the case of ordinary language the requisite analysis was supplied by P. F. Strawson in his now classical work *Individuals: An Essay in Descriptive Metaphysics*.[9] We will outline the pertinent aspects of his argument.

The question that serves as Strawson's point of departure is, How can a person use language to identify particulars, especially absent particulars, in such a way that others may recognize the particular referred to? Some particulars are recognized by relation to others, i.e., the assassin of Abraham Lincoln. To obviate an infinite regress, however, there must be some basic particulars which can be identified in a nonrelative way. Even these cannot serve as the needed basic particulars unless they can be identified and reidentified as the same individuals previously referred to.

This analysis, which I am simply outlining without any supporting detail, leads to both objective and subjective conditions for the possibility of unambiguous indentification of basic particulars through language. The objective condition involves what Strawson labels 'descriptive metaphysics'. The only type of particulars that can play the role of being basic, rather than relative, particulars, and of being identified and reidentified unambiguously through language are things, or relatively perduring objects with identifiable characteristics. Strawson concludes that a fundamental condition for the possibility of unambiguous communication is that the conceptual core of any language contains a representation of physical reality as an interrelated set of things with characterizing properties and that these things exist in a common space-time framework. This is *descriptive* metaphysics in the sense that Strawson is describing the conceptualization of reality which, he argues, must be present in any ordinary language.

The subjective aspect of Strawson's analysis concerns the subject-object distinction and the concept of 'person'. A description of an object together with its characterizing features does not, in general, suffice for unambiguous reference. It leaves open the theoretical possibility of massive reduplication. The only way in which unambiguous identification of individuals can be secured is through a common space-time framework, which the speaker shares with his hearers. My bodily presence anchors the space-time framework for me and supplies my ultimate reference point for the spatial and temporal location of other objects.

This, in turn, presupposes a sharp distinction between myself as the conscious subject of experience and the objects referred to. Yet, the terms I must use to refer to myself and my experiences are terms I learn through the use others make of them. Any self-ascription of mentalistic predicates is logically incoherent unless it also entails other-ascription. Descriptive metaphysics necessarily accords the community of speakers a distinctive ontological status.

From one point of view there is nothing novel about this conclusion. The basic content of descriptive metaphysics is the simple prescientific common-sense view of reality commonly attributed to the man on the street. The content of descriptive metaphysics is not novel. The way in which it is presented and interpreted is. It is *not* presented as a theory of what reality is in some objective sense. It is, rather, the conclusion of a transcendental argument. From the undoubted fact that absent particulars can be unambiguously identified through language Strawson argues to the condition of the possibility of this fact. This condition is a conceptual core which, Strawson argues, must be present in any language which serves as a vehicle for interpersonal communication.

Before building on Strawson's argument, two aspects of it require some critical comment. First, he concludes that this descriptive metaphysics must be the conceptual core of all spoken language, i.e., of English or Swahili, but not of FORTRAN or COBOL. Many have criticized the universality of this claim.[10] I find the arguments supporting this claim of universality unconvincing. What Strawson's argument really leads to is not a conceptual core which all spoken languages *must* possess, but a set of interrelated functions which must be fulfilled through the use of language. This set of functions radically underdetermines the conceptual structures that may be imposed on it. However, this rather Quinean objection need not be elaborated here, since it is not relevant to the use we are making of Strawson's argument. Regardless of how universal it may be, the descriptive metaphysics explicit in Strawson's account—and implicit in much of what Aristotle says—represents a conceptual core shared by the Indo-European languages. These supplied the matrix from which physics emerged.

The second point requiring some comment is that descriptive metaphysics is not a *theory* about reality. When roughly similar positions are presented as theories, they may be believed, disbelieved, criticized, defended, or simply ignored. Strawson's descriptive metaphysics is presented as a conceptual core already implicit and operative in the language we speak. It is a core in the sense that it is at the nub of a network of conceptual entailments. To rely on this, however implicitly, as a basis for material inference while explicitly contradicting its fundamental aspects is to become entangled in a peculiar sort of contradiction. This is not an overt contradiction between conflicting propositions. It is more like the sign in an American restaurant window reading: "Ici on no parle pas français." There is a conflict between the content of the communication and the act of communicating it. It is easy, and not uncommon, to avoid the possibility of such conceptual clashes by acting as if only ordinary language usage presents philosophically interesting problems for conceptual analysis. This expedient was not available to Bohr. His was the problem of showing how the conditions of unambiguous communication, implicit and operative in ordinary language usage, may be extended to a domain of discourse which appears to be radically incompatible with the descriptive metaphysics just outlined.

My purpose here is not to defend either Wittgenstein or Strawson. It is, rather, to present a framework within which Bohr's highly informal semantic analysis becomes intelligible to philosophers. Bohr never presented a systematic analysis of meaning as use in ordinary language. However, his attempts to explain how ordinary language concepts may be extended to new frameworks presuppose that the meanings of these is determined by their use in ordinary language. Bohr never used, and probably never encountered, the term 'descriptive metaphysics'. However, as the historical chapters have shown, he had a lifelong concern with the problem of how to give a descriptive account of physical reality. In the course of his career his emphasis gradually shifted from the reality described to the conditions of the possibility of using language to give unambiguous descriptions. If we accept the doctrines developed by Wittgenstein and Strawson as supplying the foundation that Bohr required, then Bohr's position achieves an overall coherence. We are accordingly presupposing these doctrines as a basis for making Bohr's doctrine intelligible. The problem then is one of explaining how to get from this basis to further conceptual frameworks. To the extent possible we will develop this by letting Bohr speak for himself.

Bohr's method of concept extension can best be explained as a series of successive modifications of the descriptive metaphysics just considered. Each modification, to be acceptable, must ultimately be presented in such a way as to preserve the conditions of the possibility of unambiguous communication. This, in turn, requires a sharp subject-object distinction.

As might be expected, Bohr did not develop this as an abstract doctrine, but through an explanation of the particular conceptual frameworks that function in physics. Thus, Bohr's frequent quasi-historical accounts of the development of physics are of interest, not for their history—Bohr was no historian—but for the epistemological principles which serve as Bohr's interpretative base. Conceptual frameworks, as he saw it, supply a basis for explaining both particular changes and ununderlying continuity:

> The main point to realize is that all knowledge presents in-self within a conceptual framework adapted to account for previous experience and that any such frame may prove too narrow to comprehend new experiences. . . .
> When speaking of a conceptual framework, we refer merely to the unambiguous logical representation of relations between experience. . . .
> The necessity, even within this comparatively simple theme, of paying constant attention to the problem of objective description has deeply influenced the attitude of philosophical schools through the ages. In our day, the exploration of new fields of experience has disclosed unsuspected presuppositions for the unambiguous application of some of our most elementary concepts and thereby given us an epistemological lesson with bearings on problems far beyond the domain of physical science.[11]

Within this developmental perspective, Bohr interpreted the physics stemming from Galileo and Newton as the first significant step beyond the representation of reality implicit in ordinary language. It did this by focusing on measurable quantities and by interpreting inanimate matter as a determi-

nistic system governed by casual laws. The conditions of objectivity were
preserved and even intensified by an absolute subject-object distinction and
by a representation of space and time as absolutes.[12]

The physicists of the Enlightenment era tended to stress the break
between the new physics and earlier views. Bohr's interpretation emphasized the perduring influence of the representation of reality implicit in
ordinary language. Heisenberg, whose position developed through protracted discussions with Bohr, expressed this quite clearly:

> Furthermore, one of the most important features of the
> development and analysis of modern physics is the experience
> that the concepts of natural language, vaguely defined as they
> are, seem to be more stable in the expansion of knowledge
> than the precise terms of scientific language, derived as an
> idealization from only limited groups of phenomena. This is in
> fact not surprising since the concepts of natural language are
> formed by the immediate contact with reality; they represent
> reality.[13]

The next major change in conceptual frameworks, as Bohr interpreted
the history of physics, was the development of field theory by Faraday and
Maxwell. This Bohr saw as sharpening the relationship between causal
determinism and space-time descriptions. It represents in this sense the
completion of classical physics. The theory of relativity represents the next
major advance in conceptual frameworks. Where Einstein saw this as the
decisive break with Newtonian physics, Bohr stressed the underlying continuity linking relativity with classical physics. Here again, the subject-
object distinction and the observer's reliance on relating himself to objects
through an ordinary space-time framework are seen as the basis for objectivity, i.e., the unambiguous communication of information concerning
physical objects:

> The situation in such respects [i.e., the determinism of
> classical physics] was not essentially changed by the recogni-
> tion, embodied in the notion of *relativity*, of the extent to
> which the description of physical phenomena depends on the
> reference frame chosen by the observer. We are here con-
> cerned with a most fruitful development which has made it
> possible to formulate physical laws common to all observers
> and to link phenomena which hitherto appeared uncorrelated.
> Although in this formulation use is made of mathematical ab-
> straction such as a four-dimensional non-Euclidean metric, the
> physical interpretation for each observer rests on the usual
> separation between space and time, and maintains the determi-
> nistic character of this description.[14]

Quantum physics constituted the distinctive break with classical physics precisely because it showed the inadequacy of the ordinary language-cum-classical physics representation of reality: "Indeed, it became clear that the pictorial description of classical physical theories represents an idealization valid only for phenomena in the analysis of which all actions involved are sufficiently large to permit the neglect of the quantum."[15] The way in which quantum physics goes beyond the classical fusion of causal determinism and space-time descriptions has already been sufficiently considered. The problem to be considered now is how in Bohr's view it was possible to preserve objectivity in quantum physics. Here it is of crucial importance to attend to the significance Bohr accorded the basic terms involved. When he spoke of 'objectivity', he did *not* refer to a correspondence between the concepts used in physics and objects as they exist independently of our knowledge of them. For Bohr, as the recent citations indicate, both classical and quantum physics are concerned with *phenomena*, in a very Kantian sense of the term. Hence, the meaning of such terms as 'reality' and 'objectivity' cannot be considered to be determined by what physical objects really are. This correspondence was the idea behind the EPR criterion of reality, which Bohr criticized as being hopelessly ambiguous. The meaning of these terms is determined by their ordinary language usage.

In ordinary language usage, 'objectivity' is meaningfully used in contrast to 'subjectivity'. An experience is subjective. A report of what is experienced is objective if others, hearing the report, understand it in the same sense. The term 'objectivity' accordingly is only meaningfully employed in a context that presupposes a sharp subject-object distinction and in which the speaker and listener relate themselves to objects through the common space-time framework implicit in the language used for interpersonal communication. This is not an attempt to impose Strawson's doctrine on Bohr. Bohr himself is quite clear on this point:

> The answer to this question [of the completeness of quantum physics as an exhaustive description of experience] evidently calls for a closer examination of the conditions for the unambiguous use of the concepts of classical physics in the analysis of atomic phenomena. The decisive point is to recognize that the description of the experimental arrangement and the recording of observations must be given in plain language, suitably refined by the usual physical terminology. This is a simple logical demand, since by the word 'experiment' we can only mean a procedure regarding which we are able to communicate to others what we have done and what we have learnt. . . . As regards all such points, the observation problem in quantum physics in no way differs from the classical physical approach.[16]

The features that distinguishes classical from quantum physics with respect to measurements is not the method of measurement as such but the way in which one distinguishes between the measuring apparatus and the system measured. Bohr's way of handling this is best seen by contrasting it with a different approach stemming from von Neumann. Since the difference concerns the presuppositions guiding interpretation, we may skip the formalism and consider a simple example.[17] An electron entering a Geiger counter illustrates some distinctive features of atomic measurement. The Geiger counter is in a metastable state, so that an atomic event triggers off a macroscopic response, a click which may be heard and recorded. In practice, there is no difficulty in using classical concepts and applying classical physics to treat clicks of Geiger counters, tracks in a cloud chamber or bubble chamber, spots on a photographic plate, or other reductions of metastable states triggered by atomic events.

Suppose that one interprets quantum theory as supplying a descriptive account of things as they exist independent of the conditions which make our knowledge of them possible, i.e., of noumena. Then, since any recording device is made of atoms, its interaction with the object measured is governed by the superposition principle. This gives a set of states with probability coefficients rather than a unique outcome. Even the act of observing the results involves an interaction between impinging photons and atoms in the eye. This introduces further superpositions, as does each interaction in the nervous system. Yet, the final observation is a unique outcome rather than a probabilistic superposition of possible outcomes. The question that emerges from this approach is one of how the act of observation reduces the wave packet or projects a unique result.

If the formalism of quantum mechanics is interpreted as supplying, at least in principle, an exhaustive description of all objects including the knower, the reduction of the wave packet, or von Neumann's projection postulate, emerges as a truly formidable problem. It seems to require drastic expedients. Thus, H. Everett's many-world interpretation holds that every process of observation involves a splitting of the universe into separate noninteracting systems, or parallel universes.[18] Or, with Wigner, one might defend a radical ontological dualism, attributing to consciousness some mysterious power to reduce wave packets.[19]

One presupposition implicit in such reasoning is that the basis for interpreting quantum theory is the correlation between the Ψ function, or a state vector, and some aspect of reality corresponding to it. This Bohr rejects. In spite of its mathematical sophistication, it rests on naive epistemological presuppositions. Such presuppositions are inevitable whenever one takes the mathematical formalism as the basis for interpretation. This formalism, Bohr insists, is essentially an inference mechanism. Its applicability presupposes rather than controls a proper interpretative basis:

> Strictly speaking, the mathematical formalism of quantum
> mechanics and electrodynamics merely offers rules of calcula-
> tion for the deduction of expectations about observations
> obtained under well-defined experimental conditions specified
> by classical physical concepts. The exhaustive character of this
> description depends not only on the freedom, offered by the
> formalism, of choosing these conditions in any manner con-
> ceivable, but equally on the fact that the very definition of the
> phenomena under consideration for their completion implies
> an element of irreversibility in the observational process em-
> phasizing the fundamentally irreversible character of the con-
> cept of observation.[20]

Many others were persuaded or at least intimidated by the von
Neumann formalism. Bohr never took it seriously.[21] As he saw it, a proper
understanding of the role of measurement in quantum mechanics hinges on
a clarification of the concept of observation and of the subject-object distinction. His analysis of the concept of observation involves two aspects, only
one of which is clearly treated. The clear aspect is the idea that an observation is objective when it accords with the conditions of the possibility of
unambiguous communication:

> The notion of complementarity does not imply any renun-
> ciation of detailed analysis limiting the scope of our inquiry,
> but simply stresses the character of objective description, inde-
> pendent of subjective judgment, in any field of experience
> where unambiguous communication essentially involves regard
> to the circumstances in which evidence is obtained.[22]

At issue here is not merely the act of observation, but the *concept*
'observation', a concept that must be learned, another classical concept.
The objectivity of quantum mechanical reports, in Bohr's sense of 'objective', is guaranteed by the classical nature of the measuring apparatus and by
the principle of epistemological irreducibility. The total experimental situation, including a measuring apparatus described in classical concepts, has a
wholeness unique to quantum mechanics. The phenomenon it defines does
not admit of further conceptual subdivision. Any attempt to determine what
happens to the atoms constituting the bubbles in the chamber, the eye, or
the neural pathways necessarily involves a new and different experiment.
This constitutes its own phenomena.

The aspect that is not so clearly treated concerns the nature of the
subject-object distinction. On this point I think that Bohr was quite clear in
his own mind on what he held. The lack of clarity enters with his failure to
make a proper distinction between the presuppositions of discourse and the
foundations of the life sciences. The reason for this failure seems to have
been Bohr's zeal for conceptual reform. He argued that his analysis of the

proper use of such concepts as 'life', 'consciousness', 'observe', and 'voluntary' should clarify the conceptual foundations requisite for the life sciences.[23] Only by considering what Bohr had to say on these issues can we understand his position on the nature of the subject-object distinction and its relation to the concept 'observe'.

In the past what the life sciences tended to borrow from physics was usually a program of mechanistic reductionism. Now that the limits of such a mode of explanation are more clearly realized, one can see that the life sciences would profit more from the type of presuppositional analysis quantum mechanics has generated than from any further programs of reductionism. Thus, in an address to the Danish Medical Society, Bohr drew this lesson from quantum physics: "This situation, novel in physical science, has demanded a renewed analysis of the presuppositions for the application of concepts used for orientation in our surroundings." Then he went on to apply this lesson to the life sciences:

> Compared to the extention of the mechanical mode of description demanded by the account of the individuality of atomic phenomena, the integrity of the organism and the unity of the personality confront us of course with a further generalization of the frame for the rational use of our means of communication. In this respect, it must be emphasized that the distinction between subject and object, necessary for unambiguous description, is retained in the way that in every communication containing a reference to ourselves we, so-to-speak, introduce a subject which does not appear as part of the content of the communication.[24]

This emphasis, it should be noted, does not exclude the use of molecular methods in studying organisms. Bohr's doctrine, as a matter of fact, played a significant role in the path leading to the discovery of the structure of DNA and the cracking of the genetic code.[25] Bohr's point is, rather, that one cannot consistently study the behavior of organisms while denying the presuppositions necessary to recognize behavior as distinct from sheer activity. These presuppositions are implicitly present and operative in the terms used to identify and describe differing behavioral activities. Such presuppositions as life, consciousness, purposefulness, and free choice are accordingly *objective*, not subjective. They are necessary presuppositions for unambiguous communication in many fields of biology and psychology.

Subjectivity enters only in a polar relation to objectivity. When an experimenter—or any one else—uses such terms 'I see', 'I observe', or 'I remember', he is implicitly positing a subject which has these experiences. Such positing, Bohr insists, is not to be interpreted ontologically. Whenever we attempt to communicate something of our own state, the dividing line between subject and content varies as a function of the content communicated. The tool Bohr found most apt for illustrating this point was an exegesis of the difficulties the young licentiate in Poul Martin Møller's *A Danish Student's Adventure* had in sorting out the progressive number of selves generated by reflecting on experience, by reflecting on one's reflecting on experience, . . . and so on for an inexhaustible series.[26] The point of this analysis of subjectivity was not to introduce subjectivism into physics. It was, on the contrary, an attempt to bring out the conditions of objectivity: "It is evident, however, that all search for an ultimate subject is at variance with the aim of objective description, which demands the contraposition of subject and object."[27]

In the light of these considerations one may, I believe, comprehend the reason for one of Bohr's most disturbing and disputed claims, the contention that the Copenhagen interpretation must remain a permanent part of physics. It will not be overturned by any future advances. In making such a claim, Bohr is not referring to the mathematical formalism. He would see no difficulty in going beyond the Schrödinger wave equation or the Dirac representation of states of a system by vectors in a Hilbert space. Such formalisms in Bohr's rather extreme opinion are essentially algorithms based on an idealization of ordinary language. Their function is to give numerical results and to ensure an overall consistency, but not to supply a basis for interpretation.

Any transformation or extension of present quantum theory must, as a condition of acceptability, meet two requirements. First, it must fulfill the conditions of the possibility of the unambiguous communication of information concerning observed systems. Second, it must accept the quantum of action. The first condition stems, not from quantum physics, but from the fundamental fact that human communication is grounded in the lived world. Until the computers take over, this is a constraint all scientific discourse must observe. The second condition is essentially pragmatic. It must be accepted to make sense of modern physics and its resounding success in solving problems. Quantum physics in the Copenhagen interpretation represents an adequate fulfillment of these conditions. It preserves an underlying continuity with prequantum physics, so that quantum theory is a rational generalization of classical physics. Any further advances must, to be intelligible and acceptable, be interpretable as rational generalizations, in Bohr's sense, of established physics. From these premises, Bohr concludes that the complementarity interpretation of quantum mechanics will remain a permanent part of physics. If this is so, then the development of this theory certainly constitutes Bohr's greatest contribution to physics.

Bohr and Kant

Bohr's philosophical position as interpreted here bears a deep resemblance to Kant's philosophical position as interpreted earlier. What philosophical education Bohr had occurred at a time when there was a widely influential revival of Kantianism in philosophy. Yet, Bohr never cites Kant or averts to the resemblance between the two positions. This might seem to be a formidable argument against the validity of the interpretation presented here.[28] This objection can, I believe, be countered. However, it imposes limits on any comparison. Bohr did not consciously draw on Kant, nor was he influenced by Kant's writings. The real resemblance comes, not from any conscious borrowing, but from the way each man solved the interpretative problems he faced. For this reason, it is best to bring out the resemblance through strategy arguments, something more concerned with ways of doing philosophy than with the doctrines developed. After developing this, I will return to the original question and present a conjectural account of possible reasons why Bohr does not refer to Kant.

In comparing the two, one must remain aware of the obvious temporal difference and also of the difference in perspective. Kant was a professional philosopher and, except in geography, an amateur scientist. Bohr was a professional scientist and an amateur philosopher. Yet, both men effectively took as basic the question, How can man come to know a world in which he is a part, a knower, and a moral agent? Also, both accepted the advances made by science as the only rationally acceptable basis for an account of what the world is. This, in both cases, leads to the question, How is the representation of the world given by science to be interpreted as a form of human knowledge?

In both cases the distinctive point of departure in interpreting science is the *primacy of pragmatics*. Each accepted the theories that had proved successful in science. They are accepted on the basis of their success in explaining physical reality, not because of their conformity to previously accepted norms of intelligibility. In the Leibniz-Wolff tradition, in which Kant was trained, Newtonian physics was seen as resting on the unintelligible doctrine of action at a distance. According to the ideals of classical physics, in which Bohr was trained, the quantum of action represents an element of irrationality at the heart of physics. Confronted with a clash between a priori norms of intelligibility and a successful physics, both men sooner or later accepted the physics that worked and then revised their norms of intelligibility.

The primacy of pragmatics suggests a methodology, one which each man developed only after various false starts and inadequate attempts. The method is the one succinctly summarized in the old Scholastic slogan "Ab esse ad posse valet illatio." The science that works must be possible. One should accordingly analyze the conditions of its possibility. No such analysis can be considered acceptable unless it somehow renders intelligible the element of irrationality that provided the original stumbling block. Science supplies a representation of the world. Yet, there is no way to interpret such a representation as a description of things as they exist objectively without introducing a radical incoherence. To handle this incoherence, each man concluded, one must examine what is involved in forming a representation of the world.

Here there is a substantive agreement and a methodological difference. The substantive agreement is on the point that one must consider the knowing subject as distinct from the content of his representation. This content is a world of objects. The subject is not one more object. Rather, it is the knowing subject who constitutes other things *as objects of knowledge.* Both are antireductionist in insisting that the subject must be accorded such nonphysicalistic attributes as life, consciousness, and freedom. Yet, both insist, this attribution must not be interpreted as ascribing properties to an object. In both cases, the argument is that the presuppositions implicit in meaningful discourse about objects are different from and complementary to the presuppositions implicit in meaningful discourse about human knowers and moral agents. Any argument that in some ultimate analysis human beings really are nothing but ordered collections of atoms necessarily presupposes a type of knowledge of noumena (Kant) or of atoms apart from the conditions that make an experiment possible (Bohr). Such an outsider's view of physical reality transcends the limits of human knowledge (Kant) or of meaningful discourse (Bohr). For both men a complementarity between a conceptual framework in which one speaks about the properties of physical objects and a conceptual framework centered on the subject of this knowing activity serves as a *via media* between vitalism and reductionism. In both cases the conflict is dissolved rather than solved, and this is done by a transformation of ontological into epistemological issues.

Here there is also a significant methodological divergence. Kant's analysis of how we constitute objects of knowledge focuses on the stages and processes involved in knowing considered as an individual cognitive act. Bohr focuses on language considered as an interpersonal means of communication. This difference, significant as it surely is, does not play a decisive role in the type of argumentation we are now considering. The reason for this may be put rather crudely. Kant effectively smuggles in the social and interpersonal aspects which Bohr, like many other contemporary philosophers, attributes to language sharing and the lived world. He does this initially by his method of determining the analytic (in Kant's sense) meaning of concepts through their use in commonly accepted judgments. He does this subsequently by making the assumptions that all human knowers have the same cognitive apparatus. For all its lack of development, Bohr's language-centered approach must be considered a methodological advance over Kant. This, of course, is more properly viewed as an advance our century has made over his.

The representation of the world is, to be sure, significantly different as one goes from Kant to Bohr. At the present level of strategy arguments, however, there are two similarities that play a crucial role in the way that scientific knowledge is interpreted. On both points each man differed from the currently prevailing philosophies of science. The first point was the primacy accorded conceptual analysis over logical deduction. We represent the world as a collection of interrelated objects with characteristic properties and activities. The properties accepted as basic and the way in which the conceptual ordering of these properties is understood are subject to alteration, as is shown by the transition from the Aristotelian to the Galilean to the Cartesian conceptualization of primary and secondary qualities. The supervening principles in terms of which the universe is understood as an ordered whole obviously vary as one goes from a theological to a scientific perspective or from classical to quantum physics.

Such differences concern content. The epistemologically significant point is that the concepts through which we represent objects, properties, and activities are embedded in conceptual frameworks whose essential features derive from ordinary language usage. The presuppositions of this framework ground the type of conceptual entailments allowed. Within this loose enveloping overall framework, there are complex networks of conceptual entailments. One working *within* a particular conceptual framework may not need to undertake the difficult task of analyzing foundational presuppositions. One attempting to explain how a conceptual framework may be altered or how to go from one conceptual framework to another must make some kind of presuppositional analysis, or an analysis of hermeneutical principles.

The Wolffians in Kant's time and the logical positivists in Bohr's time insisted that general logic is the apt tool for an elucidation of scientific explanation. Though the logics they relied on differed, both were formal in the sense that the validity of the inferential rules is considered to be independent of the content to which these rules are applied. Both Kant and Bohr emphatically rejected this as a basis for clarifying the nature of scientific reasoning. Kant could not use any formal logic to elucidate the issue of whether the concept 'body' contains the concepts 'extension' and 'weight'. Bohr could not use formal logic to explain whether the concept 'position of a body' accords with the possible experimental conditions requisite to make statements about position meaningful. Kant developed a doctrine of synthetic a priori propositions to explain foundational features necessarily present in any conceptual framework in which one speaks about objects. Bohr adumbrated presuppositions which must obtain as a condition of the possibility of unambiguous communication concerning physical objects. Both were involved in the task of explaining foundational presuppositions grounding the conceptual entailments we routinely make. Both utilized transcendental arguments proceeding from the actual to the conditions of its possibility.

Kant's doctrine of ultimate corpuscles and force fields differs from Bohr's doctrine of atomic structure to such an extent that there is little to be gained from a comparison of the contents. Here again, however, they exhibited a methodological similarity which set each man apart from the prevailing philosophies of science. By differing routes, each reached the conclusion that there is no conceptually consistent way of representing macroscopic bodies as assemblages of particles having the same basic properties as these macroscopic bodies. Yet, each insisted, the only epistemologically justifiable basis for discussing the properties of ultimate particles is our common conceptualization of the properties of macroscopic bodies. Even when new properties are postulated for the ultimate constituents, something each man did, the meaning of the terms referring to these properties cannot be explained as a relation between the terms introduced and the property referred to. Rather, the new meaning must be parasitic upon the meanings of terms already in use.

This is the type of nonvicious circularity often treated in terms of the hermeneutic principle in the Continental tradition, or through Quine's ideas on radical translation and analytic principles in the Anglo-American tradition. Kant's way of resolving the particular type of circularity considered above was to fashion a doctrine of *hypothetical* atoms. He was not clear on this when he wrote the *Monadologia*; he was when he wrote the *Metaphysical Foundations*, some thirty years later. To call atoms hypothetical does not mean that they should be classified as phenomenal in contrast to tables and chairs. The phenomenon-noumenon distinction is not operative within physics. Atoms were presented as hypothetical in the sense that they were not considered empirically real bodies. Rather, they were hypothetical entities which, properly interpreted, give Newtonian physics an intelligible and epistemologically consistent foundation.

When Bohr spoke of atoms or electrons, he was speaking of real, not hypothetical, entities. The issue that came to perplex him was the ground of meaning of the statements attributing properties to such entities. Some such statements are relatively unproblematic, such as statements attributing to the electron a definite rest mass or charge. The problematic issues usually concern the attribution of classical property concepts in the context of a conceptual framework, quantum theory, whose foundations are incompatible with the presuppositions of classical physics. Bohr's solution to this circularity problem hinged on the idea of preserving classical meanings while restricting the usage of classical terms in quantum contexts.

Finally a point that still occasions much misunderstanding. Both men were empirical realists. Neither was willing to admit any source of information concerning physical reality other than perception and the systematic extension of knowledge based on perception and language. Both thought the goal of physical science is to come to know physical reality as profoundly as humanly possible. Yet, on the basis of epistemological analysis, each had become critically aware of the limitations of doctrinaire realism.

In Kant's case, this awareness found expression in the phenomenon-noumenon distinction. What is not always properly appreciated is that this distinction is epistemologically significant only in a broadly theological context. *Noumena* are things as known by God, or by beings with intellectual intuition, if such there be. *Phenomena* are objects of knowledge constructed by the human mind from a sensory input and various forms of cognition. Since humans lack intellectual intuition, this is the only knowledge of objects humanly available. Within the context of an analysis of properly human knowledge, this distinction does one imply either idealism or any renunciation of the attempt to come to know physical reality as thoroughly as possible. It simply expresses a recognition of the fact that human knowing is a constructive process.

Bohr occasionally expressed some interest in the epistemological problems involved in discourse about God. This was usually in the context of countering objections to complementarity or indeterminacy on the grounds that since God knows things as they are, for him there is no indeterminacy principle. In his own epistemological analysis, however, no theological perspective ever played a role. Accordingly, he neither had nor needed any distinction between noumena and phenomena, for this is not a distinction made within physics. The problem of realism Bohr faced had taken a linguistic turn. 'Real' is a term whose usage is learned in an ordinary language context: Horses are real; centaurs are not. Experiences are real; illusions are not. The extension of this term from the conceptual framework of ordinary language to that of quantum theory must be guided by the same type of correspondence principle analysis as with the extension of other ordinary language terms. This empirical realism can be interpreted as a disguised idealism only when contrasted with a more dogmatic realism based on something other than a critical analysis of the advance of scientific knowledge, based, for example, on Thomas Aquinas's theory of creation or Lenin's opposition to empiriocriticism.

On a more personal note, both men concluded their public careers with a plaintive expression of awareness that even the best professional philosophers had failed to understand the real significance of the epistemology each developed. Kant's final public statement on the critical philosophy was contained in his open letter to Fichte:

> For the pure theory of science [Fichte's Wissenschafts-lehre] is nothing more or less than mere logic, and the principles of logic cannot lead to any material knowledge. Since logic, that is to say, *pure logic*, abstracts from the content of knowledge, the attempt to cull a real object out of logic is a vain effort and therefore a thing that no one has ever done.[29]

The prevailing philosophers of science Bohr encountered developed the idea that pure logic supplies normative forms for the axiomatic reconstruction of scientific theories. The contents of particular theories can best be interpreted by relating them to such idealized reconstructions. Few philosophers of whatever persuasion seemed willing to learn from Bohr techniques, not based on the primacy of logic, of interpreting the nature of scientific explanation. Bohr's final appraisal of this rejection came in an interview with Thomas Kuhn the day before Bohr's death:

> I felt . . . that philosophers were very odd people who really were lost, because they have not that instinct that it is important to learn something and that we must be prepared to learn something of very great importance. . . . There are all kinds of people, but I think it would be reasonable to say that no man who is called a philosopher really understands what one means by the complementary description. . . . They did not see that it was an objective description, and that it was the only possible objective description.[30]

The correspondence between Bohr's thought and Kant's was not a superficial resemblance. It was present at the deepest level of analysis each had achieved. Yet, Bohr never averted to this correspondence, nor referred to Kant's writings in support of his own doctrines. Why? I can offer two answers to this question. One is relatively superficial; the other is more philosophically interesting.

The relatively superficial answer is that Bohr did not seem to know very much about Kant's philosophy. Bohr knew the textbook Kantianism based on a simplification of Kant's arguments for the rejection of metaphysics, but not Kant's positive treatment of scientific knowledge. That is, he knew only the type of Kantianism which presents the axioms of Euclidean geometry and Newtonian mechanics as examples of principles whose truths are determined by a priori reasoning. This was the type of view that Bohr opposed during the conflict over causality in the early twenties. I know of no evidence indicating that he studied Kant after that period.

This reason I find relatively superficial. Bohr must have known at least enough Kantianism to realize that his later use of 'phenomenon' was quite Kantian. More pertinently, he developed his position through almost interminable discussions. Some of those he talked to, such as Rosenfeld, von Weizsacker, Pauli, Heisenberg at least in his later years, and others, had more familiarity with the philosophy of Kant. The resemblance must have come up in discussion.[31] Bohr could have explored the correspondence between his views and Kant's, had he thought it worthwhile to do so. He did not.

The more philosophically interesting reason is properly appreciated when one notes that Bohr did not refer to *any* philosophers when presenting his own philosophical position. Bohr's way of developing philosophy was much like Wittgenstein's. He went over and over the same basic points concerning the meaning and proper usage of concepts until he finally reached a position that he thought both correct and coherent. The result was, in his view, the only possible position on the interpretation of quantum theory. To cite authorities in support of this position or any of its parts would only confuse the issue. Bohr's purpose was to get others to understand that his doctrine of complementarity was the only interpretation that showed quantum theory as a rational generalization of classical physics. The only way Bohr knew of for achieving this was to try to get others to think as he thought. Though he sometimes succeeded with physicists, he never succeeded in bringing any philosophers around to his way of thinking. He never attempted to accommodate his perspective to theirs. In spite of the widespread acceptance of the Copenhagen interpretation, Bohr's distinctive epistemology was rarely understood and even more rarely accepted.

Einstein's Views on Scientific Knowledge

The context of scientific discovery differs from the context of scientific justification. These differing contexts suggest differing criteria of scientific rationality. This difference was rather strikingly illustrated by Heisenberg. Reflecting on the reasons why he, rather than Pauli, had succeeded in making the decisive breakthrough in quantum mechanics, he concluded that Pauli's difficulty was that he was too concerned with consistency.[32] A theorist confronted with an intractable problem will often entertain hypotheses incompatible with the currently accepted principles of physics. His goal is to discover what obtains in reality, not what accepted theories predict should obtain. The achievement of an overall consistency may be relegated to a later cleanup operation.

Einstein was a discoverer, certainly one of the greatest theoretical discoverers in human history. He tended to view the goal of scientific explanation as identical with the thrust of discovery: to come to know physical reality as it exists objectively. In the context of Einstein's epistemology, 'objectively' means independent of any human observer. Bohr, from the beginning to the end of his career, was acutely concerned with achieving a consistency between the methods through which scientific information is obtained and the way in which scientific knowledge is interpreted. Einstein, the discoverer, preferred to tolerate short-range inconsistencies rather than abandon or compromise the long-range goal of scientific inquiry, the knowledge of physical reality as it exists objectively.

This insistence accounts, I believe, for the long-range shift in Einstein's interpretative perspective. In the early days of his career, he was influenced by Mach, Hume, and a tradition of empiricist epistemology. This tradition's stress on the primacy of sensation seemed to support the young Einstein by supplying a rationale for reinterpreting simultaneity in terms of what can be observed and measured. Mach's critical undermining of the foundations of Newtonian physics helped to stimulate and justify Einstein's attempt to go beyond the limits set by Newtonian physics. The iconoclastic style of Hume and Mach fitted Einstein's image of himself as a rebel, a Bohemian nonconformist in dress and life-style, a socialist in a capitalist world, a scientific revolutionary.

Gradually, as noted in chapter 9, the discrepancy between the epistemology Einstein professed and the science he practiced forced him to modify his position. Even then he never achieved the type of overall consistency Bohr did. Einstein's theory of scientific explanation was never adequate even to his own practice, much less to quantum physics. Such inadequacies notwithstanding, Einstein's views on scientific explanation in atomic physics are eminently worth considering, both as a complement and as a corrective to Bohr's views. Bohr preserves what he can explain in a rational and coherent fashion. Einstein, reflecting the poet's sentiment that a man's reach should exceed his grasp, claimed, "Feeling and longing are the motive force behind all human endeavor and human creation, in however exalted a guise the latter may present themselves to us."[33]

I have struggled for a long time with, I fear, only a limited success to think myself into Einstein's perspective. Other interpreters undergoing the same struggle have attributed Einstein's sustained rejection of Copenhagen physics to an innate conservatism, to the importance he attached to determinism, or even to his mystical bent. It seems to me that if Einstein's ontic realism is accepted as basic, then the other elements fall into place. His version of ontic realism was most succinctly expressed in the opening sentence of a 1931 commemorative paper on Maxwell: "The belief in an external world independent of the perceiving subject is the basis of all natural science."[34]

Einstein was an ontic realist. Statements, whether in ordinary language or scientific theories, are true when they report or describe how things actually are. Yet, naive realism, in whatever guise, ceased to represent a viable option for Einstein. This was not simply due to his reading of Hume, Mach, Russell, and other critics of metaphysics. It was also due to his own work in physics. For years he fought for the truth of the light-quantum hypothesis against formidable opposition. Yet, he knew that it could not be interpreted as a picture of what obtains in reality, for no such picture ever worked. He was so convinced of the truth of the theory of general relativity that when his assistant Ilse Rosenthal-Schneider asked Einstein, who had just received Eddington's cable confirming the crucial prediction, what he would have done had there been no confirmation, Einstein replied, "Then I would have been sorry for the dear Lord—the theory *is* correct."[35]

Einstein was sure this theory, his greatest creation, was true. Yet, the correspondence between this theory and reality could not be explained through any simple idea of terms denoting things in such a way that propositions using these terms picture the relations between things. The laws of general relativity are expressed in a form that is covariant with respect to general transformations. If the truth of the theory resides essentially in such laws, then the meaning of truth requires a more abstract formulation.

In chapter 9, the development of Einstein's interpretative perspective was presented historically. Now, as with Bohr, we wish to consider his final position more systematically. Einstein was influenced by Kant to the degree that he accepted from Kant a sharp distinction between percepts and concepts as two quite different sorts of things. He thought of himself as anti-Kantian, however, in rejecting any doctrine of a priori principles.[36] Einstein's most comprehensive statement of his epistemological position was contained in an article written for the *Franklin Institute Journal* just after the debate on the EPR paper.[37] In its systematization of knowledge, though not in its ontic commitments, it might be described as precognitive plagarism of Quine's *Word & Object*. Knowledge is seen as stratified in a series of levels with the given of sensible experience on the lowest level and the ordering principles of theories on the highest level, much like the gradual transition from the periphery to the core of Quine's great sphere of knowledge. The emphasis from beginning to end is on the status of concepts. Though, as Einstein notes, the man of science may not be trained to analyze concepts, he it is who knows where he shoe pinches: "In looking for a new foundation, he must try to make clear in his own mind just how far the concepts which he uses are justified and are necessities."[38]

The lowest level of knowledge is the level of sense experience plus representations. The nature of these representations and their relation to language was treated in a bit more detail in a radio address Einstein gave in 1941.[39] The origins of language occurred when people (or primates) learned to link acoustical signs to sense impressions. A further development, language properly speaking, is reached when further signs are introduced whose function is to link the established signs rather than to refer to sense impressions directly. Language so conceived is partially independent of the background of impressions. The first representation of the real external world is that of a world of bodily objects of varying kinds. The concepts used to refer to such objects are not identical either with individual sense impressions or with the totality of sense impressions. Concepts are free creations of the human mind. The meaning and justification of a concept, however, depend on the cluster of sense impressions associated with it.

Einstein did not study linguistics or make any systematic study of language. His views on the origin and structuring of language are similar to, and probably dependent upon, the account of language acquisition which Bertrand Russell described as proper to everyone except Patagonians, Thomas Carlyle, and Lord Macaulay.[40] Where Bohr insists on ordinary language usage geared to the lived world as the ground of meaning, Einstein, like Russell, interprets the advancement of knowledge as an attempt to get beyond the primitive commonsense view of reality reflected in ordinary language. His concern is not with meaning, but with the fact that this descriptive metaphysics is utterly lacking in logical unity.

The second stage, the first significant step beyond the formation of bodily concepts of varying kinds, comes in attributing to the concept of a bodily object a significance which is to a high degree independent of the sense impressions which originally gave rise to it. This, Einstein claims, is what we mean when we attribute to a bodily object a 'real existence'. Such a position, as Einstein points out, is explicitly Kantian in one respect. The superposition of a conceptual ordering on sense impressions makes possible our representation of a real external world:

> One may say 'the eternal mystery of the world is its comprehensibility.' It is one of the great realizations of Immanuel Kant that the postulation of a real external world would be senseless without this comprehensibility.[41]

His position is, however, very anti-Kantian concerning the role of a priori principles, something Einstein takes as basic to Kant's doctrine: "In my opinion, nothing can be said a priori concerning the manner in which the concepts are to be formed and connected, and how we are to coordinate them to sense experience."[42] This, as he explicitly notes, makes the formation of concepts a mystery science cannot treat, and the comprehensibility of the world of sense experiences a miracle. After surveying his more systematic epistemology, we will return to his method of accommodating such mysteries and miracles.

Further stages in the development of knowledge are systematized in terms of the stratification of concepts. Primary concepts are directly and

intuitively connected with typical complexes of sense experiences. All other
notions have meanings only insofar as they are connected by theorems with
primary concepts. These higher levels achieve an increasingly greater degree of logical unity:

> Thus, the story goes on until we have arrived at a system
> of the greatest conceivable unity, and of the greatest poverty
> of concepts of the logical foundations, which is still compatible
> with the observations made by our senses. We do not know
> whether or not this ambition will ever result in a definitive sys-
> tem. If one is asked for his opinion, he is inclined to answer
> no. While wrestling with the problem, however, one will never
> give up the hope that this greatest of all aims can really be
> attained to a very high degree.[43]

One point Einstein repeatedly emphasizes is that concepts, at every
level in the process of stratification, are free constructions of the human
mind. They cannot be explained by any process of induction or abstraction
from human experience. He expressed the logical independence of concepts
and sense experiences through a metaphorical language that was simple, but
to the point: "The relation is not analogous to that of soup to beef but rather
of check number to overcoat."[44]

After sketching this idea of human knowledge as stratified in levels
where successive sematic impoverishment leads to gradually increasing
logical unity, Einstein interpreted the advancement of physical knowledge
as the gradual, though yet only partial, realization of this ever-increasing
logical unification through the successive introduction of freely created
concepts. Einstein was no more a historian of science than Bohr. In both
cases, the historical summaries are of interest primarily as a revelation of the
way in which each man understood how the different parts of physics form a
coherent whole.[45]

Einstein's interpretations of the history of science invariably take the
work of the late Renaissance as a starting point: "Physics really begins with
the invention of mass, force, and an inertial system. These concepts are all
free creations."[46] With these concepts it was possible to fashion a new view
of the universe, one which departed radically from the commonsense view:
"Human thought creates an ever-changing picture of the universe. Galileo's
contribution was to destroy the intuitive view and replace it by a new one."[47]
This new view culminated in the mechanistic universe. In discussing this,
Einstein regularly attributed to Newton a view which corresponds more
closely to that of Euler and the French Newtonians than to Newton himself.
For our present purposes such historical discrepancies may be ignored.
'Newton' and 'mechanism' can be considered code names for the first

stratification of knowledge beyond the descriptive metaphysics implicit in our ordinary language.

The mechanistic view of reality presupposes a concept of objective time and also presupposes that Euclidean geometry supplies a valid basis for the description of space.[48] Newton's greatest contributions to science were his construction of the concept of mass and his realization, more acute than any of his followers, of the limitations of his system. The completed system can be understood in terms of four basic concepts and the laws that give these new or refurbished concepts a precise expression. These are: material point, the law of inertia, the law of motion, and the law of force. Logically considered, this new conceptual framework leads to atomism, where atoms are conceived of as the ultimate constituents obeying these mechanical laws.

Atoms, however, were unknown. Mechanism accordingly had to be redeveloped as a phenomenological physics: "It is characteristic of this kind of physics that it makes as much use as possible of concepts which are close to experience but, for this reason, has to give up, to a large extent, unity in the foundations."[49] In phenomenological physics, heat, light, and electricity are treated as separate branches of science, each having its own foundational concepts. In each case the key mathematical tool is the differential equation, which expresses a continual causal interaction. Such phenomenological physics supports the type of inductivism developed by Mill and Mach, an understanding of science based more on ordinary language concepts than on the new concepts which should play a foundational role.

Kinetic theory can be interpreted as the proper completion of classical mechanism considered as a foundational science. Instead of a phenomenological physics, one assumes ultimate constituents which are understood in terms of the foundational concepts of mechanics. These, in turn, supply a unified basis for the proper understanding of heat and thermodynamics. The classical atomism of kinetic theory reaches its culmination in Einstein's explanation of Brownian motion.

Classical mechanics postulated, but did not explain, forces. In lieu of a proper explanation it relied on the notion of action at a distance. The concept of a field, as developed by Faraday and Maxwell, represents a profound transformation of classical physics. Here one has a simpler concept supporting a more elaborate mathematics. This concept and the associated mathematics supply the basis for a higher-level stratification. On this level 'field' is accepted as expressing the most basic reality, while Maxwell's equations are interpreted as describing the structure of this field.[50] As Einstein saw it, something of the same developmental sequence obtained here as on the lower, mechanical, level of stratification. Maxwell attempted to explain 'field' mechanically, i.e., by reducing the new concepts to the already established stratum. When this appeared implausible, Hertz took the next logical step and explained fields as part of a phenomenological physics.

The special and general theories of relativity may be seen as the partial completion of this stratum. In these theories the concept 'field' is taken as foundational, rather than as either a derivative or a phenomenological notion. The transmission of action is then given its proper space-time description. In general relativity, this notion of a field is extended from electromagnetic to gravitational interactions.

In spite of these advances, the present state of field theory cannot be considered either definitive or final. Instead of one foundational concept, there are two, 'field' and 'particle'. Maxwell's equations describe the structure of fields in empty space, not their sources. Lorentz reformulated classical electromagnetic theory to include electrons. In his theory, however, there is an irreducible dualism between fields and particles. Even the general theory of relativity did not remove this dualism. Yet, if the goal of human striving is a logical unification based on deductive ordering from as simple a basis as possible, this dualism must go:

> What appears certain to me, however, is that in the foundations of any consistent field theory, the particle concept must not appear in addition to the field concept. The whole theory must be based solely on partial differential equations and their singularity-free solutions.[51]

This was the goal Einstein never achieved and never relinquished.

This general approach can also supply an interpretative framework for quantum theory. After 1935, Einstein accepted both quantum theory and its statistical interpretation as self-consistent and empirically adequate. What he steadfastly rejected was the idea that quantum mechanics, together with the Copenhagen interpretation, can be considered the final or definitive theory in this domain:

> Probably never before has a theory been evolved which has given a key to the interpretation and calculation of such a heterogeneous group of phenomena of experience as has quantum theory. In spite of this, however, I believe that the theory is apt to beguile us into error, in our search for a uniform basis for physics, because, in my belief, it is an *incomplete* representation of real things, although it is the only one which can be built out of the fundamental concepts of force and material points (quantum corrections to classical mechanics).[52]

One point, sometimes confused in the literature, requires clarification. It was Einstein, not Bohr, who championed the ensemble interpretation of quantum mechanics. Bohr insisted that quantum mechanics does supply a basis for describing individual atomic interactions. He did so on the episte-mological grounds that one could neither rationally require nor unambig-uously interpret more of a description than quantum mechanics already supplies. This clearly did not satisfy Einstein:

> It seems to be clear, therefore, that Born's statistical inter-pretation of quantum theory is the only possible one. The ψ-function does not in any way describe a state which could be that of a single system; it relates rather to many systems, to an 'ensemble of systems' in the sense of statistical mechanics. If, except for certain special cases, the ψ-function furnishes only *statistical* data concerning measurable magnitudes, the reason lies not only in the fact that the *operation of measuring* intro-duces unknown elements, which can be grasped only statisti-cally, but because of the very fact that the ψ-function does not in any sense, describe the state of *one* system.[53]

Quantum mechanics, so interpreted, fits into the dialectical develop-ment Einstein thought proper to physics. The initial attempts to give the new quantum theory a realistic interpretation, through de Broglie's various wave-particle models and Schrödinger's successive interpretations of the ψ-function, all failed. As is their wont, physicists rely on phenomenological physics as a fallback position. By a 'phenomenological interpretation' Ein-stein does not mean reducing theoretical concepts to observational terms. He means explaining the concepts proper to a new stratum by reducing it to a previously established stratum. Quantum mechanics can be interpreted as a statistical treatment of ensembles, a treatment in which individual interac-tions are not properly described. The complementarity interpretation of this physics may be considered as rules for the proper extension and restriction of classical concepts to this new strata. Quantum theory so interpreted constitutes phenomenological physics. Einstein is not arguing that quantum theory goes too far. His contention is that, in spite of its success, it does not go far enough. What he wanted was, not a return to classical physics, but a more radical break with it. He anticipated a distinctively new stratum having its own, rather than borrowed and modified, foundational concepts. These should supply a basis for a deductive unification, one that should replace the phenomenological physics that is present quantum theory.

Einstein seems to have thought of the situations as analogous to that which obtained in relativity earlier. To most physicists the remarkable success of Newtonian physics was the surest warranty of that theory's validity. Yet, the development of relativity showed that the foundational concepts of classical mechanics, the concepts of space, time, mass, and motion, were all incorrect. The new theory, both special and general relativ-ity, reproduced the numerical results of the old theory within the limits where the old theory gave correct results. Yet, the new theory rested on a foundation of radically different concepts. Similarly, the theory that should eventually replace quantum theory should reproduce all the correct results obtained by quantum theory. Yet, the new to-be-developed theory should be grounded in a foundation of concepts radically different from those basic to quantum theory. It should also employ a deductive methodology quite different from the phenomenological method proper to the Copenhagen interpretation of quantum mechanics. For these reasons one cannot hope to develop the new theory by extending the concepts and methods of the present quantum theory. Rather, one must abandon these concepts and methods and start afresh. The goal to be sought is a totally new theory, not an extension or modification of the present theory.[54]

If such a unified basis could be found, then physics would achieve the ideal toward which it is innately directed. The article we have been summarizing concludes with one of the clearest statements Einstein ever made on the ideal of a unified physical science. It is worth quoting in full:

> Physics constitutes a logical system of thought which is in a state of evolution, whose basis cannot be distilled, as it were, from experience by an inductive method, but can only be arrived at by free invention. The justification (truth content) of the system rests in the verification of the derived propositions by sense experiences, whereby the relations of the latter to the former can only be comprehended intuitively. Evolution is proceeding in the direction of increasing simplicity of the logical basis. In order further to approach this goal, we must resign ourselves to the fact that the logical basis departs more and more from the facts of experience, and that the path of our thought from the fundamental basis to those derived propositions, which correlate with sense experiences, becomes continually harder and longer. . . .
>
> I try to demonstrate, furthermore, why in my opinion, quantum theory does not seem capable to furnish an adequate foundation for physics: one becomes involved in contradictions if one tries to consider the theoretical quantum description as a *complete* description of the individual physical system or event.
>
> On the other hand, the field theory is as yet unable to explain the molecular structure of matter and of quantum phenomena. It is shown, however, that the conviction of the inability of field theory to solve these problems by its methods rests upon prejudice.[55]

After this paper, Einstein continued his efforts to make 'field' the foundational concept for physics. Even when the prospect looked hopeless, he would not abandon it. He once told Otto Stern, "I have thought a hundred times as much about the quantum problems as I have about general relativity theory."[56] To accord quantum theory and the complementarity interpretation a foundational role in scientific explanation deeply offended Einstein's aesthetic sensibilities. He simply did not want to be part of a universe that could be explained by Bohr's type of physics.

> Some physicists, among them myself, cannot believe that we must abandon, actually and forever, the idea of direct representation of physical reality in space and time; or that we must accept the view that events in nature are analogous to a game of chance. It is open to every man to choose the direction of his striving; and also every man may draw comfort from Lessing's fine saying, that the search for truth is more precious than its possession.[57]

In this summary, I have attempted to present Einstein's interpretative perspective without introducing any adverse criticism. Yet, his epistemology presents one difficulty so basic that it cannot be ignored without leaving his systematization radically incomplete. Einstein accepted from Kant the idea that sense experiences and concepts are radically different in their nature and function. Sense experiences are determined by external bodies causally interacting with our sensory apparatus. Concepts are free creations of the human mind, introduced to order the blooming buzzing confusion of the world of sense experiences. The ultimate order of our conceptual systems is determined, not by the given of experience, but by rational ideals of conceptual simplicity, aesthetic satisfaction, and deductive unification. What ground is there for believing that this conceptual order or its constitutive concepts corresponds to something objectively real?

To appreciate the significance of this, it may be of some initial help to compare Einstein's position with other treatments of the same basic difficulty. Kant countered it by two strategic moves that Einstein would reject. The first is the acceptance of innate categories which play a constitutive role in determining an object of knowledge. Einstein insisted that all concepts are free creations of the human mind. Those that should ultimately prove constitutive, far from being innate, have not even been created yet. The second point was one in which Kant differed from his young followers as well as from Einstein. It was the primacy accorded pragmatics. One accepts the science that works on the ground of its success. Then one searches for the conditions of its possibility. One does not use rationalistic ideals to predetermine what science must be. Einstein's position was based on a rejection, rather than an acceptance, of the most successful physics available. He rejected it in the sense of refusing to accept it as something whose ultimacy and necessity require explanation.

Einstein would not accept Kant's strategy. Neither would he accept the views held, especially by earlier epistemological empiricists, that concepts can be explained by some process of induction or abstraction from sense experiences. Nor would he accept the view of Bohr, Wittgenstein, and many philosophers loosely dubbed 'analysts' that the meaning of basic concepts is determined by the way terms expressing these concepts function in language. In Einstein's view, foundational concepts are given meaning by the creative minds that invent them. Later linguistic usage may come to serve as a repository of shared meanings, but never as a creator of new concepts. It seems that Einstein never seriously considered the older rationalistic idea that sensations are some sort of confused concepts.

Concepts are free creations imposed to order sense experiences. They are not caused by or abstracted from reality as experienced, though sense experiences supply the ultimate grounds for testing conclusions deduced from statements involving basic concepts. In spite of this radical dichotomy between sense experiences, which are causally determined by external physical objects, and concepts, which are free creations of the human mind, Einstein also insisted that physical reality, as it exists independently of any observer, is only adequately represented by concepts and conceptual structures far removed from the given of experience. Thus, in his "Notes on the Origin of the General Theory of Relativity," Einstein indicates that he began with a simple physical interpretation of coordinates as signifying direct results of measurements. The use of nonlinear transformations, however, prohibited any consistent reliance on such a direct physical interpretation. Physical content had to be found on a more abstract level: "A physical significance attaches not to the differentials of the coordinates but only to the Riemannian metric corresponding to them. A workable basis had now been found for the general theory of relativity."[58] This was the foremost of many scientific experiences in Einstein's career that seemed to support a general conclusion, "Our experience hitherto justifies us in believing that nature is the realization of the simplest conceivable mathematical ideals."[59]

Concepts and conceptual structures, though free creations of the human mind, correspond to physical reality as it exists objectively. How can this be explained? Einstein's answer to this question is quite clear. This connection cannot be explained; it must simply be accepted. Thus, in commenting on the statement that "the eternal mystery of the world is its comprehensibility," Einstein declared: "It is in this sense that the world of sense experiences is comprehensible. The fact that it is comprehensible is a miracle."[60] When an epistemological analysis leads to a foundation of positions accepted on faith, to mysteries, and to miracles, there seems to be only one way out. One must seek a theological escape route.

Einstein's theological interjections are usually passed over in a some-what embarrassed silence. Yet, these reflect an integral part of his overall perspective. One can no more understand how Einstein's interpretative framework hangs together while excluding his theology than one can understand Bohr's while neglecting his position on language or on the subject-object distinction. Einstein's theological perspective was unique. In an attempt to understand its significance we will attempt to trace the way it developed and indicate the role it played in his theory of scientific explanation.

Einstein was raised in a secularized Jewish family, one that rejected traditional religious beliefs as superstitious yet respected the ethical traditions of Judaism. Einstein's earliest religious training came from his first five years in a Catholic grammar school, where he took the same catechism lessons as the Catholic students. When he attended the Luitpold Gymnasium, he encountered rigid rules and a restrictive regime. This served as a catalyst, transforming the young Einstein into a self-conscious rebel opposed to organized authority. While there he also received instructions in the Jewish religion and passed through a brief period of what he later described as deep religiosity. This juvenile fervor ended rather abruptly at the age of twelve:

> Through the reading of popular scientific books I soon reached the conviction that much in the stories of the Bible could not be true. The consequence was a positively fanatic [orgy of] freethinking coupled with the impression that youth is intentionally being deceived by the state through lies; it was a crushing impression.[61]

Science, rather than religion, emerged as the path to truth and emotional security. The five-year-old Einstein had felt a sense of wonder at seeing his father's pocket compass, an experience he repeatedly referred to in his later years. This sense of wonder about the mysteries of nature became emotionally linked to Einstein's wonder at his own power of unraveling mysteries when he found, at the age of thirteen, that he could easily work through all the problems in Spieker's *Lehrbuch der ebenen Geometrie*.[62] Einstein seems to have transferred to science, and to his own ability to unravel the mysteries of science, the feelings that he had attached to religion. Contemplation of the mysteries hidden in nature seemed to open a path to a different sort of paradise: "The road to this paradise was not as comfortable and alluring as the road to the religious paradise; but it has proved itself as trustworthy, and I have never regretted having chosen it."[63]

This may sound like youthful enthusiasm; it was written when Einstein was sixty-seven. Einstein's subsequent position seems to have been a consistent outgrowth of this early development. He emphatically rejected the religious traditions of the West and any notion of an anthropomorphic God, one who is concerned with human affairs. Yet, he insisted throughout his career that a cosmic religious feeling, supporting a belief in the intrinsic rationality of the universe, is the strongest motive for scientific research. In 1929, he replied to the questions of a Japanese scholar, "Certain it is that a conviction, akin to religious feeling, of the rationality or intelligibility of the universe lies behind all scientific work of a higher order."[64] In an essay written a year later for the *New York Times Magazine* he claimed:

> I maintain that the cosmic religious feeling is the strongest and noblest motive for scientific research. . . . What a deep conviction of the rationality of the universe and what a yearning to understand, were it but a feeble reflection of the mind revealed in this world, Kepler and Newton must have had to enable them to spend years of solitary labor in disentangling the principles of celestial mechanics.[65]

In a 1939 address to the Princeton Theological Seminary he claimed:

> But science can only be created by those who are thoroughly imbued with the aspiration toward truth and understanding. This source of feeling, however, springs from the sphere of religion. To this there also belongs the faith in the possibility that the regularities valid for the world of existence are rational, that is comprehensible to reason. I cannot conceive of a genuine scientist without this profound faith.[66]

Einstein repeatedly claimed that a profound religious feeling is at the base of all truly creative scientific work. He would have been hard pressed to prove this by induction from the example of his contemporaries, Planck excepted. The basis of this claim is obviously Einstein's strongly felt convictions. The relation of this theological perspective to the interpretation of science was most clearly spelled out in a letter the aged Einstein wrote in 1952 to his old friend Maurice Solovine, who had asked what Einstein meant by the comprehensibility of the world. Einstein answered with a distinction. One can order chaotic events in arbitrary or conventional ways. This is quite different from discovering an order already present. The success of science presupposes a high degree of order in the objective world. The internal rationality grounding this comprehensibility is the miracle which grows ever clearer with the development of our knowledge of nature: "It is here where one finds the weak point of the positivists and professional atheists, who feel happy because they are aware not only of having, with great success, stripped the world of gods, but also of having stripped off the miracles."[67]

The goal of human striving, Einstein insisted, was a unification of human knowledge based on deductive unfolding from the simplest and purest concepts the mind of man can create. When pressed on his religious commitments Einstein replied, "I believe in Spinoza's God who reveals himself in the orderly harmony of what exists, not in a God who concerns himself with fates and actions of human beings."[68] Spinoza's God is, as is well known, the center and substance of a pantheistic system. Unlike many, Einstein seems to have gone on to read book 2 of Spinoza's *Ethics*. This is concerned with the three levels of human knowledge. Knowledge of the first kind, opinion or imagination, is the knowledge one has by being acted upon. One can advance to the second level of scientific knowledge, which is more active than passive. Activity is shown primarily through deductive logical ordering. One moves from confused and inadequate ideas toward adequate ideas which are simpler. When ideas are truly adequate, the order and connection of ideas is the order and connection of things. The third level is an intuitive immediate knowledge of the infinite essence of God. This is a knowledge which man has in principle, since the human mind is part of God's infinite intellect, but which is confused with an admixture of imaginative and common notions.[69]

This is both the God and the ideal of human knowledge Einstein accepted. A God whose eternal essence is manifested by a deductive unfolding of the concept 'substance' is not a God who can leave anything to chance. Man, with his inadequate ideas and partial knowledge, may need to rely on statistical reasoning. But *Gott würfelt nicht*. He does not play dice. Our present ideas only represent a feeble approximation to the true rational order that is objectively present. Yet, this order is discoverable, for the God who is pure rationality immanent in the universe cannot misdirect the unfolding of reason. Over Einstein's desk in Princeton hung a framed version of his famous slogan, "Raffiniert ist der Herr Gott aber boshaft is er nicht." God is subtle, but he is not malicious.

An epistemological position that relies on a natural affinity between human rationality modeled on a deductive system and immanent cosmic rationality couched in pantheistic terms will inevitably be unacceptable to most for a variety of scientific, epistemological, and theological reasons. It must, however, be admitted that those of us who reject it do not have to live with the miracle of Einstein's creativity. He did. This miracle—he certainly thought of it in these terms—drove him into a position peculiarly reminiscent of Newton's feeling of a special affinity to God and of a willingness to invoke God to close the gaps in his system.

If this theological escape route is disallowed, then Einstein's epistemological position remains incoherent and inadequate to explain even his own achievements. Yet, it brings out dimensions lacking in Bohr's more coherent, but more limited, account. Einstein took theories as cognitive units seriously. Bohr's analysis of meaning focused on key ordinary language terms and the proper modes of restricting and extending them. Einstein interpreted general scientific laws as expressions of, or approximations to, relations obtaining in reality. Bohr repeatedly manifested a willingness to sacrifice such general principles as energy conservation. Einstein interpreted the history of science as a series of conceptual revolutions. Bohr interpreted the same breakthroughs as rational generalizations of earlier stages of development. Even those who accept the general lines of the Kant-Bohr interpretation of knowledge must eventually come to grips with the difficulties Einstein's position was developed to counter.

After this detailed survey of the points on which Bohr and Einstein disagreed, it might be helpful to conclude with a point of shared agreement. Both men realized that the issues dividing them concerned the foundational concepts of quantum theory and the proper goals of scientific explanation. They did not at all concern the mathematical formalism of quantum mechanics. For this reason both assumed as obvious the idea that a rigorous axiomatic reformulation of this formalism would contribute nothing whatsoever toward a resolution of the problematic they shared.

The burgeoning tradition loosely labeled 'quantum logic' presupposes, as a point that hardly requires discussion, that such a formal logical redevelopment is the only effective means of clarifying the foundations of quantum theory. This is obviously not a disagreement of technical details. It represents a radical divergence between the way scientists and philosophers of science interpret the nature of scientific explanation. Any attempt at resolving such a fundamental divergence requires a method quite different from the type of historical survey presented in this volume.

Notes

Frequently Cited Sources

We list here abbreviations for a few sources which either supply background material on the history of atomism or are frequently cited in the notes.

Abbreviation	Source
APHK	Bohr, Niels. *Atomic Physics and Human Knowledge*. New York: Wiley, 1958.
ATDN	Bohr, Niels. *Atomic Theory and the Description of Nature*. Cambridge: Cambridge University Press, 1934.
Besso Correspondence	Speziale, Pierre, ed. *Albert Einstein: Michele Besso. Correspondence, 1903–1955*. Paris: Hermann, 1972.
Bohr, *Works*	Rosenfeld, L., et al., eds. *Niels Bohr: Collected Works*. Amsterdam: North-Holland, 1972 and later.
Boorse and Motz	Boorse, Henry A., and Motz, Lloyd. *The World of the Atom*. New York: Basic Books, 1966.
Born Correspondence	Born, Max, ed. *The Born-Einstein Letters*. Trans. Irene Born. New York: Walker & Co., 1971.
Brush, *Kind of Motion*	Brush, Stephen G. *The Kind of Motion We Call Heat: A History of the Kinetic Theory of Gases in the 19th Century*. Amsterdam: North-Holland, 1976.
Brush, *Kinetic Theory*	Brush, Stephen G. *Kinetic Theory*, 1, *The Nature of Gases and Heat*. Oxford: Pergamon Press, 1965.
BSC	Bohr, Scientific Correspondence (see SHQP)
HSPS	*Historical Studies in the Physical Sciences*.
Hund	Hund, Friedrich. *The History of Quantum Theory*. Trans. Gordon Reece. New York: Barnes & Noble, 1974.
Ideas and Opinions	Einstein, Albert. *Ideas and Opinions by Albert Einstein*. Trans. Sonja Bargmann. New York: Crown Publishers, 1954.
Jammer, *History*	Jammer, Max. *The Conceptual Development of Quantum Mechanics*. New York: McGraw-Hill, 1966.
Jammer, *Philosophy*	Jammer, Max. *The Philosophy of Quantum Mechanics: The Interpretation of Quantum Mechanics in Historical Perspective*. New York: Wiley, 1974.
Pauli	Pauli, Wolfgang, ed. *Niels Bohr and the Development of Physics*. New York: Pergamon Press, 1955.
Pauli Briefwechsel	*Wolfgang Pauli: Wissenschaftlicher Briefwechsel mit Bohr, Einstein, Heisenberg U.A. Band I: 1919–1929*. Ed. A. Hermann, K. v. Meyenn, and V. F. Weisskopf. New York: Springer-Verlag, 1979.
Pauli, *Papers*	Kronig, R., and Weisskopf, V. F., eds. *Collected Scientific Papers by Wolfgang Pauli*. New York: Interscience, 1964.

PSA 1980	*PSA 1980: Proceedings of the 1980 Biennial Meeting of the Philosophy of Science Association.* Ed. P. Asquith and R. Giere. East Lansing, Mich.: Philosophy of Science Association, 1980.
Rozental	Rozental, S., ed. *Niels Bohr: His Life and Work as Seen by His Friends and Colleagues.* Amsterdam: North-Holland, 1967.
Schilpp	Schilpp, P. A., ed. *Albert Einstein: Philosopher-Scientist.* The Library of Living Philosophers, New York: Tudor, 1949.
SHQP	*Sources for the History of Quantum Physics.* Interviews are cited by typescript folder number and page; correspondence by microfilm number. Thus, BSC 14 stands for Bohr, Scientific Correspondence, reel 14.
Solovine Correspondence	Einstein, Albert. *Lettres à Maurice Solovine.* Trans. M. Solovine. Paris: Gauthier-Villars, 1956.
Sommerfeld Correspondence	Albert Einstein/Arnold Sommerfeld, *Briefwechsel.* Basel: Schwabe & Co., 1968.
van der Waerden	van der Waerden, B.L. *Sources of Quantum Mechanics.* New York: Dover, 1967.
Whittaker, *History*	Whittaker, Sir Edmund. *A History of the Theories of Aether and Electricity.* New York: Harper, 1960.

Introduction

1. The best overall survey of this development is Frederick Suppe's "The Search for Philosophical Understanding of Scientific Theories," in F. Suppe, ed., *The Structure of Scientific Theories* (Urbana: University of Illinois Press, 1974), pp. 3–241.

2. A more detailed evaluation may be found in my "Scientific Realism: The New Debates," *Phil. Sc.* 46 (1979): 501–32.

3. This stage of scientific development is discussed by Dudley Shapere in "Scientific Theories and Their Domains" in Suppe, *Structure of Scientific Theories*, pp. 518–99.

4. These are fragments 591 and 593 in G. S. Kirk and J. E. Raven, *The Presocratic Philosophers* (Cambridge University Press, 1962), pp. 423–44.

5. I am indebted to George Farre for sending me a copy of his preprint "The Role of Functions in Natural and in Mathematical Philosophy," which traces the changes in the concept of a function.

6. Alfred North Whitehead, *Process and Reality: An Essay in Cosmology* (New York: Macmillan, 1929), p. 4.

7. I have tried to indicate the relation between a rational reconstruction of quantum theory and Bohr's type of informal semantic analysis in my "Jeffrey Bub, *The Interpretation of Quantum Mechanics*: A Critical Evaluation," *Philosophia* 10 (1981): 89–124.

8. Chapter 2 on Kant was completed before the appearance of W. H. Werkmeister's *Kant: The Architectonic and Development of His Philosophy*, which brings out the relevance of Kant to issues in contemporary philosophy of science.

Chapter One

1. G. S. Kirk and J. E. Raven, *The Presocratic Philosophers* (Cambridge: Cambridge University Press, 1962), p. 405.

2. Ibid., fragment 589, p. 422.

3. Lucretius, *On the Nature of the Universe*, trans. R. E. Latham (Baltimore: Penguin, 1951), p. 36.

4. Ibid., pp. 63–64.

5. Kirk and Raven, *Presocratic Philosophers*, fragment 593, p. 424.

6. The classical source for the early history of atomism is Kurd Lasswitz, *Geschichte der Atomistik: Vom Mittelalter Bis Newton*, 2 vols. (Hamburg: L. Voss, 1890). Some other general references utilized in the present survey are A. C. Crombie, *Medieval and Early Modern Science*, 2 vols., 2d rev. ed. (Garden City, N.Y.: Doubleday Anchor, 1959); E. J. Dijksterhuis, *The Mechanization of the World Picture*, trans. C. Kikshoorn (Oxford: Clarendon Press, 1961); W. B. Fleischmann, *Lucretius and English Literature: 1680–1740* (Paris: A. G. Nizet, 1964); Max Jammer, *Concepts of Force: A Study in the Foundations of Dynamics* (New York: Harper Torchbook, 1962); *Concepts of Mass in Classical and Modern Physics* (New York: Harper Torchbook, 1964); Robert Hugh Kargon, *Atomism in England from Hariot to Newton* (Oxford: Clarendon Press, 1966); Ernan McMullin, ed., *The Concept of Matter* Notre Dame, Ind.: University of Notre Dame Press, 1963); Andrew G. van Melsen, *From Atomos to Atom: The History of the Concept Atom*, trans. H. J. Koren (New York: Harper Torchbook, 1960); Olaf Pederson and Mogens Pihl, *Early Physics and Astronomy: A Historical Introduction* (London: MacDonald & Janes, 1974); and Lancelot Law Whyte, *Essay on Atomism: From Democritus to 1960* (Middletown, Conn.: Wesleyan University Press, 1961).

7. See Kargon, *Atomism in England*, chap. 7.

8. In this historical survey I am refraining from imposing any normative terminology. Such terms as 'quality', 'quantity', 'property', 'attribute', 'cause', 'effect', 'essence', and 'nature' will, as much as possible, simply reflect the usage proper to the sources cited.

9. See Ernan McMullin, "Matter as Principle," in McMullin, *Concept of Matter*, pp. 169–208; and Joan Kung, "Aristotle on Essence and Explanation," paper presented at the meeting of the American Philosophical Association, Pacific Division, in Berkeley, Calif. 27 March 1976.

10. Aquinas's most detailed treatment of this point is contained in his commentary *In Librum Boethii de Trinitate*, ed. Paul Wyser, O.P (Fribourg: Société Philosophique, 1948). Part of this commentary has been translated and republished as *The Division and Methods of the Sciences*, trans. A. Maurer, C.S.B., 2d rev. ed. (Toronto: Pontifical Institute of Mediaeval Studies, 1958).

11. St. Thomas Aquinas, *Summa Theologiae*, 1, q. 42, a. 1, ad 1; *Quaestio Disputata de Virtute in Commune*, a. 11, ad 10, in *Quaestiones Disputatae*, 8th rev. ed., vol. 2 (Turin: Marietti, 1949), p. 741.

12. St. Thomas Aquinas, *In Metaph.* X, lect. 2, ed. M. R. Cathala and R. M. Spiazzi (Turin: Marietti, 1950), n. 1938.

13. St. Thomas Aquinas, 3, *Summa Theologiae*, q. 7, a. 9.

14. The discussion of these two issues is based chiefly on Dijksterhuis, *Mechanization*, pp. 186–209.

15. See Marshall Clagett, *The Science of Mechanics in the Middle Ages* (Madison: University of Wisconsin Press, 1959); and Pederson and Pihl, *Early Physics*, chaps. 15–17.

16. From Clagett, *Science of Mechanics*, p. 292. He gives an analysis of this argument on pp. 294–97. Surprisingly, Swinesherd and his contemporaries did not apply the idea of uniform acceleration to falling bodies. This was first done by Domingo de Soto. On this point, see William A. Wallace, "The Enigma of Domingo De Soto: *Uniformiter difformis* and Falling Bodies in Late Medieval Physics," *Isis* 59 (1968): 348–401.

17. Marshall Clagett, *Nicole Oresme and the Medieval Geometry of Qualities and Motions* (Madison: University of Wisconsin Press, 1968), p. 165. Oresme's work was entitled *Tractatus de configurationibus qualitatum et motuum* (Treatise on the Uniformity and Difformity of Intensities).

18. The development of a new conceptual language which made discussions of measurement possible and popular is treated by John Murdoch in "Philosophy and the Enterprise of Science in the Later Middle Ages," in Y. Elkana, ed., *The Interaction between Science and Philosophy* (Atlantic Highlands, N.J.: Humanities Press, 1974), pp. 51–74.

19. "On the Heavens," book 3, chap. 2 (302a16–19).

20. "On Generation and Corruption," book 1, chap. 10 (327b–328a).

21. See van Melsen, *From Atomos to Atom*, pp. 58–77; and Dijksterhuis, *Mechanization*, pp. 205–9.

22. This school is treated by Lewis White Beck in his *Early German Philosophy: Kant and His Predecessors* (Cambridge, Mass.: Harvard University Press, 1969), pp. 172–79.

23. See W. R. Ong, S.J., *Ramus, Method, and the Decay of Dialogue* (Cambridge, Mass.: Harvard University Press, 1958).

24. Most accounts of Bacon's philosophy confused his position on atomism because they did not take into account the chronological order in which his works were written rather than published. This confusion is straightened out in Kargon, *Atomism in England*, chap. 5.

25. Maurice Clavelin, *The Natural Philosophy of Galileo: Essay on the Origins and Formation of Classical Mechanics*, trans. A. J. Pomerans (Cambridge, Mass.: MIT Press, 1974), p. 458.

26. The influence of the writings of Descartes and Boyle on the young Newton is discussed in A. Rupert Hall and Marie Boas Hall, *Unpublished Scientific Papers of Isaac Newton: A Selection from the Portsmouth Collection in the University Library, Cambridge* (Cambridge: Cambridge University Press, 1962), p. 187; and in John Herivel, *The Background to Newton's "Principia": A Study of Newton's Dynamical Researches in the Years 1664–84* (Oxford: Clarendon Press, 1965), pp. 42–53.

27. The version of Descartes's *Principes de la philosophie* used is that given in Charles Adam and Paul Tannery, *Oeuvres de Descartes*, part IX, sec. 2 (Paris: Vrin, 1964). The citation is from part IV, sec. 202 (p. 320).

28. Ibid., part II, secs. 16–20 (pp. 71–74). The argument from the unintelligibility of a real void to the incoherence of the atomistic doctrine goes back to Aristotle.

29. Ibid., part II, secs. 19–20 (pp. 73–74).

30. Ibid., part II, sec. 4 (p. 65).

31. Carl B. Boyer, *A History of Mathematics* (New York: Wiley, 1968), pp. 370–80.

32. Parts III and IV of the *Principes*, effectively Descartes's system of the world, contain geometrical arguments but never rely on numerical values.

33. Christian Huygens, *Treatise on Light*, trans. S. P. Thompson, in *Great Books of the Western World*, vol. 34 (Chicago: Encyclopaedia Britannica, 1952), pp. 551–659. The citation refers to chap. 1.

34. Marie Boas Hall, *Robert Boyle on Natural Philosophy* (Bloomington: Indiana University Press, 1965), collects Boyle's various accounts of the mechanical explanation of qualities on pp. 231–75. In the interpretation given here I have followed Kargon, *Atomism in England*, chap. 9, rather than M. Hall in presenting Boyle more as a Baconian experimentalist than a Cartesian theorist. The interpretation of Boyle's method presented here was also influenced by J. E. McGuire, "Boyle's Conception of Nature," *Journal of the History of Ideas* 33 (1972): 523–42; and Frederick J. O'Toole, "Qualities and Powers in the Corpuscular Philosophy of Robert Boyle," *Journal of the History of Philosophy* 12 (1974): 295–315.

35. Harry Frankfurt, "Descartes on the Creation of Eternal Truths," *Phil. Rev.* 86 (1977): 36–57, has argued convincingly that Descartes held an extreme form of voluntarism, that the truths we intuit as a priori, e.g., the laws of logic, are those Gods wills us to hold, but need not be true for God or radically different minds. This interpretation would make Descartes into a proto-Kantian, holding that correct deduction from clear and distinct ideas yields a science of phenomena, while the noumena are, conceivably, totally different. This would not change the present account, which is concerned with Descartes's public influence (or the phenomenal Descartes).

36. This aspect of Boyle's thought is developed in detail by McGuire, "Boyle's Conception of Nature," pp. 523–42.

37. On the development of Newton's atomism, in addition to the works cited, see I. B. Cohen, *Introduction to Newton's Principia* (Cambridge, Mass.: Harvard University Press, 1971); Alexandre Koyré, *Newtonian Studies* (Chicago: University of Chicago Press, 1965); A. R. Hall and M. B. Hall, "Newton and the Theory of Matter," in Robert Palter, ed., *The Annus Mirabilis of Sir Isaac Newton, 1666–1966* (Cambridge, Mass.: MIT Press, 1970), pp. 54–68; C. Truesdell, *Essays in the History of Mechanics* (New York: Springer-Verlag, 1968); and Richard Westfall, *Force in Newton's Physics*: The Science of Dynamics in the Seventeenth Century (London: Macdonald, 1971), and *The Construction of Modern Science*: Mechanisms and Mechanics (New York: Wiley, 1971).

Citations will generally be from the Motte-Cajori translation of the *Principia* (2 vols. [Berkeley: University of California Press, 1962]). This is a rather interpretative translation. Whenever precise wording is important, I will give my own, more literal translation from the critical text given in A. Koyré and I. Bernard Cohen, *Isaac Newton's Philosophiae Naturalis Principia Mathematica*, 2 vols. (Cambridge, Mass.: Harvard University Press, 1972). This will be cited as the *Variorum edition*.

38. Since I am confining myself to conceptual problems concerning atomism, I will not enter into the complex and controverted issues concerning Newton's doctrine of forces. An excellent treatment of this is given in Ernan McMullin, *Newton on Matter and Activity* (Notre Dame, Ind.: University of Notre Dame Press, 1978). His position differs from Westfall's.

39. This aspect of Newton's thought is given a detailed analysis in J. E. McGuire, "Atoms and the 'Analogy of Nature': Newton's Third Rule of Philosophizing," *Stud. Hist. Phil. Sc.* 1 (1970): 3–58.

40. McMullin, *Newton on Matter*.

41. Westfall, *Force*.

42. Herivel, *Background to Newton*, pp. 128–82, reproduces the writings on dynamics contained in *The Waste Book*.

43. This is developed in detail in McMullin, *Newton on Matter*. A textual basis for the three levels of explanation is contained in *Principia, book* 1, Prop. 69, "Scholium."

44. The draft preface and conclusion intended for the first edition are given in the Halls' book *Unpublished Papers*, sec. IV. The citation is from p. 305.

45. Ibid, pp. 333–35.

46. Ibid., p. 186.

47. From I Bernard Cohen, ed., *Isaac Newton's Papers and Letters on Natural Philosophy: And Related Documents* (Cambridge, Mass.: Harvard University Press, 1958), p. 408.

48. *Principia* book 3, Prop. VII.

49. Ibid., book 2, Props. XIX and XXXII.

50. Ibid., Prop. XXIII.

51. Richard S. Westfall, "Newton and the Fudge Factor," *Science* 179 (1973): 751–58.

52. See, for example, book 2, Prop. XL, p. 355.

53. *Principia*, p. 383.

54. Ibid., p. 7.

55. The text is given in the Halls' *Unpublished Papers*, pp. 259–64 (Latin), 265–68 (English).

56. According to A. R. Hall, *The Scientific Revolution: 1500–1800, the Formation of the Modern Scientific Attitude* (Boston: Beacon, 1956), p. 238, liquid thermometers were introduced about the middle of the seventeenth century. The first systematic calibration using two fixed points was in 1665. Fahrenheit's scale was introduced in 1714.

57. *Principia*, p. 38.

58. This is developed in Newton's "Tractatus de Quadrature Curvarum," first published in 1704. It is translated in Derek T. Whiteside, ed., *The Mathematical Works of Isaac Newton*, vol. 1 (New York: Johnson Reprint Co., 1964), pp. 141–60.

59. *Principia*, p. 29.

60. This summary is derived from Morris Kline, *Mathematical Thought from Ancient to Modern Times* (New York: Oxford University Press, 1972), pp. 356–70.

61. *Principia*, p. 39.

62. Ibid., p. 38.

63. Ibid., p. 196.

64. Ibid., p. 302.

65. *Variorum ed.*, vol. 2, p. 583, note on lines 31-36. In his introduction, pp. 154–56, Cohen discusses the probable reasons for the deletion of this sentence from later editions.

66. René Dugas, *Mechanics in the Seventeenth Century*, trans. F. Jacquot (Neuchâtel, Switzerland: Griffon, 1958), summarizes the arguments given by Huygens (pp. 439–50) and Leibniz (pp. 483–90) to show that Newton's theory of gravity is unintelligible.

67. *Variorum ed.*, vol. 2, p. 551.

68. See Cohen's *Introduction*, pp. 23–26 and plates 1–3 (after p. 48), for photographs of the rewritten manuscripts.

69. *Variorum ed.*, vol. 2, pp. 552–54.

70. The first edition of the *Opticks* was published in 1704. The second Latin edition added seven new queries. The account of atoms contained in query 23 in the second Latin edition became query 31 in the third (2d English) edition of 1718. The quotations cited here are from the Dover (1952) reprint of the fourth (1730) edition, pp. 400–405.

71. The historical development of the method of analysis and synthesis is treated in Jaakko Hintikka and Unto Remes, *The Method of Analysis* (Dordrecht: Reidel, 1974), The Cartesian and Newtonian adaptions are treated on pp. 104–17.

72. *Opticks*, p. 405.

73. This interpretation draws heavily on McGuire, "Atoms."

74. McMullin, *Newton on Matter*.

75. Yehuda Elkana, "Scientific and Metaphysical Problems: Euler and Kant," in R. Cohen and M. Wartofsky, eds., *Boston Studies in the Philosophy of Science*, vol. 14 (Dordrecht: Reidel, 1974), pp. 277–305.

76. References to Euler will be to the *Opera Omnia* published by the Swiss Society of Natural Sciences. The *Mechanics* is in series II, vol. 1, citation from p. 38 (my translation).

77. Ibid., p. 51.

78. L. Euler, "Recherches physiques sur la nature des moindre parties de la matière," *Opera Omnia*, series III, vol. 1, pp. 1–15.

79. Euler, "Anleitung zur Naturlehre," ibid., pp. 16–178. Since this was not published until 1962, it had no contemporary influence. In the opinion of the editor of this series (F. Rudio, *Vorwart*, p. viii), this is the most important of Euler's treatises on the property of matter.

80. Ibid., p. 80.

81. Newton speculated on this in his unpublished treatise, "De Gravitatione et Aequipondi Fluidorum," in the Halls' *Unpublished Papers*, pp. 75–156. On p. 140 Newton lists the conditions which God would have to impose on space to produce the bodies we experience. The extensions of space should be mobile, impenetrable, and able to excite sense perceptions. Newton's basic purpose in this obscure early tract seems to have been one of refuting Descartes by showing that motion rather than space is absolute.

82. L. Euler, "Principes generaux de l'état d'equilibre des fluids," in *Opera Omnia*, series II, vol. 12, pp. 2–53, citation from p. 3.

83. Ibid., Prologue, pp. ix–cxxv, contains a history of rational fluid mechanics from 1687–1765. The citation is from p. 1xxxi.

84. L. Euler, "Lettres à une princess d'Allemagne," *Opera Omnia*, series III, vol. II. A bibliography of the various editions and translations is given on pp. 1xi–1xxi.

85. Ibid., letters 69–70, pp. 149–52.

86. Ibid., letters 126, 127, pp. 297–301.

87. For the popularization of Newtonian physics in France, see Dugas, *Mechanics*, pp. 577–91; and Leonard M. Marsak, "Bernard de Fontenelle: The Idea of Science in the French Enlightenment," *Transactions of the American Philosophical Society* 49 (1959): 3–64.

88. J. d'Alembert, "Discours préliminaire de L'Encyclopédie," in *Oeuvres complètes de d'Alembert*, vol. 1 (Geneva: Slatkin, 1967), pp. 17–99.

89. Jean d'Alembert, *Traité de dynamique*: A Reprint of the Second Edition, Paris, 1758, with a New Introduction and Bibliography by Thomas L. Hankins (New York: Johnson Reprint Co., 1968).

90. D'Alembert's classification of knowledge is given in his "Explication détaillées du système des connaissances humaines," in d'Alembert, "Discours," pp. 99–109, and p. 114 for a schematic outline of the interrelation of all branches of knowledge.

91. Joseph L. Lagrange, *Mécanique Analitique*, in J.-A. Serret and G. Darboux, eds., *Oeuvres de Lagrange*, vols. 11 and 12 (Paris, 1788).

92. Ibid., pp. 1–27, for the principles of statics; pp. 237–62, for the principles of dynamics.

93. Ibid., p. 85.

94. This interpretation and the arguments supporting it are developed in Maurice Mandelbaum, *Philosophy, Science, and Sense Perception: Historical and Critical Studies* (Baltimore: Johns Hopkins University Press, 1964).

95. J. Lagrange, "Sur l'objet du calcul des fonctions et sur les fonctions en general," in *Oeuvres*, vol. 10, pp. 7–12, citation from p. 9.

96. "The Analyst; or A Discourse Addressed to an Infidel Mathematician," in A. A. Luce and T. E. Jessup, eds., *The Works of George Berkeley, Bishop of Cloyne*, vol. 3 (London: Nelson, 1951), pp. 55–102, citation from p. 67.

Chapter Two

1. A comprehensive account of the interrelations of science and philosophy in the Kantian and pre-Kantian eras is contained in Gerd Buchdahl, *Metaphysics and the Philosophy of Science. The Classical Origins: Descartes to Kant* (Cambridge, Mass.: MIT Press, 1969). The survey of the development of Kant's thought that I found most helpful is H. J. de Vleeschauwer, *The Development of Kantian Thought*, trans. A. R. C. Duncan (London: Thomas Nelson, 1962). In spite of its datedness and its attempt to cast Kant's development in a Hegelian mould, the old two-volume work of Edward Caird, *The Critical Philosophy of Immanuel Kant* (Glasgow: James Maclehose & Sons, 1899), gives a good survey of Kant's early writings. John Herman Randall's *The Career of Philosophy*, vol. 2 (New York: Columbia University Press, 1965), and Frederick Copleston's *A History of Philosophy*, vol. 6 (Garden City, N.Y.: Doubleday Image Book, 1960), both supply the background presupposed in the present account. Two helpful surveys of Kantian scholarship are H. J. de Vleeschauwer, "A Survey of Kantian Philosophy," *Rev. of Metaphysics* 11 (1957): 122–42; and M. J. Scott-Taggart, "Recent Work on the Philosophy of Kant," *American Philosophical Quarterly* 3 (1966): 171–209.

2. Immanuel Kant, "Gedanken von der wahren Schätzung der lebendigen Kräfte . . . "in *Kant's gesammelte Schriften*, herausgegeben von der Köiglichen Preussischen Akademie der Wissenschaften, vol. 1, pp. 1–181. Hereafter, this edition will simply be referred to as *Schriften*.

3. This point was developed in a paper "Kant and the Dynamical Tradition: A Case Study of the Role of Matter in Explanation," which Barbara Jill Buroker presented at the October 1972 Philosophy of Science Association meeting.

4. "Gedanken," pp. 22–23.

5. Ibid., pp. 40–42.

6. Ibid., pp. 144–45.

7. On the complex history involved in this acceptance, see Carolyn Iltis, "D'Alembert and the *Vis Viva* Controversy," *History and Philosophy of Science* 1 (1970): 135–44.

8. Jaako Hintikka and Unto Remes, *The Method of Analysis: Its Geometrical Origin and Its General Significance* (Dordrecht: Reidel, 1974).

9. I. Kant, "Untersuchung der Frage, Ob die Erde in ihren Umdrehung . . ." *Schriften*, vol. 1, pp. 183–91. Others, especially A. C. Clairaut, had explained the apparent slowing down of eclipses by postulating a lunar recession.

10. Instead of the dipole effect of the moon on the tidal bulge and the inverse cube force which Kant considered, one must consider the difference between the two tidal bulges. This gives an effect proportional to the inverse sixth power of the distance.

Kant's lack of facility in mathematical calculations might have been a subjective factor contributing to his persistent concern with explaining why mathematical laws fit physical reality. Physicists with a facility for mathematical reasoning rarely seem to find the applicability of mathematics to nature an anomaly requiring explanation.

11. Kant, "Allgemeine Naturgeschichte und Theorie des Himmels," *Schriften*, vol. 1, pp. 215–368. Part of this is translated in *Kant's Cosmology*, trans. W. Hastie, revised by Willy Ley (New York: Greenwood, 1968).

12. The analogy is cited in the subtitle to part 3, p. 349. Kant guesses that degrees of intelligence depend inversely on distance from the sun so that we are probably smarter than Venusians but inferior to Saturians.

13. I. Kant, "Meditationum quardundam de igne succinta delineatio," *Schriften*, vol. 1, pp. 369–84.

14. I. Kant, "Neue Anmerkungen zur Erläuterung der Theorie der Winde," ibid., pp. 489–503.

15. I. Kant, "Von den Ursachen der Erderschütterungen . . . ," ibid., pp. 417–28.

16. I. Kant, "Principiorum Primorum Cognitionis Metaphysicae Nova, Dilucidatio," ibid., pp. 385–416. This is translated in an appendix to F. E. England, *Kant's Conception of God* (London: George Allen & Unwin, 1929), pp. 211–52.

17. I. Kant, "Versuch einiger Betrachtungen über den Optimismus," *Schriften*, vol. 2, pp. 27–35.

18. Christian Wolff, *Preliminary Discourse on Philosophy in General*, trans. Richard Blackwell (Indianapolis: Bobbs-Merrill, 1963), p. 6.

19. John A. Reuscher, "A Clarification and Critique of Kant's *Principiorum Primorum Nova Dilucidatio*," *Kant-Studien* 68 (1977): 18–32.

20. Citation from England, *Kant's Conception of God*, pp. 214–42.

21. I. Kant, "Metaphysicae cum geometria iunctae usus in philosophia naturali, cuius specimen I. continet monadologiam physicam," *Schriften*, vol. 1, pp. 473–87.

22. As a dynamist Kant speaks of bodies as having differing *vires inertiae* rather than different masses. His argument is more simply understood when recast in terms of masses. If a monad's attractive force is Gm/r^2 and its repulsive force is Km/r^3, its effective sphere of impenetrability is defined by $r = K/G$, which is independent of m. However, Kant insists (theorem 12) that specific differences in bodily densities must be due to diversities in the inertia (i.e., mass) of the elemental parts.

23. I. Kant, "Die falsche Spitzfindigkeit der vier syllogistischen Figuren erwiesen," *Schriften*, vol. 2, pp. 45–61, citation from p. 49.

24. I. Kant, "Versuch den Begriff der negativen Grössen in die Weltweisheit einzuführen," ibid., pp. 165–204.

25. I. Kant, "An Inquiry into the Distinctness of the Principles of Natural Theology and Morals," in L. W. Beck, ed. and trans., *Immanuel Kant: Critique of Practical Reason and Other Writings in Moral Philosophy* (Chicago: University of Chicago Press, 1949), pp. 261–85.

26. Ibid., p. 269.

27. Ibid., p. 271.

28. Ibid., p. 271.

29. Ibid., pp. 272–73.

30. I. Kant, "Dissertation on the Form and Principles of the Sensible and Intelligible World," trans. John Handyside, in *Kant's Inaugural Dissertation and Early Writings on Space* (Chicago: Open Court, 1929), pp. 35–85, citation from p. 47. A good account of this dissertation and its bearing on the first *Critique* is contained in Robert Paul Wolff, *Kant's Theory of Mental Activity* (Cambridge, Mass.: Harvard University Press, 1963), pp. 11–22.

31. *Inaugural Dissertation*, p. 79.

32. Kant's first mention of this is in a letter of 21 February 1772 to Marcus Herz, where he announced his new key to the whole secret of metaphysics: "I asked myself: What is the ground of the relation of that in use which we called 'representation' to the object?" This is from Arnulf Zweig, *Kant: Philosophical Correspondence, 1759–99* (Chicago: University of Chicago Press, 1967). The letter to Herz is on pp. 70–76, citation from p. 71. Some of the ideas basic to Kant's critical revolution were anticipated in a letter Lambert sent Kant commenting on his *Inaugural Dissertation*. Inter alia Lambert interpreted Kant's distinction between a sensible and intelligible world as "the old *phaenomenon* and *noumenon* distinction" (letter in Zweig, *Kant*, pp. 60–67, citation from p. 61). In the *Prolegomena* Kant claimed that it was the memory of Hume that aroused him from his dogmatic slumber. Much later, in a latter to Christian Garve, 21 September 1798 (in Zweig, *Kant*, pp. 250–52), Kant claimed that the antinomies first aroused him from his dogmatic slumber.

33. From David Hume's *An Inquiry concerning Human Understanding*, ed. C. Hendel (Indianapolis: Bobbs-Merrill, 1955), p. 45. Buchdahl, *Metaphysics and the Philosophy of Science*, has a discussion of Hume on pp. 325–87 that is particularly good on clarifying the difference between Hume's acceptance of ordinary scientific reasoning and his profound skepticism concerning theoretical justifications of the principles employed.

34. I. Kant, *Prolegomena to Any Future Metaphysics*, L. W. Beck's revision of the Carus trans. (Indianapolis: Bobbs-Merrill, 1950), p. 8. The way in which this work probably came to influence Kant is analyzed in R. P. Wolff's "Kant's Debt to Hume via Beattie," *Journal of the History of Ideas* 21 (1960): 117–23.

35. The letter to Mendelssohn is in Zweig, *Kant*, pp. 54–57, citation from p. 55.

36. In "Der einzig mögliche Beweisgrund zu einer Demonstration des Daseins Gottes," *Schriften*, vol. 2, pp. 63–163, Kant concluded, in his examination of the ontological argument for God's existence, that existence is *not* a predicate. From this it follows that conceptual analysis can never give real existence.

37. From Carl J. Friedrich, ed., *The Philosophy of Kant: Immanuel Kant's Moral and Political Writings* (New York: Modern Library, 1949), p. 19.

38. *Inaugural Dissertation*, pp. 66–71.

39. See de Vleeschauwer, *Development*, pp. 138–76, for a historical survey of such criticisms.

40. See Kenneth T. Gallagher, "Kant and Husserl on the Synthetic A Priori," *Kant-Studien* 63 (1972): 342–52.

41. See P. F. Strawson, *The Bounds of Sense: An Essay on Kant's Critique of Pure Reason* (London: Methuen, 1966); and Jonathan Bennett, *Kant's Analytic* (Cambridge: Cambridge University Press, 1966). The interpretation of the first *Critique* presented in these two works, especially Bennett's, is totally at variance with the interpretation presented here. Both authors effectively argue that the valid aspects of the *Critique* are best determined by ignoring both Kant's concern with science and the historical background of the *Critique* and then attempting to see which of Kant's agruments can be reconstructed in accord with the methods of ordinary language analysis.

42. See esp. Wilfrid Sellars, "Is There a Synthetic A Priori?" in his *Science, Perception,*

and Reality (London: Routledge & Kegan Paul, 1963), pp. 298–320; and also Hilary Putnam, "The Analytic and the Synthetic," in H. Feigl and G. Maxwell, eds., *Minnesota Studies in the Philosophy of Science*, vol. 3 (Minneapolis: University of Minnesota Press, 1962), pp. 358–97. The dismissal of such psychological considerations as whether thinking a subject involves thinking a predicate is more a transferral than a dismissal. The banished psychology tends to reemerge whenever one attempts to explain how language is learned.

43. From *Schriften*, vol. 2, p. 202.

44. See de Vleeschauwer, *Development*, pp. 82–88; Randall, *Career*, pp. 83–88; and Giorgio Tonelli's article on Tetens in the *Encyclopedia of Philosophia*.

45. See de Vleeschauwer, *Development*, pp. 82–114.

46. I. Kant, *Logic*, trans. Robert Hartmann and Wolfgang Schwarz (Indianapolis: Bobbs-Merrill, 1974). At Kant's request, Jäsche prepared the original edition from Kant's notes for his lecture of 1782, the year after the *Critique* was published. The long introduction by the translators contains an excellent account of Kant's doctrine of analytic and synthetic methods, concepts, and judgments and of the way these are interrelated in both general and transcendental logic. The relation of Kant's *Logic* to his overall philosophy is discussed by James Collins in "Kant's Logic as a Critical Aid," *Review of Metaphsics* 30 (1977): 440–61.

47. I. Kant, *The Critique of Pure Reason*, trans. Norman Kemp Smith (London: Macmillan, 1963), A 709 = B 737—A 739 = B 767.

48. *Logic*, pp. 67–70.

49. Ibid., p. 64.

50. Ibid., p. 70. See also *Critique*, A 725 = B 735.

51. *Critique*, A 727 = B 755.

52. *Logic*, p. 75.

53. This is cited from the translators' introduction to Kant's *Logic*, p. 1xvi (note). Kant's crucial distinction between real and merely logical connections seems to have been influenced by Lambert's distinction between formal (logical) and material knowledge. In his first letter to Kant, 13 November 1765 (in Zweig, *Kant*, pp. 43–47), Lambert insisted that form gives only the ideal, but not real knowledge. In his next letter, 3 February 1766 (Zweig, *Kant*, pp. 50–53), Lambert concentrated on the relation of this material-formal distinction to the concepts and axioms of science. His general idea, that analysis of concepts does not yield real existence, was independently developed by quite a few others. On this point, see de Vleeschauwer, *Development*, pp. 19–33.

54. *Critique*, A 372.

55. Ibid., A 226 = B 273.

56. Kant summarized his position on this question in the *Prolegomena* (Note 34), pp. 42–74.

57. Kant's treatment of these schemata is A 137 = B 176—A 147 = B 187. Two points require some comment. First, Kant's treatment of the foundations of mathematics is best interpreted in a modern context as dealing with applied mathematics. Second, the role of imagination is one of the most underdeveloped topics in the first *Critique*. This inadequacy supplies the point of departure for Martin Heidegger's *Kant and the Problem of Metaphysics*, trans. J. S. Churchill (Bloomington: Indiana University Press, 1962). See esp. pp. 93–118 for a discussion of the interrelation of images, concepts, and schemata. The apparent conflict between the demands of Kant's architectonic and the development of the schemata is presented in the harshest form in Normal Kemp Smith, *A Commentary to Kant's Critique of Pure Reason* (New York: Humanities Press, 1962), pp. 334–42. This criticism is partially answered by R. P. Wolff, *Kant's Theory of Mental Activity*, pp. 206–18, on the grounds that in the actual working out of the Transcendental Analytic Kant effectively treats concepts as rules for the synthesis of a manifold of intuitions. This is the significance which is needed for Kant's treatment of the foundations of science.

58. *Critique*, B 202. Whenever there is a difference in formulation between the A and B editions I will cite the B edition, on the grounds that the later formulation is less dependent on the psychology Kant used and more in accord with the doctrine of the *Foundations*.

59. *Critique*, B 207.

60. *Prolegomena* (my note 34), p. 56, note 7: "Quantitas qualitatis est gradus."

61. *Critique*, B 201.

62. Ibid., A 162 = B 202.

63. Ibid., B 218.

64. Ibid., B 224.

65. "All appearances contain the permanent (substance) as the object itself, and the transitory as its mere determination, that is, as a way in which the object exists." A 182, NKS trans., p. 212. In his *Commentary* (see my note 57), p. 358, Kemp Smith brings out the difference between the two editions and refers to the second edition doctrine as weaker. It is, I believe, much more defensible.

66. *Critique*, B 233.

67. *Inaugural Dissertation*, p. 71.

68. *Critique*, B 257.

69. *Logic*, pp. 25–30.

70. *Critique*, B 266.

71. The interpretation of the Antinomies followed here is essentially the one developed by W. H. Walsh in *Kant's Criticism of Metaphysics* (Chicago: University of Chicago Press, 1975), pp. 195–204.

72. *Critique*, A 434–5 = B 462–3.

73. This is what Kant treats in his section 7–9 explaining the Antinomies, NKS trans., pp. 443–64.

74. I am using the translation by James Ellington (Indianapolis: Bobbs-Merrill, 1970), which also contains a useful survey, "The Unity of Kant's Thought in His Philosophy of Corporeal Nature," pp. 135–218. Another survey of Kant's philosophy of nature that attempts to relate this philosophy directly to modern science, rather than, as in the present study, to the problem of conceptual foundations, is W. H. Werkmeister, "Kant's Philosophy and Modern Science," *Kant-Studien* 66 (1975): 35–57.

75. *Foundations*, p. 3.

76. Kant attempted to bridge the gap between the metaphysics of nature and the metaphysics of morals chiefly by suggesting possible ways in which man as a natural species might have evolved into man as a moral agent. He also explicitly considered the possibility of evolution of species and of the evolution of man from monkeys. This, however, he rejected, rather reluctantly it seems, on the grounds that such a doctrine of evolution had not received empirical support. His writings on this topic have been collected by L. W. Beck in *Kant on History* (Indianapolis: Bobbs-Merrill, 1963).

77. *Foundations*, p. 4. This judgment on chemistry as not yet a science should be seen in historical context. It occurred just three years after Lavoisier had burned the books of Stahl and his followers to signal his own attempt to turn chemistry into a science.

78. I. Kant, *The Critique of Judgment*, trans. J. C. Meredith (Oxford: Clarendon Press, 1952).

79. In 1940 the International Union of Pure and Applied Chemistry established a Committee on the Reform of Inorganic Nomenclature to settle such issues. See Aaron J. Ihde, *The Development of Modern Chemistry* (New York: Harper & Row, 1964), p. 587.

80. *Schriften*, vol. 2, p. 32. The translation used is that by John Manolesco, *Dreams of a Spirit Seer* (New York: Vantage Press, 1969), p. 33 (note).

81. Since Kant makes so few references to sources, it is difficult to determine influence by counting citations. To check this I used the Insel edition, *Kant: Werke* (Wiesbaden: Insel

Verlag, 1960), which has a helpful index. In his treatise on negative magnitudes, vol. 1, p. 780, Kant cites Euler's Memoire to the Prussian Academy, discussed in chap. 1. In his *Inaugural Dissertation*, trans. p. 85, and elsewhere he cites Euler's *Letters to a German Princess*. Both of these works treat the concept of matter. In "Aus Sömmering, über das Organ der Seele," Insel ed., vol. 6, p. 257, Kant refers to Euler's theory of fluids. There are other references to 'the well-known Euler', where no explicit citation is given. Though he must have been familiar with it, Kant never refers to Euler's *Mechanics*. Then, as now, philosophers are reluctant to cite introductory physics texts.

What is much more significant than the number of citations is Kant's willingness to accept Euler's position on disputed issues, even when Euler disagrees with Newton. Thus, in *Foundations*, p. 72 (note), and in the third *Critique*, Meredith trans., p. 66, Kant accepts Euler's wave theory of light and colors rather than Newton's particle theory, which was still the established theory. He also uses Euler's ideas on point-masses, the ether, and even his attack on the doctrine of physical monads, though Kant made the most significant contribution to this doctrine. I know of only one place where he cites Euler to disagree with him, indicating his unwillingness to accept Euler's a posteriori proof of absolute space. This is in his "Concerning the Ultimate Foundation of the Differentiation of Regions in Space," in *Kant: Selected pre-Critical Writings and Correspondence with Beck*, trans. G. B. Kerford and D. E. Walford (Manchester: Manchester University Press, 1968), p. 57.

82. In spite of his rewriting of this section after completing the *Foundations*, Kant gave the same order of characteristics in B 12 as in A 8.

83. Manolesco, *Dreams*, p. 83. In the beginning of this work Kant clarified the notion of 'body' as a counterpoise to a possible notion of spirit. The key difference is that spirits, if there are such things, could not be impenetrable. If they were, they would be bodies for all practical purposes.

84. *Foundations*, p. 18.

85. Ibid., p. 40.

86. Ibid., pp. 64–66.

87. Euler, *Mechanics*, p. 38.

88. Ibid., p. 83.

89. Though Kant revised the analysis of body given in the first edition of the first *Critique*, he retained the idea that weight is *not* an analytic property of 'body'. That is, for the *analytic*, or meaning-determined-by-use, concept, the judgment "A body has weight" is a synthetic a priori judgment. For the *synthetic*, or meaning-determined-by-postulation-and-construction, concept, the same judgment is analytic. This is a point repeatedly confused by critics who consider systhetic a priori judgments without also considering the distinction between synthetic and analytic methods and concepts.

90. *Foundations*, p. 80.

91. Ibid., p. 81.

92. Ibid., p. 95.

93. Ibid., p. 99.

94. Ibid., pp. 98–99.

95. Ibid., p. 102.

96. Ibid., p. 104.

97. Ibid., p. 106.

98. Ibid., p. 118.

99. From Zweig, *Kant*, p. 253.

100. My account of the *Opus Postumum* is based chiefly on de Vleeschauwer, *Development*, pp. 177–98; Smith, *Commentary*, pp. 607–41; and W. H. Werkmeister, "The Critique of Pure Reason and Physics," *Kant-Studien* 68 (1977): 33–45.

101. Smith, *Commentary*, p. 618.

102. See L. Pearce Williams, "Kant, Naturphilosophie, and Scientific Method," in R. Giere and R. Westfall, eds., *Foundations of Scientific Method: The Nineteenth Century* (Bloomington: Indiana University Press, 1973), pp. 3–22.

103. Trans. from *Johann Georg Hamann, Sämtliche Werke*, vol. 3 (Vienna: Verlag Herder, 1951), p. 286. The background to Hamann's criticism is explained in W. M. Alexander, *Johann Georg Hamann: Philosophy and Faith* (The Hague: M. Nijhoff, 1966), pp. 99–119. J. G. Herder, another former student of Kant, also wrote a *Metakritik* arguing that Kant's reliance on faculty psychology is incompatible with an adequate account of how language is learned. His position is summarized in Isaiah Berlin's *Vico and Herder* (New York: Viking Press, 1976), pp. 167–76.

104. Wilfrid Sellars, "Empiricism and the Philosophy of Mind," in his *Science, Perception, and Reality* (London: Routledge & Kegan Paul, 1963), pp. 127–96; and in *Science and Metaphysics: Variations on Kantian Themes* (New York: Humanities Press). His position on this point is critized in my article-length review of the latter book in *Philosophical Forum* 1 (1969): 509–45.

Chapter Three

1. My interpretation of the philosophical significance of this historical pattern will be presented elsewhere. Two differing interpretations of this pattern are presented in William Wallace, *Causality and Scientific Explanation*, 2 vols. (Ann Arbor: University of Michigan Press, 1974), who accords the philosophy of nature and the type of inferential reasoning it sanctions a much more normative role than the present account; and Stephen Toulmin, *Human Understanding*, vol. 1 (Princeton: Princeton University Press, 1972), who develops a detailed, provocative, but somewhat fanciful, analogy between the emergence of a scientific discipline and the emergence of a biological species.

2. Roger Boscovich, *A Theory of Natural Philosophy*, trans. J. M. Child, paperback ed. (Cambridge, Mass.: MIT Press, 1966).

3. Robert E. Schofield, *Mechanism and Materialism: British Natural Philosophy in an Age of Reason* (Princeton: Princeton University Press, 1970). Arnold Thackery, *Atoms and Powers* (Cambridge, Mass.: Harvard University Press, 1970), covers much of the same ground but concentrates more on the concept of matter than on natural philosophy as such. He effectively records a similar shift with the gradual abandonment of the Newtonian (and later Boscovichean) insistence on the inertial homogeneity of ultimate matter. Thackery's survey covers Continental as well as British developments.

4. This is from Tiberius Cavallo, *The Elements of Natural or Experimental Philosophy*, 4 vols. (London, 1803), citation from vol. 1, p. 3. Earlier textbooks with a similar orientation are Richard Helsham, *A Course of Lectures in Natural Philosophy* (London, 1777) and William Nicholson, *An Introduction to Natural Philosophy*, 2 vols. (London, 1790).

5. I. Bernard Cohen, *Franklin and Newton* (Philadelphia: American Philosophical Society, 1956), pp. 118–19.

6. Ibid., p. 147.

7. Wilson Scott, *The Conflict between Atomism and Conservation Theory: 1644–1860* (London: Macdonald, 1970).

8. Francis Bacon, *The New Organon*, ed. F. H. Anderson (Indianapolis: Bobbs-Merrill, 1960), book 2, secs. x–xxi, pp. 130–62.

9. Isaac Newton, *Mathematical Principles of Natural Philosophy*, Motte-Cajori trans. (Berkeley: University of California Press, 1962), book 2, Prop. XXIII, "Scholion," p. 302.

10. This historical account is adapted from Robert Fox, *The Caloric Theory of Gases: From Lavoisier to Regnault* (Oxford: Clarendon Press, 1971); and D. S. L. Cardwell, *From Watt to Clausis: The Rise of Thermodynamics in the Early Industrial Age* (Ithaca: Cornell University Press, 1971).

404 Notes to Pages 92–95

11. Lavoisier treated this in his "Du passage des corps solides à l'état liquide, par l'action du calorique," in *Oeuvre de Lavoisier* (Paris: Imprimerie Impériale, 1864), vol. 2, pp. 765–72. His account of the basic properties of caloric appears in his *Traité élémentaire de chimie*, ibid. vol. 1, pp. 19–31. The influence of Lavoisier's caloric theory on his chemical doctrines is treated in Robert J. Morris's "Lavoisier and the Caloric Theory," *British Journal for the History of Science* 6 (1972): 1–38.

12. A translation of Laplace's "Velocity of Sound in Air and Water" is contained in R. Bruce Lindsay, ed., *Acoustics: Historical and Philosophical Development* (Stroudsburg, Pa.: Dowden, Hutchinson & Ross, no date), pp. 181–82.

13. See Stephan J. Goldfarb, "Rumford's Theory of Heat: A Reassessment," *British Journal for the History of Science* 10 (1977): 25–36.

14. These accounts may be found in William Magie, *A Source Book in Physics* (Cambridge, Mass.: Harvard University Press, 1935), pp. 151–61 (Rumford), pp. 161–65 (Davy).

15. Cardwell, *Watt to Clausius*, p. 113.

16. These are summarized in Fox, *Caloric Theory*, pp. 276ff.

17. The pertinent passages from Bernoulli's *Hydrodynamica* (1738) are reproduced in H. A. Boorse and L. Motz, *The World of the Atom* (New York: Basic Books, 1966), vol. 1, pp. 109–18; and in Stephen Brush's *Kinetic Theory* (Oxford: Pergamon Press, 1965), pp. 57–65. Stephen Brush, *The Kind of Motion We Call Heat: A History of the Kinetic Theory of Gases in the 19th Century*, 2 vols. (Amsterdam: North-Holland, 1976), supplies the best overall guide to the study of kinetic theory. See esp. his introductory survey, pp. 3–103.

18. Their papers are reproduced in Boorse and Motz, *World of the Atom*, pp. 196–205 (Herapath), pp. 213–35 (Waterston).

19. The key papers of Mayer, Joule, and von Helmholtz are translated in Brush, *Kinetic Theory*, pp. 71–110. Peter Clark, "Atomism versus Thermodynamics," in Colin Howison, ed., *Method and Appraisal in the Physical Sciences: The Critical Background to Modern Science, 1800–1905* (Cambridge: Cambridge University Press, 1976), pp. 41–105, interprets kinetic theory and thermodynamics as competing research programs. The interpretation has some plausibility for the period after 1880, but not for the early stage. Thermodynamics, a general method of treating heat, did not involve molecular assumptions. Kinetic theory applied these general principles to treatments of gases which involved molecular models of a gas.

20. This paper is reproduced in Brush, *Kinetic Theory*, pp. 111–34.

21. Ibid., pp. 135–47.

22. Maxwell's statement of his attitude was given in a letter to Stokes excerpted in Brush, *Kinetic Theory*, pp. 26–27.

23. *The Scientific Papers of James Clerk Maxwell*, ed. W. D. Niven (New York: Dover, no date), vol. 2, pp. 1–26. For an overview of Maxwell's development, see C. W. F. Everitt, *James Clerk Maxwell: Physicist and Natural Philosopher* (New York: Charles Scribner's Sons, 1975).

24. Lavoisier, *Traité*, p. 7.

25. The background and development of Dalton's position are explained in Arnold Thackery, *John Dalton: Critical Assessments of His Life and Science* (Cambridge, Mass.: Harvard University Press, 1972), and in Alan J. Rocke, "Atoms and Equivalents: The Early Development of the Chemical Atomic Theory," HSPS, vol. 9, pp. 225–63. A more polemical account of the conflicting accounts given by Dalton, Gay-Lussac, Avagadro, and Berzelius is presented in terms of the methodology of research programs in Martin Frické, "The Rejection of Avagadro's Hypothesis," in Howison, *Method and Appraisal*, pp. 277–307. Surveys of the development of chemical atomism through the end of the nineteenth century may be found in D. P. Mellor, *The Evolution of the Atomic Theory* (Amsterdam: Elsevier, 1971); and David M. Knight, *Atoms and Elements: A Study of the Theory of Matter in England in the Nineteenth Century* (London: Hutchinson, 1967). An older but still helpful survey is Leonard K. Nash's "The Atomic and Molecular Theory," in James Bryant Conant, ed., *Harvard Case Histories in*

Experimental Sciences, vol. 1, pp. 217–32. Most of the papers cited here are reproduced in D. L. Hurd and J. J. Kipling, *The Origin and Growth of Physical Science*, vol. 2 (Middlesex, England: Penguin, 1964), pp. 19–100, and in Boorse and Motz, *World of the Atom*, pp. 157–71 (Gay-Lussac), pp. 171–81 (Avagadro), pp. 277–97 (Cannizzaro). A general survey of the development of chemistry in this period may be found in Aaron J. Ihde, *The Development of Modern Chemistry* (New York: Harper & Row, 1964), pp. 57–258.

26. John Dalton, *A New System of Chemical Philosophy*, 2 vols. (London: R. Bickerstaff, 1808), esp. vol. 1, pp. 211–16.

27. The debates on this issue are summarized in Knight, *Atoms and Elements*, chap. 6, Rocke, "Atoms and Equivalents," argues that even the chemists who talked in terms of equivalents really relied on chemical atomism, but were generally unwilling to identify chemical atoms with real atoms, i.e., the postulated ultimate indivisible units.

28. J. W. van Sponsen, *The Periodic System of Chemical Elements* (Amsterdam: Elsevier, 1969), chap. 5, claims that the basic idea of the periodic system was discovered independently by A. de Chancourtois (1862, Paris), J. Newlands (1864, London), W. Olding (1864, London), G. Hinrichs (1867, Iowa City), L. Mayer (1868, Breslau), and D. Mendeleev (1869, St. Petersburg). After the Karlsruhe conference of 1860, the idea was so in the air that it could emerge with relative ease from a careful comparison of relative atomic weights and chemical properties. Mendeleev's chart is recognized as the most basic because of its choice of properties and because it left gaps for as yet undiscovered elements.

29. Maxwell, *Scientific Papers*, vol. 2, p. 456.

30. The development of spectroscopy is covered in William McGucken, *Nineteenth-Century Spectroscopy: Development of the Understanding of Spectra, 1802–97* (Baltimore: Johns Hopkins University Press, 1969); and in Sir Edmund Whittaker, *A History of the Theories of Aether and Electricity* (New York: Harper Reprint, 1951), vol. 1, chap. 12.

31. This paper is reproduced in Magie, *Source Book*, pp. 354–60.

32. These and other theories are covered in McGucken, *Nineteenth-Century Spectroscopy*, pp. 73–101.

33. Stanley L. Jaki, *The Relevance of Physics* (Chicago: University of Chicago Press, 1966), pp. 106–7.

34. Balmer's paper is given in Magie, *Source Book*, pp. 360–65.

35. A survey of these developments and of the physics involved is given in H.E. White, *Introduction to Atomic Spectra* (New York: McGraw-Hill, 1934), chap. 1.

36. John Theodore Merz, *A History of European Scientific Thought in the Nineteenth Century* (New York: Dover, 1965), vol. 1, pp. 419–20.

37. Maxwell, *Scientific Papers*, vol. 2, p. 378. See also Brush, *Kind of Motion*, p. 32.

38. Barbara Giusti Doran, "Origins and Consolidation of Field Theory in Nineteenth-Century Britain: From the Mechanical to the Electromagnetic View of Nature," HSPS, vol. 6, pp. 133–260.

39. Faraday's research and reasoning on this issue are summarized in Joseph Agassi's *Faraday as a Natural Philosopher* (Chicago: University of Chicago Press, 1971), chap. 8. Agassi interprets Faraday as a theorist deeply frustrated at the fact that his theories were not taken seriously. A more conventional interpretation of Faraday as a careful experimenter and cautious inductivist who never really departed from the Boscovichean atom is given in L. Pearce Williams, *Michael Faraday: A Biography* (London: Chapman & Hall, 1965), chap. 7. A general survey of some of the difficulties involved in developing consistent atomic models is given by Elizabeth Garber, "Molecular Science in Late-Nineteenth-Century Britain," HSPS, Vol. 9, pp. 265–97.

40. This is from his article "On the Size of Atoms" reproduced as an appendix in William Thomson (Lord Kelvin) and Peter Guthrie Tait, *Treatise on Natural Philosophy*, rev. ed. (Cambridge: Cambridge University Press, 1879), vol. 2, citation from p. 495.

41. Sir William Thomson, Baron Kelvin, *Mathematical and Physical Papers*, ed. J. Larmor (Cambridge: Cambridge University Press, 1910), vol. 4, pp. 1–12, 13–68.

42. See Mary Hesse, *The Structure of Scientific Inference* (Berkeley: University of California Press, 1974), chap. 11; Robert Kargon, "Model and Analogy in Victorian Science: Maxwell and the French Physicists," *Journal of the History of Ideas* 30 (1969): 423–36; and Joseph Turner, "Maxwell on the Method of Physical Analogy," *British Journal for the Philosophy of Science* 6 (1955): 226–38. For Maxwell's contributions to kinetic theory, see Brush, *Kind of Motion*, chap. 5.

43. The address is in Maxwell *Scientific Papers*, vol. 2, pp. 418–38, citation from p. 433.

44. Ibid., pp. 445–84, esp. p. 471.

45. See McGucken, *Nineteenth-Century Spectroscopy*, pp. 176ff., and Garber, "Molecular Science," pp. 272–88.

46. J. C. Maxwell, review of Henry Watson, *The Kinetic Theory of Gases, Nature* 16 (1877): 242–46 (not in Maxwell's *Scientific Papers*).

47. Ibid., p. 245.

48. This citation is from Silvanus P. Thompson, *The Life of William Thomson, Baron Kelvin of Largs* (London: Macmillan, 1910), vol. 2, pp. 984. I have consistently used the familiar name 'Kelvin' rather than 'William Thomson', so that it would not be necessary to use different names and titles when this most honored scientist of all time became a Knight (of different countries), a Baron, an Honorary Colonel, an Honorary Liveryman of the Clothworker's Company, etc. S. P. Thompson gives a list of such honors on pp. 1215–22 and, on pp. 1223–74, a list of Kelvin's 661 publications. It seems likely that only Euler had a larger number of technical publications.

49. Ibid., p. 1047, note 1.

50. "Aepinus Atomized," appendix E, pp. 541–68, in Lord Kelvin, *Baltimore Lectures on Molecular Dynamics and the Wave Theory of Light* (London: C. J. Clay, 1904).

51. Kelvin used the terms 'quantum' and 'quantum number' to refer to the number of electrons in an atom. As late as 1909 Max Planck was still using these terms in this sense. Thus, he said, "The elementary quantum of electricity, or the free electric charge of a monovalent ion or electron, in electrostatic measure is: $e = 4.69 \times 10^{-10}$." This is from Max Planck, *Eight Lectures on Theoretical Physics: Delivered at Columbia University in 1909*, trans. A. P. Wills (New York: Columbia University Press, 1915), p. 95.

52. These models are given in Kelvin's *Papers*, vol. 6, pp. 204–43.

53. Surveys of Boltzmann's contributions to kinetic theory may be found in Brush, *Kind of Motion*, chap. 6; in Martin J. Klein, *Paul Ehrenfest*, vol. 1, *The Making of a Theoretical Physicist* (Amsterdam: North-Holland, 1970), pp. 94–128; and in Yehuda Elkana, "Boltzmann's Scientific Research Program and Its Alternatives," in Y. Elkana, ed., *The Interaction between Science and Philosophy* (Atlantic Highlands, N.Y.: Humanities Press, 1974), pp. 243–79. The interpretation of Boltzmann's thought presented in the text is chiefly based on my reading of Boltzmann's *Lectures on Gas Theory*, trans. Stephen G. Brush (Berkeley: University of California Press, 1964); and Boltzmann's *Theoretical Physics and Philosophical Problems: Selected Writings*, ed. Brian McGuinnes (Dordrecht: Reidel, 1974). These will be referred to as *Gas Theory* and *Writings*.

54. Boltzmann, *Gas Theory*, p. 272.

55. This is the characterization Boltzmann gave in *Writings*, p. 42. This movement is separate from the slightly later phenomenology developed by E. Husserl.

56. This is from Ernst Mach, *The Science of Mechanics* (1883), trans. T. J. McCormick (La Salle, Ill.: Open Court, 1942). See also Mach's *Popular Scientific Lectures*, trans. T. J. McCormick (Chicago: Open Court, 1895). This includes his lecture "The Economical Nature of Physical Inquiry," which has been included in many philosophy of science anthologies. *Synthese*, vol. 18 (1968) has a collection of articles on the philosophy of Mach. Of these, the one

most pertinent to our present concern, S. Brush's study of the development of Mach's thought from his early support of atomism to his later rejection, is revised and reproduced in Brush, *Kind of Motion*, chap. 8.

57. See Erwin N. Hiebert, "The Energetic Controversy and the New Thermodynamics," in Duane H. D. Roller, ed., *Perspectives in the History of Science and Technology* Norman: University of Oklahoma Press, 1971), pp. 67–86.

58. Max Planck, *Scientific Autobiography and Other Papers*, trans. Frank Gaynor (London: William & Norgate, 1950), pp. 32–33, discusses Planck's original unwillingness to accept Boltzmann's stress on atomism and his interpretation of entropy. After the development of the quantum theory, Planck came to accept Boltzmann's statistical interpretation of thermodynamics. After Boltzmann's death, Planck effectively succeeded to the position of Mach's leading opponent. The key papers in the Planck-Mach controversy are reproduced in S. Toulmin, ed., *Physical Reality* (New York: Harper Torchbook, 1970), pp. 1–52.

59. See B. G. Doran, "Field Theory."

60. J. Clerk Maxwell, *Matter and Motion* (1877) (New York: Dover, no date), uses point-masses as a limit notion and countenances action at a distance (e.g. on p. 42), but rejects absolute space and time.

61. Maxwell, *Scientific Papers*, vol. 2, p. 418.

62. Hertz's paper is reproduced in Hurd and Kipling, *Origin and Growth*, pp. 254–66.

63. In addition to B. G. Doran "Field Theory," see William T. Scott, "Resource Letter FC-1 on the Evolution of the Electromagnetic Field Concept," *Amer. J. of Phys.* 31 (1963): 1–8.

64. This development is traced in Salvo d'Agostino's "Hertz's Researches on Electromagnetic Waves," HSPS, vol. 6, pp. 261–323.

65. A survey of this development is given in Whittaker, *History*, chap. 11.

66. His paper is reproduced in Hurd and Kipling, *Origin*, pp. 293–306.

67. This paper is also reproduced in Hurd and Kipling, *Origin*, pp. 339–62.

68. As a young man Whittaker made some contributions to the Maxwell-Lorentz theory. The summary of classical radiation theory given in Whittaker *History*, vol. 1, chaps. xii, xiii, remains the basic account of these developments.

69. H. A. Lorentz, "Sur la théorie de la réflexion et de la réfraction de la lumière" (his 1875 Leiden dissertation), in *H. A. Lorentz: Collected Papers*, vol. 1 (The Hague: M. Nijhoff, 1935), pp. 193–383, citation from p. 383.

70. The brief account that follows is essentially a summary of material given in Russell McCormmach, "H. A. Lorentz and the Electromagnetic View of Nature," Isis 61 (1970): 459–70, and in James T. Cushing, "Electromagnetic Mass, Relativity, and the Kaufmann Experiments," preprint.

71. Boltzmann, *Writings*, p. 258.

72. Ibid., p. 227.

73. Ibid., p. 258.

74. See especially his essays "On the Indispensability of Atomism in Natural Science," *Writings*, pp. 41–53, and "On the Question of the Objective Existence of Processes in Inanimate Nature," Ibid., pp. 57–76.

75. Heinrich Hertz, *The Principles of Mechanics: Presented in a New Form*, trans. D. E. Jones and J. T. Walley (New York: Dover, 1956), p. 1.

76. This statement was made by Lord Salisbury. Boltzmann cited and strongly endorsed it in an article written for *Nature*, in *Writings*, pp. 201–9, citation from p. 201.

77. Boltzmann, *Writings*, p. 64.

78. See Boltzmann, *Writings*, pp. 93, 258.

79. Ibid., p. 258.

80. Ibid., p. 259.

81. From the preface to J. Willard Gibbs, *Elementary Principles in Statistical Mechanics* (1902) (New York: Dover, 1960), p. ix.

Chapter Four

1. In addition to the general surveys given in Whittaker, Jammer, and Hund three studies of this period are D. ter Haar, *The Old Quantum Theory* (London: Pergamon Press, 1967), which translates some of the basic articles; A. Hermann, *The Genesis of Quantum Theory (1899–1913)*, trans. C. W. Nash (Cambridge, Mass.: MIT Press, 1971); and Thomas S. Kuhn, *Black-Body Theory and the Quantum Discontinuity, 1894–1912* (Oxford: Clarendon Press, 1978), which will be discussed when we consider Planck's work.

2. See "The Law of Partition of Kinetic Energy," in *Scientific Papers by Lord Rayleigh (John William Strutt)* (New York: Dover, 1964), vol. 4, pp. 433–51.

3. "Remarks upon the Law of Complete Radiation," Ibid., pp. 483–85.

4. "The Dynamical Theory of Gases and of Radiation," Ibid., vol. 5, pp. 248–53.

5. In Sir Joseph Larmor, *Mathematical and Physical Papers* (Cambridge: Cambridge University Press, 1929), vol. 2, pp. 396–415.

6. Pierre Duhem, *The Aim and Structure of Physical Theory*, trans. P. P. Wiener (Princeton: Princeton University Press, 1954), part I, chap. 4.

7. A. Einstein, "Folgerungen aus den Capillaritätserscheinungen," *Ann. der Phys.* 4 (1901): 513–23.

8. A. Einstein, "Ueber die thermodynamische Theorie der Potentialdifferenz zwischen Metallen und vollständig dissociierten Lösungen ihre Salze, und über eine elektrische Methode zur Erforschung der Molekularkräfte," *Ann. der Phys.* 8 (1902): 798–814.

9. A. Einstein, "Eine Theorie der Grundlagen der Thermodynamik," *Ann. der Phys.* 11 (1903): 170–87. Summaries of Einstein's early work in statistical mechanics may be found in Kuhn, *Black-Body Theory*, and in Max Born, "Einstein's Statistical Theories," in *Albert Einstein: Philosopher-Scientist*, ed. P. A. Schilpp (New York: Tudor, 1949), pp. 163–77.

10. A. Einstein, "Autobiographical Notes," ibid., p. 47. Hereafter this book will be referred to as Schilpp. These earliest papers are discussed in M. Klein's "Thermodynamics in Einstein's Thought," *Science* 157 (1967): 509–16.

11. A. Einstein, "Über die von der molekularkinetischen Theorie der Wärme geforderte Bewegung von in ruhenden Flüssigkeiten suspendierten Teilchen," *Ann. der Phys.* 17 (1905): 549–60. This, plus four later articles, is translated in R. Furth, ed., *Investigations on the Theory of Brownian Motion*, trans. H. D. Cowper (New York: Dover, 1956), pp. 1–18. This will be cited as *Investigations*.

12. Furth, *Investigations*, p. 18.

13. A general survey of the development of the theory of Brownian motion together with a detailed account of Perrin's work is given in Mary Joe Nye, *Molecular Reality: A Perspective on the Scientific Work of Jean Perrin* (London: Macdonald, 1972). The early history is also treated in Milton Kerker, "Brownian Movement and Molecular Reality Prior to 1900," *Journal of Chemical Education* 51 (1974): 764–68. Svedberg's work is analyzed in Milton Kerker's "The Svedberg and Molecular Reality," *Isis* 67 (1976): 190–216.

14. Jean Perrin, *Atoms* (1913), trans. D. Hammick from the 11th French ed. (London: Constable, 1923), pp. 215–16.

15. Martin Klein's principal articles on Planck's work are "Max Planck and the Beginnings of the Quantum Theory," *Archives Hist. Exact. Sc.* 1 (1962): 459–79; "Einstein and the Wave-Particle Duality," *Natural Philosopher* 3 (1964): 1–49; "Thermodynamics and Quanta in Planck's Work," *Physics Today* 19 (November 1966): 23–32; and "Einstein, Specific Heats, and the Early Quantum Theory," *Science* 148 (9 April 1965): 173–80. For the text of Planck's papers, we are using the critical edition of H. Kangro, *Planck's Original Papers in Quantum*

Physics, German and English ed., trans. D. ter Haar and S. Brush (London: Taylor & Francis, 1972).

16. Kuhn, *Black-Body Theory*. Though Hans Kangro's article "Max Planck," in the *Dictionary of Scientific Biography*, vol. 11, pp. 7–17, was written before this work of Kuhn's was published, Kangro had discussed the central thesis with Kuhn. In note 27 he lists the sources in Planck's *Collected Works* where Planck seems to associate energy discreteness with individual resonators.

17. Max Planck, *Scientific Autobiography and Other Papers*, trans. Frank Gaynor (London: Williams & Norgate, 1950); and in his Nobel Prize address, "The Genesis and Present State of Development of the Quantum Theory," in *Nobel Lectures: Physics, 1901–1921* (Amsterdam: Elsevier, 1967), pp. 408–18. Here, p. 413, he claims that the crucial question after the derivation of his radiation formula was whether the new conception of a quantum of action was fictional or foundational. Einstein's theorizing and experimental confirmation supported the second alternative.

18. Planck, *Autobiography*, p. 32.

19. Ibid., p. 41.

20. Planck's introduction of this concept is discussed in Kuhn, *Black-Body Theory*, pp. 82–84, 120–25. Planck stressed the analogy in his December 1900 paper. See Kangro ed., *Planck's Original*, p. 7 (39). We will refer to this critical edition by citing the German (English) pages.

21. Planck's original derivation of this crucial formula is summarized in Jammer, *History*, appendix A, pp. 383–85.

22. Hans Kangro, *Vorgeschichte des Planckschen Strahlungsgesetzes* (Wiesbaden: Steiner, 1970), pp. 195–206, where the experimental results are summarized.

23. Kangro ed., *Planck's Original*, p. 4 (36).

24. Planck, *Autobiography*, p. 40.

25. Kangro ed., *Planck's Original*, p. 5 (37).

26. Planck, *Autobiography*, p. 41.

27. This is Planck's own statement of what he is doing, Kangro ed., *Planck's Original*, p. 7 (39).

28. Max Planck, "Ueber das Gesetz der Energieverteilung im Normalspectrum," *Ann. der Phys.* 4 (1901): 553–64.

29. Though Boltzmann developed the relation between entropy and probability, he never used the proportionality constant k. This was Planck's contribution. It is rather ironic that the Boltzmann memorial in Vienna is a large bust of Boltzmann with the single formula $S = k \log W$.

30. Planck's statement of this is in his December address, Kangro ed., *Planck's original*, p. 8 (40), and in the *Annalen* paper, p. 556. This remains the principal basis of the claim that Planck introduced the concept of energy quantization.

31. Stirling's formula in first approximation is $N! = 2\pi N (N/e)^N$, where e is the basis of the natural logarithms. Even at relatively low values of N the discrepancy between Planck's and the true approximation is extreme. Thus, for $N = 50$, $N! = 3.041 \times 10^{64}$. The true first approximation to Stirling's formula gives 3.036×10^{64}, which is off by less than 0.2 percent. Planck's formula $N! = N^N$ gives 8.88×10^{84}, which is off by a factor of 2×10^{20}. This is about the ratio of one millisecond to the estimated age of the universe, 17 billion years. This extreme discrepancy, however, does not introduce a significant error. On the assumptions $P \gg N \gg 1$, formula (4.21) becomes $R = 1/2\pi N(N + P)^{N+P}/N^N P^P$. When one takes the logarithm, the square root term contributes to the (then unknown) additive constant and may be disregarded.

32. This is in the December address, Kangro ed., *Planck's Original*, pp. 12–14 (44–45), and in an article appended to his 1901 derivation "Ueber die Elementarquanta der Materie und

der Elektricität," *Ann. der Phys.* 4 (1901): 564–66. Here, as in his other writings prior to the 1911 Solvay Conference, Planck used the term 'quantum' for elementary bits of matter or electricity.

33. Max Planck, *Eight Lectures on Theoretical Physics: Delivered at Columbia University in 1909*, trans. A. P. Wills (New York: Columbia University Press, 1915), p. 95.

34. Cornelius Lanczos, *The Einstein Decade (1905–1915)* (New York: Academic Press, 1974), contains synopses of Einstein's papers during this period. Of the sixty-seven works outlined, twenty-three are concerned with statistical reasoning.

35. A. Einstein, "On a Heuristic Point of View about the Creation and Conversion of Light," trans. ter Haar, *Quantum Theory*, pp. 91–107. A survey of competing theories is given in Roger Stuewer's "Non-Einsteinian Interpretations of the Photoelectric Effect," *Minnesota Studies in the Philosophy of Science*, vol. 5, pp. 246–63.

36. A. Einstein, "Die Plancksche Theorie der Strahlung und die Theorie der spezifischen Wärme," *Ann. der Phys.* 22 (1907): 180–90.

37. Ibid., p. 184.

38. A. Einstein, "Zum gegenwärtigen Stand des Strahlungsproblems," *Phys. Zeit.* 10 (1909): 185–93.

39. See Roger H. Stuewer, *The Compton Effect: Turning Point in Physics* (New York: Science History Publications, 1975), chaps. 1, 2; and J. L. Heilbron, *H. G. S. Mosley: The Life and Letters of an English Physicist, 1887–1915* (Berkeley: University of California Press, 1974), chap. 5, for surveys of these developments.

40. An outline of the work Einstein did on this problem may be found in Hermann, *Genesis of Quantum Theory*, pp. 59–64.

41. A. Einstein, "Autobiographical Notes," in Schilpp, p. 45.

42. See Hermann, *Genesis of Quantum Theory*, chap. 2. The letter to Wien is on p. 36. See also Kuhn, *Black-Body Theory*, pp. 188–96.

43. Hermann devotes a chapter to each of these men. A more detailed survey may be found in John Lewis Heilbron, "A History of the Problem of Atomic Structure from the Discovery of the Electron to the Beginning of Quantum Mechanics," Ph.D. dissertation, University of California, Berkeley, 1964.

44. A detailed account of this proof, developed while Ehrenfest was effectively exiled in Russia, may be found in Martin J. Klein, *Paul Ehrenfest*, vol. 1, *The Making of a Theoretical Physicist* (New York: American Elsevier, 1970), chap. 10.

45. See Kurt Mendelssohn, *The World of Walther Nernst* (Pittsburgh: University of Pittsburgh Press, 1973). These developments and the 1911 Solvay Conference are discussed in M. Klein's "Einstein, Specific Heats, and the Early Quantum Theory," *Science* 148 (1965): 173–80.

46. The papers and discussions are in P. Langevin and M. de Broglie, eds., *La théorie du rayonnement et les quanta: Rapports et discussions* (Paris: Gauthier-Villars, 1912). This will be cited as Solvay, 1911. A survey of the Solvay Conferences is given in Jagdish Mehra, *The Solvay Conferences on Physics* (Dordrecht: Reidel, 1975). The citation from Einstein's letter is taken from Mehra, p. xiv (note).

47. Solvay, 1911, pp. 53–73.

48. Ibid., p. 77.

49. Ibid., pp. 78–86.

50. Ibid., pp. 93–114.

51. Ibid., p. 436.

52. Ibid., p. 293.

53. H. Poincaré, "Sur la théorie des quanta," *Journal de Physique* 2 (1912): 5–34.

54. J. H. Jeans, *Report on Radiation and the Quantum Theory* (London: Physical Society

of London, 1914). In the light of Jeans's strong opposition to Planck's formula, this is a surprisingly good survey of all the basic developments.

55. Wilhelm Wien, "On the Laws of Thermal Radiation," *Nobel Lectures: Physics, 1901–1921* (Amsterdam: Elsevier, 1967), pp. 275–86, citation p. 286.

Chapter Five

1. SHQP, interview 5, p. 3.

2. Jammer, *History*, pp. 172–00.

3. Gerald Holton, "The Roots of Complementarity," in his *Thematic Origins of Scientific Thought* (Cambridge, Mass.: Harvard University Press, 1973), pp. 115–61.

4. L. Rosenfeld, introduction, in his edition of Bohr's *On the Constitution of Atoms and Molecules* (Copenhagen: Munksgaard, 1963), pp. xi–liv.

5. John L. Heilbron and Thomas S. Kuhn, "The Genesis of the Bohr Atom," HSPS, vol. 1, pp. 211–90.

6. Bohr, *Works*, vol. 1, p. 503.

7. This is Bohr's own interpretation of the history as presented in his M.A. paper, "Electron Theory of Metals," trans. in Bohr, *Works*, vol. 1, pp. 131–61.

8. Ibid., p. 158.

9. "Studies on the Electron Theory of Metals," in Bohr, *Works*, vol. 1, pp. 167–290 (Danish); pp. 291–395 (English); citation from pp. 293–99.

10. J. H. Van Vleck, "Quantum Mechanics: The Key to Understanding Magnetism," *Science* 201 (July 1978): 113–20.

11. Ibid., p. 113.

12. In a letter to his brother, Harald, on 23 December 1912, Bohr uses the term 'classical state of the atom' (*klassiske Tilstand af Atomerne*); in Bohr, *Works*, vol. 1, p. 563.

13. Ibid., p. 300.

14. Ibid., p. 339.

15. Ibid., p. 378.

16. Ibid., pp. 380–83, esp. Bohr's footnote on pp. 382–83.

17. Ibid., pp. 383–92, for the problems just summarized.

18. Ibid., p. 389. This conclusion is drawn in the type of understatement that characterized most of Bohr's writings.

19. Bohr's handwritten copy of this address is photographically reproduced in *Works*, vol. 1, pp. 413–20, citation from p. 419.

20. P. Zeeman's original paper, "On the Influence of Magnetism on the Nature of the Light Emitted by a Substance," *Phil. Mag.* 43 (1897): 226–36, contains a summary of Lorentz's explanation, pp. 232–36. Lorentz's more complete account is contained in his *Theory of Electrons and Its Application to the Phenomena of Light and Radiant Heat* (Dover reprint, 1952, of 2d ed., 1915), pp. 98–103.

21. J. J. Thomson, "On the Structure of the Atom," *Phil. Mag.* 7 (1904): 237–65.

22. Lord Rayleigh, "On Electrical Vibrations and the Constitution of the Atom," *Phil. Mag.* 11 (1906); reproduced in *Scientific Papers*, vol. 5, pp. 287–91.

23. SHQP, interview 1, pp. 7–8.

24. In addition to the references cited in chap. 4, note 32, surveys of the then current atomic models are given in G. K. T. Conn and H. D. Turner, *The Evolution of the Nuclear Atom* (London and New York: American Elesvier, 1965), chaps. 4–7; and in Carl E. Behrens, "Atomic Theory from 1904 to 1913," *Amer. J. of Phys.*, 11 (1943): 60–78, and "The Early Development of the Bohr Atom," Ibid., pp. 135–47, 272–81.

25. Heilbron and Kuhn, *Genesis*, pp. 230–31.

26. Some discussion of the possible reasons for this neglect is contained in E. N. da C. Andrade, *Rutherford and the Nature of the Atom* (New York: Doubleday Anchor, 1964), pp. 109–22. Andrade worked with Rutherford in the period prior to World War I.

27. Bohr, *Works*, vol. 1, p. 559.

28. This Rutherford memorial is reproduced, except for some unessential omissions, in Rosenfeld, *Constitution of Atoms*, pp. xii–xxviii. This memorial is primarily concerned with the conditions of mechanical stability; radiative instability is not discussed.

29. Ibid., p. xxiii.

30. In a letter to C. W. Oseen, 1 December 1911, in Bohr, *Works*, vol. 1, p. 427. Similarly, in SHQP, interview 1, p. 11, Bohr claimed that he thought Nicholson was just playing with numbers.

31. An analysis of Nicholson's work is given in Russell McCormmach, "The Atomic Theory of John William Nicholson," *Archives Hist. Exact Sc.* 3 (1966): 160–84.

32. In Bohr, *Works*, vol. 1, p. 563. In SHQP, interview 3, p. 1, Bohr explained his reaction to Nicholson's work: "Nicholson seems to have got some agreement with something, so let's look for how it could be."

33. SHQP, interview 3, p. 11.

34. Heilbron and Kuhn, *Genesis*, pp. 262–66.

35. Ibid., p. 266.

36. Moseley was already familiar with the idea introduced by A. van den Broek some six months earlier that atomic number rather than mass is the key parameter for a systematic ordering of the elements, but Moseley did nothing to test or develop it. Bohr visited Manchester toward the end of the first week in July 1913 and had his first long conversation with Moseley. During this visit Bohr gave his first public presentation of his new model of the atom. After the discussion Moseley declared, as Hevesy recalled it, "We shall see what quantity determines the X-ray spectra." This is from J. L. Heilbron, *H. G. J. Moseley: The Life and Letters of an English Physicist* (Berkeley: University of California Press, 1974), p. 84.

37. N. Bohr, "On the Constitution of Atoms and Molecules," *Phil. Mag.* 26 (1913): 1–25, 476–502, 857–75. These were published in book form by Rosenfeld, *Constitution of Atoms*.

38. Ibid.

39. Schilpp, pp. 46–47.

40. BSC 3.

41. A historical account of these developments may be found in L. Rosenfeld and E. Rüdinger, "The Decisive Years," in S. Rozental, ed., *Niels Bohr: His Life and Work as Seen by His Friends and Colleagues* (Amsterdam: North-Holland, 1967), pp. 38–73.

42. N. Bohr, "On the Effects of Electrical and Magnetic Fields on Spectral Lines," *Phil. Mag.* 27 (1914): 506–24.

43. N. Bohr, "On the Quantum Theory of Radiation and the Structure of the Atom," *Phil. Mag.* 30 (1915): 394–415. The assumptions cited are on pp. 396–98.

44. J. Franck and G. Hertz, "On the Excitation of the 2536 Å Mercury Resonance Line by Electron Collisions," trans. in ter Haar, *Old Quantum Theory*, pp. 160–66.

45. A survey of the difficulties and changes involved in interpreting this experiment may be found in George L. Trigg, *Crucial Experiments in Modern Physics* (New York: Van Nostrand, 1971), pp. 69–75.

46. The chronological development is traced out in detail in Bohr, *Works*, vol. 3, pp. 1–46.

47. A. Sommerfeld, "Application de la théorie de l'élement d'action aux phénomènes moléculaires non périodiques," Solvay, 1911, pp. 313–72; rule cited is on page 316.

48. A. Sommerfeld, "Zur Theorie der Balmerschen Serie," *Sitzungsb. der Kgl. Bayer* (1915), pp. 425–58. The key idea of quantizing action integrals was developed independently by W. Wilson and J. Ishiwara. Their work is discussed in Jammer, *History*, pp. 89–109.

49. Paul Forman summarized Sommerfeld's work and the interpretative difficulties he encountered in his "Alfred Landé and the Anomalous Zeeman Effect, 1919–21," HSPS, vol. 2, pp. 153–261, esp. pp. 179–221.

50. A detailed summary of this work may be found in George Birtwistle, *The Quantum Theory of the Atom* (Cambridge: Cambridge University Press, 1929) chaps. 5–9.

51. P. Ehrenfest published accounts of his theorem in English, German, and Dutch. The English version is in van der Waerden, *Sources*, pp. 79–93. A general account of this development is given in Martin J. Klein, *Paul Ehrenfest*, vol. 1, *The Making of a Theoretical Physicist* (New York: American Elsevier, 1970), chap. 11.

52. N. Bohr, "On the Quantum Theory of Line-Spectra," *Dan. Vid. Selsk. Skrifter, naturvid.-mat. Afd.* (1918), pp. 1–100; reprinted in Bohr, *Works*, vol. 3, pp. 67–166. The summary given here slightly simplifies Bohr's account to avoid discussing Hamilton-Jacobi equations and perturbation theory.

53. Klaus Michael Meyer-Abich, *Korrespondenz, Individualität und Komplementarität: Eine Studie zur Geistesgeschichte der Quantentheorie in den Beiträgen Niels Bohr* (Wiesbaden: Franz Steiner Verlag, 1965), trans. from p. 86.

54. A summary of the results obtained by using Bohr's correspondence principle is given in A. Sommerfeld, *Atomic Structure and Spectral Lines*, trans. from the 3d (1922) German ed. by H. L. Brose (London: Methuen, 1923), chap. 6.

55. N. Bohr, "On the Series Spectra of the Elements," in Bohr, *Works*, vol. 3, pp. 242–82, citation from pp. 245–46. Later (p. 281) the article states that the general idea applies to cases such as the anomalous Zeeman effect for which an adequate descriptive account has not yet been given.

56. The gradual changes in Bohr's position on these points is traced out in the introduction to Bohr, *Works*, vol. 4.

57. Kossel's work and its influence are analyzed in J. L. Heilbron, "The Kossel-Sommerfeld Theory and the Ring Atom," *Isis* 58 (1967): 451–85.

58. Forman, "Alfred Landé," pp. 160–78, summarizes their work.

59. This dating comes from a letter of Bohr's to Irving Langmuir, 3 December 1920; in BSC 8, sec. 4.

60. This summary is based on Irving Langmuir's "The Arrangement of Electrons in Atoms and Molecules," *Journal of the American Chemical Society* 41 (1919): 868–92, 931–32; reprinted in R. Westfall and V. Thoren, *Steps in the Scientific Tradition* (New York: Wiley, 1968), pp. 525–44.

61. Ibid., p. 525.

62. N. R. Campbell, "Atomic Structure," *Nature* 106 (25 November 1920): 408–9.

63. N. Bohr, "Atomic Structure," *Nature* 107 (24 March 1921): 104–7. This letter also contains a preliminary account of Bohr's explanation of the periodic table. The citation is from p. 104. Bohr's distinctive use of the term 'formal' may have been suggested by Sommerfeld. In a letter of 5 February 1919, Sommerfeld referred to Bohr's correspondence principle as 'Ihr formales Analogie-Princip' (Bohr, *Works*, vol. 3, p. 688).

64. N. Bohr, "The Structure of the Atom and the Physical and Chemical Properties of the Elements." We are following the English translation given in Bohr's *Theory of Spectra and Atomic Constitution*, 2d ed. (Cambridge: Cambridge University Press, 1924), pp. 61–126, with an updating appendix, pp. 127–38. This is reproduced with the original pagination preserved in Bohr, *Works*, vol. 4.

65. N. Bohr, "The Structure of the Atom," trans. in Bohr, *Works*, vol. 3, pp. 467–82.

66. N. Bohr, "On the Interaction between Light and Matter," trans. in Bohr, *Works*, vol. 3, pp. 227–40, citation from p. 235.

67. Bohr, *Works*, vol. 3, p. 432.

68. Bohr, *Works*, vol. 4, p. 351.

69. N. Bohr, "Über die Anwendung der Quantentheorie auf den Atombau," *Zs. f. Phsy.* 13 (1923): 117–65, trans. in *Proceedings of the Cambridge Philosophical Society Supplement*, 1924, pp. 1–42, citation from p. 1. Bohr's intense concern with an overall coherence is most vividly manifested in a letter he wrote to Sommerfeld on 30 April 1922: "For me it is not a matter of didactic trifles but of such an inner coherence that one can attain a secure basis for futher development. I understand quite well how little the matters are clarified as yet, and how helpless I am at expressing my thoughts in easily accessible form." Translated from Bohr, *Works*, vol. 3, pp. 691–92.

70. Bohr, *Works*, vol. 3, p. 35.

71. Ibid., p. 38.

72. A. H. Compton, "A Quantum Theory of the Scattering of X-rays by Light Elements," *Phys. Rev.* 21 (1923): 483–502.

73. N. Bohr, "Zur Polarisation des Fluorescenzlichtes," *Naturwiss.* 12 (1924): 1115–17.

74. Edmund C. Stoner, "The Distribution of Electrons among Atomic Levels," *Phil. Mag.* 48 (1924): 719–36.

75. Cited from Bohr, *Works*, vol. 4, p. 41.

76. Bohr, "Structure," pp. 80–81.

77. W. Pauli, "Über den Einfluss der Geschwindikeitsabhangigkeit der Electronmasse auf den Zeemaneffekt," *Zs. f. Phys.* 31 (1925): 373–85.

78. W. Heinsenberg, "Zur Quantentheorie der Multiplettstruktur und der anomalen Zeemaneffekte," *Zs. f. Phys.* 32 (1925): 841–60. Apart from its role in the Landé g formula, Heisenberg argued, the ℓ number characterizes the ellpiticity of an orbit only when used to specify an optical electron above a closed shell.

79. The crisis that quantum theory experienced in 1923 is strikingly brought out in Paul Forman's "The Doublet Riddle and Atomic Physics *circa* 1924," *Isis* 59 (1968): 156–74.

80. A. Einstein, "Zur Quantentheorie der Strahlung," *Phys. Zeit.*, vol 18 (1917), but printed elsewhere in 1916. It is translated in van der Waerden, pp. 63–78.

81. The changes in Compton's interpretative perspective are discussed in detail in R. H. Stuewer, *The Compton Effect: Turning Point in Physics* (New York: Science History, 1975).

82. Einstein, "Zur Quantentheorie," p. 76.

83. J. C. Slater, "Radiation and Atoms," *Nature* 113 (28 January 1924): 307.

84. Slater's disagreement was brought out in a letter to van der Waerden and cited on p. 13 of his book. The disagreement is discussed by Slater in his *Solid-State and Molecular Theory: A Scientific Biography* (New York: Wiley, 1975), chap. 2, where Bohr's antagonism to the light-quantum hypothesis is stressed.

85. N. Bohr, H. A. Kramers, and J. C. Slater, "The Quantum Theory of Radiation," *Phil. Mag.* 47 (1924): 785–802, reprinted in van der Waerden, pp. 159–76; citation from p. 785. That the ideas expressed were Bohr's rather than just Kramer's seems clear from Bohr's presentation of basically the same ideas in a draft, probably written in 1923 or 1924 on "Problems of the Atomic Theory," in Bohr, *Works*, vol. 3, pp. 569–74.

86. Bohr, Kramers, and Slater, "Quantum Theory of Radiation," in Bohr, *Works*, vol. 3, p. 790.

87. Thus, in a letter to A. A. Michelson, 7 February 1924 (BSC 14, sec. 1) Bohr wrote: "It may perhaps interest you to hear that it appears to be possible for a believer in the essential reality of the quantum theory to take a view which may harmonize with the essential reality of the wave theory conception even more closely than the views I expressed during our conversation. In fact, on the basis of the correspondence principle, it seems possible to connect the discontinuous processes occurring in atoms with the continuous character of the radiation field in a somewhat more adequate way than hitherto perceived. The essentially new idea involved has been suggested by Dr. Slater."

88. Bohr, "Zur Polarisation."

89. Mara Beller, "Reality and Acausality in Quantum Physics 1919–1927: The Conservative Aspect of the Quantum Revolution," preprint.

90. M. J. Klein, "The First Phase of the Bohr-Einstein Dialogue," HSPS, vol. 2, pp. 1–41, esp. pp. 29–39.

91. W. Bothe and H. Geiger, "Über das Wesen des Comptoneffekts; ein experimenteller Beitrag aur Theorie der Strahlung," *Zs. f. Phys.* 32 (1925): 639–63.

92. A. H. Compton and A. W. Simon, "Directed Quanta of Scattered X-Rays," *Phys. Rev.* 26 (1925): 289–99.

93. In BSC 11, sec. 2.

94. Cited in Klein "First Phase," p. 36.

95. N. Bohr, "Atomic Theory and Mechanics." A revised version is contained in Bohr's *Atomic Theory and the Description of Nature* (Cambridge: Cambridge University Press, 1934), p. 34. The claim that Bohr was not yet familiar with Heisenberg's breakthrough is based on the Bohr-Heisenberg correspondence. Though Heisenberg wrote to Bohr from Helgoland on 8 June when he was developing quantum mechanics, he said little about his own work except that it was going slowly. In a letter of 31 August 1925, Heisenberg wrote that he had done something in quantum mechanics, that Kramers was familiar with it, and that he would send Bohr a copy soon. From BSC 11, sec. 2.

Chapter Six

1. Hund's appendix, "Centers of Research," pp. 245–50, presents a general survey.

2. For a general outline of Planck's philosophy of science, see Stanley Goldberg, "Max Planck's Philosophy of Nature and His Elaboration of the Special Theory of Relativity," in HSPS, vol. 7, pp. 125–60; and Victor Lenzen, "Planck's Philosophy of Science," in H. M. Evans, ed., *Men and Moments in the History of Science* (Seattle: University of Washington Press, 1959), pp. 112–19. Max Planck, *The Philosophy of Physics*, trans. W. H. Johnson (London: Allen & Unwin, 1936), contains four lectures Planck gave in the thirties. His basic position seems to have changed little between the twenties and the thirties. John Heilbron, ed., *Max Planck, a Bibliography of His Non-technical Writings* (Berkeley: Office of History of Science and Technology, University of California, 1977), shows the extensive scope and presumably wide influence of Planck's quasi-philosophical writings.

3. See Paul Forman, "Weimar Culture, Causality, and Quantum Theory, 1918–1927: Adaption by German Physicists and Mathematicians to a Hostile Intellectual Environment," HSPS, vol. 3, pp. 1–115. Though Lewis S. Feuer's *Einstein and the Generation of Science* (New York: Basic Books, 1974) covers much the same period, I cannot recommend his method of interpretation.

4. My article "Heisenberg, Models, and the Rise of Matrix Mechanics," HSPS, vol. 8, pp. 137–88, gives a more detailed explanation of Heisenberg's paper initiating quantum mechanics. Though it omits the development of quantum mechanics, Daniel Serwer's "Unmechanischer Zwang: Pauli, Heisenberg, and the Rejection of the Mechanical Atom, 1923–25," HSPS, vol. 8, pp. 189–256, has some useful information on Pauli's development and on the intermediate stages of Heisenberg's development. Neither of these studies adequately explains Heisenberg's core model. The complex and confusing history of the various adjustments, expedients, rule violations, and partial successes centering around Heisenberg's adaption of the core model was finally clarified in David Cassidy's "Heisenberg's First Core Model of the Atom: The Formation of a Professional Style," HSPS, vol. 10, pp. 187–224.

5. W. Heisenberg, "Die absoluten Dimensionen der Karmanschen Wirbelbewegung," *Phys. Zeit.* 23 (1922): 363–66.

6. For a general history of the problem to 1921, see Paul Forman, "Alfred Landé and the Anomalous Zeeman Effect, 1919–1921," HSPS, vol. 2, pp. 153–261.

7. M. A. Catalan, "Series and Other Regularities in the Spectrum of Manganese," *Philosophical Transactions of the Royal Society* 233 (1923): 127–73.

8. A. Landé, "Über den anomalen Zeeman effekt," *Zs. f. Phys.* 5 (1921): 231–41; 7 (1921): 398–405. In his original account Landé gave three different *g* factors for singlets, doublets, and triplets, Half-integral quantum numbers entered as a projection of the azimuthal quantum number on the direction established by an external magnetic field.

9. W. Heisenberg, "Quantentheorie der Linienstruktur und die anomalen Zeemaneffekte," *Zs. f. Phys.* 8 (1922): 273–99.

10. This is from a letter Heisenberg wrote to Pauli, 19 November 1921, included in *Wolfgang Pauli: Wissenschaftlicher Briefwechsel mit Bohr, Einstein, Heisenberg, u.a.*, vol. 1, *1919–1929*, ed. A. Hermann, K. v. Meyenn, and V. F. Weisskopf (New York: Springer-Verlag, 1979), citation from p. 38. Henceforth this will be referred to as *Pauli Briefwechsel*.

11. *Pauli Briefwechsel*, p. 143.

12. A. Landé, "Termstruktur und Zeemaneffekt der Multiplets," *Zs. f. Phys.* 15 (1923): 189–205; and A. Landé and W. Heisenberg, "Termstruktur der Multipletts höherer Stufe," *Zs. f. Phys.* 25 (1924): 279–86.

13. "It became clearer than ever that an explanation of the anomalous Zeeman effect must bring about profound modifications in our quantum theoretical conceptions. This is especially notable in the failure of the *Aufbauprinzip* with respect to statistical weights of the atom core and electrons," W. Heisenberg, "Über ein Abänderung der formalen Regeln der Quantentheorie beim Problem der anomalen Zeemaneffekte," *Zs. f. Phys.* 26 (1924): 291–307, citation from p. 291.

14. Ibid.

15. M. Born and W. Heisenberg, "Über Phasenbeziehungen bei den Bohrschen Modellen von Atomen und Molekeln," *Zs. f. Phys.* 16 (1923): 229–43; "Zur Quantentheorie der Molekeln," *Ann. der Phys.* 74 (1924): 1–31; and "Über den Einfluss der Deformierbarkeit der Ionen auf optische und chemische Konstanten," *Zs. F. Phys.* 23 (1924): 388–410; 26 (1924): 196–204.

16. This is taken from Max Born, *The Born-Einstein Letters: Correspondence between Albert Einstein and Max and Hedwig Born from 1916 to 1955 with Commentaries by Max Born, Translated by Irene Born* (New York: Walker & Co. 1971), pp. 75–76.

17. In his retrospective comment on this letter Born said: "Heisenberg and I pursued a different objective [from Bohr's concern with the periodic table]. We had reason to doubt that Bohr's ingenious but basically incomprehensible combination of quantum rules with classical mechanics was correct. We therefore intended to carry out a thorough calculation of the two-body problem of the helium atom (nucleus with two electrons), and thus needed to use Poincaré's rigorous approximation technique. The result was quite negative, and that led us finally to turn our backs on classical mechanics and establish a new quantum mechanics," ibid., pp. 78–79. Max Born's paper "Über Quantenmechanik," *Zs. f. Phys.* 26 (1924): 379–95, translated in van der Waerden, pp. 181–98, was intended as a first step toward a new quantum mechanics.

18. Thus, in an interview with T. S. Kuhn (SHQP, interview 1, p. 4) Heisenberg said: "But the strongest impression on me at that time was that Bohr thought so differently on these problems from Sommerfeld. He never looked on the problems from the mathematical point of view, but from the physics point of view.

"I should say that I have learned more from Bohr than from anybody else that the new type of theoretical physics which was almost more experimental than theoretical. That is, you have to cover the experimental situation by means of concepts which fit. Later on you have to put the concepts into mathematical form, but that is more or less a trivial process which has to be solved. But the primary thing here is that you must find the words and concepts to describe a funny situation in physics which is very difficult to understand."

19. For the historical influence of Compton's discovery, see R. H. Stuewer, *The Compton Effect: Turning Point in Physics* (New York: Science History Publications, 1975).

20. R. Ladenburg, "Die quantentheoretische Deutung der Zahl der Dispersions-elektronen," *Zs. f. Phys.* 4 (1921): 451–68, trans. in van der Waerden, pp. 139–57. The problem of dispersion in the old quantum theory is summarized in Jammer, *History*, pp. 181–95, and in van der Waerden, pp. 9–18.

21. N. Bohr, "Über die Anwendung der Quantentheorie auf den Auombau," *Zs. f. Phys.* 13: 117–65; trans. in *Proceedings of the Cambridge Philosophical Society Supplement* (1924), pp. 1–24, see esp. p. 38.

22. A. Smekal, "Zur Quantentheorie der Dispersion," *Naturwiss.* 11 (1923): 873–75.

23. H. A. Kramers, "The Law of Dispersion and Bohr's Theory of Spectra," *Nature* 113 (1924): 673–76, reproduced in van der Waerden, pp. 177–80.

24. H. A. Kramers and W. Heisenberg, "Über die Streuung von Strahlen durch Atome," *Zs. f. Phys.* 31 (1925): 681–708, trans. in van der Waerden, pp. 223–52.

25. A more detailed explanation of this development may be found in MacKinnon, "Heisenberg, Models," pp. 149–55.

26. W. Heisenberg, "Anwendung des Korrespondenzprinzips auf die Frage nach der Polarisation Fluoreszenzlichtes," *Zs. f. Phys.* 31 (1925): 617–26, citation from p. 621.

27. Thus, in a letter to me of 12 July 1974 commenting on the first draft of my article "Heisenberg, Models, and the Rise of Matrix Mechanics," Heisenberg wrote: "I was especially glad to see that you noticed how important the paper on the polarization of fluorescent light has been for my further work on quantum mechanics. Actually in Copenhagen I felt that this paper contained the first step in which I could go beyond the views of Bohr and Kramers."

28. Heisenberg's personality and public career are treated in Armin Hermann, *Werner Heisenberg: 1901–1976*, trans. Timothy Nevill (Bonn-Bad Godesberg: Inter Nationes, 1976). Heisenberg summarized Pauli's influence on him in "Erinnerungen an die Zeit der Entwicklung der Quantenmechanik," in M. Fierz and V. Weisskopf, eds., *Theoretical Physics in the Twentieth Century: A Memorial Volume to Wolfgang Pauli* (New York: Interscience, 1960), pp. 40–47; and in "Wolfgang Pauli's Philosophical Outlook," in Heisenberg's *Across the Frontiers*, trans. Peter Heath (New York: Harper & Row, 1974), pp. 30–38.

29. This philosophical background is discussed in M. Fierz's article "Wolfgang Pauli, Jr." in *Dictionary of Scientific Biography*, vol. 10 (New York, 1974), pp. 422–25. Fierz reports the surprising fact that, though Pauli was Jewish, he was baptized Catholic with Mach as his godfather.

30. W. Heisenberg, *Der Teil und das Ganze: Gespräche im Umkriss der Atomphysik* (Munich: R. Piper, 1969), p. 56, trans. A. J. Pomerans as *Physics and Beyond: Encounters and Conversations* (New York: Harper & Row, 1971).

31. W. Pauli, Jr., "Merkurperihelbewegung und Strahlenablenkung in Weyls Gravita-tionstheorie," *Verhandlungen der Deutschen Physikalischen Gesellschaft*, vol. 21 (1919), citation from pp. 749–50. This is reproduced in R. Kronig and V. Weisskopf, eds., *Collected Scientific Papers by Wolfgang Pauli* (New York: Interscience, 1964), vol. 2, pp. 1–9.

32. W. Pauli, "Über die gesetzmässigkeiten des anomalen Zeemaneffektes," *Zs. f. Phys.* 16 1923): 155–64.

33. Paul Forman, "The Doublet Riddle and Atomic Physics *circa* 1924," *Isis* 59 (1968): 156–74.

34. *Pauli Briefwechsel*, p. 147.

35. W. Heisenberg to N. Bohr, 25 August 1924, BSC 11, sec. 2.

36. Pauli's letter to Bohr of 12 December 1924, in *Pauli Briefwechsel*, pp. 186–89. Pauli sent this with a draft of his paper on the exclusion principle. This letter summarizes arguments against the core model, arguments which are developed in detail in Pauli's "Über den Einfluss der Geschwindigkerisabhängigkeit der Elektronenmasse auf den Zeemaneffekt," *Zs. f. Phys.* 31 (1925): 373–85.

37. The exclusion principle was given in W. Pauli, "Über den Zusammenhang des Abschlusses der Elektronengruppen im Atom mit der Komplexstruktur der Spektren," *Zs. f. Phys.* 31 (1925): 765–85, trans. in D. ter Haar, *The Old Quantum Theory* (Oxford: Pergamon Press, 1967), pp. 184–203. The reminiscence is from Pauli's Nobel Prize address, contained in Boorse and Motz, *World of the Atom,* pp. 970–84, citation from p. 972.

38. Heisenberg to Pauli, 15 December 1924, citation from *Pauli Briefwechsel*, p. 192.

39. *Pauli Briefwechsel*, p. 188.

40. W. Heisenberg, "Zur Quantentheorie der Multiplettstruktur und der anomalen Zeemaneffekte," *Zs. f. Phys.* 32 (1925): 841–60.

41. MacKinnon, "Heisenberg, Models," pp. 162–88.

42. R. Kronig, the recipient, has reproduced the main part of this letter in Fierz and Weisskopf *Pauli Memorial*, pp. 23–25. Three years earlier Heisenberg had discussed this anharmonic oscillator in a letter to Pauli of 29 September 1922 (*Pauli Briefwechsel*, p. 67). There it supplied a convenient classical basis for calculating damped harmonic radiation which could approximate the discontinuous radiation proper to quantum theory.

43. A. Sommerfeld and W. Heisenberg, "Die Intensität der Mehrfachlinien und ihrer Zeemankomponenten," *Zs. F. Phys.* 11 (1922): 131–54.

44. Born, "Über Quantentheorie." In his third footnote he thanks Heisenberg for his advice and help with the calculations.

45. This calculation is worked out in George Birtwistle, *The New Quantum Mechanics* (Cambridge: Cambridge University Press, 1928), chap. 12, using matrix mechanics, and in his earler book *Quantum Theory of the Atom* (Cambridge: Cambridge University Press, 1926), chap. 17, using the earlier Born perturbation method.

46. *Pauli Briefwechsel*, p. 231.

47. M. Born and P. Jordan, "Zur Quantenmechanik," *Zs. f. Phys.* 34 (1925): 858–88, partially trans. in van der Waerden, pp. 277–306.

48. M. Born, W. Heisenberg, and P. Jordan, "Zur Quantenmechanik II," *Zs. f. Phys.* 35 (1926): 557–615, trans. in van der Waerden, pp. 321–85. Van der Waerden discussed this paper with each of the authors and, in his reconstruction (pp. 42–54), carefully explains who wrote each part.

49. This historical sketch is based on two articles in Fierz and Weisskopf, *Pauli Memorial*: R. Kronig's "The Turning Point" and B. L. van der Waerden's "Exclusion Principle and Spin" and on the historical reminiscences Samuel Goudsmit and George Uhlenbeck gave in *Physics Today* 29 (June 1976): 40–48.

50. G. E. Uhlenbeck and S. Goudsmit, "Ersetzung der Hypothese vom unmechanische Zwang durch eine Forderung bezüglich des inneren Verhaltens jedes einzelnen Elektrons," *Naturwiss.* 13 (1925): 953. This brief, one-page letter is reproduced in *Physics Today* 29 (June 1976): p. 43. They gave a more systematic account in "Spinning Electrons and the Structure of Spectra," *Nature* 117 (1926): 264-65. It is interesting to note that both Einstein and Bohr were willing to accept the idea of electron spin before either of the young radicals, Heisenberg and Pauli.

51. Pauli to Bohr, 5 February 1926, *Pauli Briefwechsel*, pp. 289–91.

52. L. H. Thomas, "The Motion of the Spinning Electron," *Nature* 177 (1926): 514.

53. Pauli, "Uber das Wasserstoffspektrum vom Standpunkt der neuen Quantenmechanik," *Zs. f. Phys.* 36 (1926): 336–63, trans. in van der Waerden, pp. 387–415.

54. Born expressed this opinion in the Born-Jordan paper and in lectures he subsequently gave at MIT, where he summarized the new developments Heisenberg has initiated: "In his brief paper the leading physical ideas are clearly stated, but only exemplified on account of the lack of appropriate mathematical equipment. The required machinery Jordan and I have developed in the matrix calculus." This is from Max Born, *Problems of Atomic Dynamics* (New York: Ungar, 1960), a reprint of the 1926 edition. Dirac expressed essentially the same opinion, accepting the physical basis of Heisenberg's new method as adequate but seeking to redevelop his mathematics along the lines of noncommutative algebra in his "Quantum Mechanics and a Preliminary Investigation of the Hydrogen Atom," *Proc. Roy. Soc. A* 110 (1926): 561–69, included in van der Waerden, pp. 417–27.

55. Heisenberg, *Der Teil*, p. 92.

56. Ibid., pp. 111–12.

Chapter Seven

1. Edward MacKinnon, "De Broglie's Thesis: A Critical Retrospective," *Amer. of Phys.* 44 (1976): 1047–55.

2. In addition to the general sources cited earlier, surveys of the development of de Broglie's thought may be found in Johannes Gerber, "Geschichte der Wellenmechanik, "*Archives Hist. Exact. Sc.* 5 (1969): 349–414; Fritz Kubli, "Louis de Broglie und die Entdeckung der Materiewellen," Ibid. 7 (1970): 26–68; and Heinrich A. Medicus, "Fifty Years of Matter Waves," *Physics Today* 27 (February 1974): 38–45; and in de Broglie's own retrospective accounts given in *Louis de Broglie, physicien et penseur*, ed. A. George (Paris: Editions Albin Michel, 1953), pp. 457–86; in his Nobel Prize address in *Nobel Lectures: Physics* (Amsterdam: Elsevier, 1965), pp. 244–56; and in his "The Beginning of Wave Mechanics," contained in William C. Price et al., eds., *Wave Mechanics: The First Fifty Years* (New York: Wiley, 1973), pp. 12–18.

3. Their collaboration resulted in a book, M. de Broglie and L. de Broglie, *Introduction à la physique des rayons X et gamma* (Paris: Gauthier-Villars, 1928).

4. Louis de Broglie, "Sur les interférences et la théorie des quanta de lumière," *C. R. Acad. Sc.* (Paris) 175 (1922): 811–13. Essentially the same derivation of Planck's law, but without de Broglie's interpretation, had been given by Planck himself in his paper at the 1911 Solvay Conference, ed. P. Langevin and M. de Broglie, *La Théorie du rayonnement et les quanta* (Paris: Gauthier-Villars, 1912), esp. p. 105.

5. L. de Broglie, "Ondes et quanta," *C. R. Acad. Sc.* (Paris) 177 (1923): 507–10.

6. The justification for this claim may be found in V. V. Raman and Paul Forman, "Why Was It Schrödinger Who Developed de Broglie's Ideas?" HSPI, vol. 1, pp. 291–314.

7. L. de Broglie, "Recherches sur la théorie des quanta," reproduced in its entirety in *Annales de Physique* 3 (1925): 22–128.

8. (S. N.) Bose, "Plancks Gesetz und Lichtquantenhypothese," *Zs. f. Phys.* 26 (1924): 178–81.

9. A. Einstein, "Quantentheorie des einatomigen idealen Gases," *Berliner Berichte* (1924), pp. 261–67 (paper 1); (1925), 3–14 (paper 2).

10. Thus, in a 1926 paper de Broglie cited Schrödinger's new developments and commented, "Up to the present I have been unable to accept this point of view; for me the associated waves have physical reality and must be expressed as functions of the three space coordinates and of the time." This is from L. de Broglie and L. Brillouin, *Selected Papers on Wave Mechanics*, trans. W. Deans (London: Blackie, 1928), p. 72.

11. This list of simultaneous developers is from E. Schrödinger, *Collected Papers on Wave Mechanics*, trans. J. Shearer and W. Deans, 2d ed. (London: Blackie, 1927), p. 126, note 1. Though his results were not published, W. Pauli also developed this equation. On this point, see

B. L. van der Waerden, "From Matrix Mechanics and Wave Mechanics to Unified Quantum Mechanics," in J. Mehra, *The Physicist's Conception of Nature* (Dordrecht: Reidel, 1973), pp. 276–96.

12. De Broglie and Brillouin, *Selected Papers*, p. 101.

13. Ibid., P. 112.

14. Pauli summarized his criticism of de Broglie's position in George, *Louis de Broglie*, pp. 33–42. Further criticism by de Broglie himself may be found in his *Introduction to the Study of Wave Mechanics*, trans. H. Flint (London: Methuen, 1930), pp. 119ff.

15. Thus, in a card he wrote to me on 27 January 1977, commenting on my article (note 1), de Broglie said: "Twenty-five years ago I reconsidered and considerably developed the ideas which were mine during the period of my doctoral thesis and the following year, ideas which I then abandoned under the influence of the Copenhagen school. I am now convinced that my initial ideas were the best and that I was wrong to abandon them." He summarized this rethinking of his initial position in his article "The Reinterpretation of Wave Mechanics," *Foundations of Physics* 1 (1970): 5–15.

16. A general survey of Schrödinger's development may be found in William T. Scott, *Erwin Schrödinger: An Introduction to His Writings* (Amherst: University of Massachusetts Press, 1967). The most detailed study of the development of Schrödinger's wave mechanics is Linda Wessels, "Schrödinger's Interpretation of Wave Mechanics" (Ph.D. dissertation, University of Indiana, 1975). This was written at the same time I was writing my article, "The Rise and Fall of the Schrödinger Interpretation," in P. Suppes, ed., *Foundations of Quantum Mechanics: The 1976 Stanford Seminar* (Lansing, Mich.: Philosophy of Science Association; 1979), pp. 1–57. The present revision of my earlier account was strongly influenced by a detailed correspondence with Professor Wessels. The account of how Schrödinger came to develop wave mechanics is based largely on her article "Schrödinger's Route to Wave Mechanics, *Stud. Hist. Phil. Sc.* 10 (1977): 311–40; and on P. A. Hanle's articles "The Coming Age of Erwin Schrödinger: His Quantum Statistics of Ideal Gases," *Archives Hist. Exact Sc.* 17 (1977): 165–92, "Erwin Schrödinger's Reaction to Louis de Broglie's Thesis on the Quantum Theory," *Isis* 68 (1977): 606–9; and "The Schrödinger-Einstein Correspondence and the Sources of Wave Mechanics," *Amer. J. of Phys.* 47 (1979): 644–48. My article in the Suppes volume was criticized by Linda Wessels in "The Intellectual Sources of Schrödinger's Interpretations" in the same volume. The present account is intended to be my synthesis of these earlier, somewhat opposed accounts.

17. From Erwin Schrödinger, *My View of the World*, trans. C. Hastings (Cambridge: Cambridge University Press, 1964), p. viii.

18. See, for example, his *What Is Life and Mind and Matter* (Cambridge: Cambridge University Press, 1967), pp. 136, 158. Schrödinger's opposition to the dualism he attributed to Kant might have predisposed him to oppose a similar dualism which he found in Bohr's doctrine of complementarity.

19. This is developed in Schrödinger's "Seek for the Road" contained in *My View*. He presents a philosophical position that is more closely related to science in his *Science and the Human Temperament*, trans. J. Murphy and W. Johnston (New York: W. W. Norton & Co., 1935). The unrevised reprint was retitled *Science, Theory, and Man* (Cambridge: Cambridge University Press, 1954). In spite of Schrödinger's professional interest in both physics and philosophy, his writings in philosophy never manifested the same analytic power that his writings in physics did.

20. This is developed in Schrödinger's *Mind and Matter*, esp. pp. 99–109.

21. Schrödinger, *My View*, preface.

22. Schrödinger's philosophical position, developed in piecemeal fashion in the works cited above, was given a coherent summary in Wessels's dissertation, "Schrödinger's Interpertation of Wave Mechanics," chap. 2.

23. This is from Schrödinger's lecture "Our Image of Matter," delivered in 1952 and included in the collection *On Modern Physics* (New York: Clarkson N. Potter, 1961), pp. 37–56, citation from p. 38.

24. From Schrödinger, *Science and the Human Temperament*, p. xiv.

25. Ibid., p. 147.

26. E. Schrödinger, "Bohrs neue Strahlungshypothese und der Energiesatz," *Naturwiss.* 12 (1924): 720–24; and his first letter to Bohr sent on 24 May 1924, BSC 16, sec. 2.

27. Schrödinger, *My View.* p. 19.

28. E. Schrödinger, "Die Energiekomponenten des Gravitationssfelds," *Phys. Zeit. 19* (1918): 4–7. A. Einstein, "Notiz zu E. Schrödingers Arbeit," *Phys. Zeit. 19* (1918): 115–16; and "Bemerkung zu Herrn Schrödingers Notiz, 'Über ein Lösungssystem der allgemeinen kovarianten Gravitationsgleichungen,'" *Phys. Zeit. 19* (1918): 165–66.

29. Schrödinger had detailed calculations of both planetary orbits and relativistic corrections in his "Die Erfüllbarkeit der Relativitatsforderung in der klassischen Mechanik," *Ann. der Phys.* 77 (1925): 325–36.

30. E. Schrödinger, "Über Eine Bemerkenswerte Eigneschaft der Quantenbahnen eines einzelnen Elektrons," *Zs. f. Phys.* 12 (1922): 170–76. The significance of this work for Schrödinger's path to wave mechanics was discussed by Ramann and Forman, "Why Schrödinger?"

31. E. Schrödinger, "Die wasserstoffähnlichen Spektrum von Standpunkt der Polarisierbarkeit des Atomrumpfes," *Ann. der Phys.* 77 (1925): 43–70.

32. The pivotal role that Schrödinger's work in gas theory played in the transition to wave mechanics was first suggested by M. Klein in "Einstein and the Wave-Particle Duality," in *Natural Philosopher* 3 (1964): 1–49, and further developed by L. Wessels and P. Hanle (see note 16).

33. E. Schrödinger, "Gasenartung und freie Weglänge," *Phys. Zeit.* 25 (1924): 41–45.

34. See note 9.

35. M. Planck, "Zur Frage der Quantelung einatomiger Gase," *Sitzungsberichte der Preussisch Akademie der Wissenschaften* (1925), pp. 49–55.

36. E. Schrödinger, "Bemerkungungen über die statistische Entropiedefinition beim idealen Gas," ibid., pp. 434–41.

37. E. Schrödinger, "Zur einsteinschen Gastheorie," *Phys. Zeit.* 27 (1926): 95–107.

38. This citation is from Hanle's *Isis* article, p. 607.

39. This citation is from Wessels's dissertation, "Schrödinger's Interpretation of Wave Mechanics," p. 45.

40. P. Debye, "Der Wahrscheinlichkeitsbegriff in der Theorie der Strahlung," *Ann. der Phys.* 33 (1910): 1427–34.

41. E. Schrödinger, *Statistical Thermodynamics*, rev. ed. (Cambridge: Cambridge University Press, 1962), chap. 7.

42. E. Schrödinger, "Quantisierung als Eigenwertproblem" (Erste Mitteilung), *Ann. der Phys.* 79 (1926): 361–76. This series was translated in *Collected Papers*, pp. 1–12. Henceforth this will be cited as WM (Wave Mechanics).

43. This was brought out in "Peter J. Debye—an Interview," with the participation of E. E. Salpeter, D. R. Corson, and Scott Bauer, *Science* 145 (1964): 554–59.

44. In an obituary notice on Schrödinger, Dirac claimed: "He immediately applied his method to the electron in the hydrogen atom, duly taking into account relativistic mechanics for the motion of the electron, as de Broglie had done. The results were not in agreement with observation. We know now that Schrödinger's method was quite correct, and the discrepancy was due solely to his not having taken the spin of the electron into account. However, electron spin was unknown at that time, and Schrödinger was very disappointed and concluded that his method was no good and abandoned it. Only after some months did he return to it, and then

noticed that if he treated the electron non-relativistically his method gave results in agreement with observation in non-relativistic approximations." This notice is in *Nature* 189 (1961): 355–56. This early relativistic work presents something of a puzzle. A former student of Schrödinger's, D. M. Denison, claimed that Schrödinger gave him a typed carbon copy of his unpublished early article on relativistic wave mechanics (in his interview with Thomas Kuhn, on 30 January, 1964, SHQP, interview 3, p. 7). Yet, no copy of this paper remains, nor is there any trace of this early work in the copious notebooks that Schrödinger has left, at least in those available in mocrofilm copy in the Berkeley Center for the History of Science.

45. This is cited in Wessels's "Schrödinger's Route," p. 57.

46. WM, p. 2. The English translation is misleading in citing the conditions Ψ must fulfill. It reads "real over the whole of co-ordinate space" where it should have "real over the whole of configuration space" as a translation of "Konfigurationenraum." The significance of this distinction will be discussed later. The later theoretical discussions of the conditions the Ψ function should meet are summarized in Jammer, *History*, pp. 267–70.

47. WM, p. 13.

48. Ibid., pp. 7–8. Because Ψ tends to zero at infinity as $1/r$, $\partial\Psi/\partial r$ tends to zero as $1/r^2$. Hence, the surface integral, involving a factor of r^2, is still of the same order at infinite as at finite distances. Allowing this would prohibit the continuous spectrum for positive total energy values. Accordingly, Schrödinger added the further postulate that the integral effectively vanishes over an infinite sphere.

49. Pascual Jordan, "Early Years of Quantum Mechanics: Some Reminiscences," in Jagdish Mehra, ed., *The Physicist's Conception of Nature* (Dordrecht: Reidel, 1973), pp. 294–99, discusses David Hilbert's Göttingen seminar on the mathematical foundations of quantum theory in which all the contributors to quantum theory except Schrödinger participated. Hilbert was convinced on general grounds that there should be eigenvalue equations having both discrete and continuous solutions and was delighted when Schrödinger found a relatively simple example of such an equation.

50. WM, p. 9.

51. "Quantisierung als Eigenwertproblem: Zweite Mitteilung," *Ann. der Phys.* 79 (1926): 489–527, trans. in WM, pp. 13–40.

52. WM, p. 16. A description of the wave front as progressive but stationary may seem contradictory. It is really Schrödinger's abbreviated way of bringing out two distinctive features of the solution. His Hamilton-Jacobi equation yields a family of surface for the action function in configuration space. It is stationary in the sense that at any earlier or later instant than the comparison instant the same group of surfaces illustrates the same action distribution. It is progressive in the sense that the action value for any particular point in configuration space is linearly dependent on time.

53. Ibid., pp. 25ff.

54. Ibid., p. 25.

55. Ibid., p. 30.

56. Ibid., p. 34.

57. E. Schrödinger, "An Undulatory Theory of the Mechanics of Atoms and Molecules," *Phys. Rev.* 28 (1926): 1049–70, citation from p. 1049.

58. The basic letters are given in K. Przibram, ed., *Letters on Wave Mechanics*, trans. M. Klein (New York: Philosophical Library, 1967). Some further letters are translated in Wessels, "Schrödinger's Interpretation of Wave Mechanics," chap. 3.

59. C. T. R. Wilson, "Investigations on X-Rays and γ-Rays by the Cloud Method," *Proc. Roy. Soc. A* 104 (1923): 1–24 and 192–212.

60. In Przibram, *Letters*, pp. 43–54, esp. p. 47.

61. Ibid., pp. 55–56.

62. E. Schrödinger, "Der stetig Übergang von der Mikro—zur Makromechanik," *Naturwiss.* 14 (1926): 664–66, trans. in WM, pp. 41–44.

63. Citation in Przibram, *Letters*, p. 61.

64. E. Schrödinger, "Über das Verhältnis der Heisenberg-Born-Jordanschen Quantenmechanik zu der meinen," *Ann. der Phys.* 79 (1926): 734–56, trans. in WM, pp. 45–61.

65. Pauli's unpublished paper on this is given and its background explained in B. L. van der Waerden, "From Matrix Mechanics and Wave Mechanics to Unified Quantum Mechanics," in Mehra, *Physicist's Conception*, pp. 276–96. On 12 April 1926 Pauli sent Jordan a letter, effectively a circular letter to the Göttingen community, in which he expressed his opinion that Schrödinger's new work ranked among the most important of modern times. Then he explained the relation between matrices and Ψ functions, $F_{nm} = \int_{-\infty}^{\infty} F(a) \, \Psi_n \, \Psi_m \, dx$. (*Pauli Briefwechsel*, pp. 317–20). He concluded this letter with the observation: "Just as in the Göttingen, so also in the Schrödinger formulation of the quantum problem essentially no spatial description of the motion of the electron in the atom is given." On 24 May 1926 Pauli sent Schrödinger a letter (ibid., pp. 324–26) in which he again explained the mathematical equivalence of the two systems, the relation which Schrödinger had discovered two months earlier. In both of these letters Pauli stated that he did not understand the real physical significance of these equations.

66. Carl Eckart, "Operator Calculus and the Solutions of Quantum Dynamics," *Phys. Rev.* 28 (1926): 711–26.

67. See James D. Watson, *The Double Helix* (New York: Atheneum Press, 1968).

68. See Nicholas Wade, "Guillemin and Schally," in *Science* 200 (1978): 279–82, 411–16, 510–13.

69. WM, p. 46, note 1.

70. Przibram, *Letters*, p. 63.

71. SHQP, microfilm 80.

72. WM, p. 59. I believe that the strongest case that could be made for an underlying continuity in Schrödinger's technical physics and his later opposition to the Copenhagen interpretation would come more from his use of the word 'intelligible' than from his philosophical reinterpretations of physics. The rationalist tradition in its variant forms has always stressed the primary role of intelligible principles in any adequate explanation of reality and tended to reject discontinuities in nature on the grounds that they are unintelligible. The empirical tradition, on the other hand, usually tends to stress the primacy of observation and is willing to accept as basic whatever laws emerge from a process of scientific investigation, inductive generalization, and hypothetical-deductive reasoning, regardless of whether these laws have an independent intelligibility or must simply be accepted as given. This criterion would put Bohr, Heisenberg, Dirac, and Pauli in the empiricist camp, while Schrödinger and Einstein, at least in his later years, would be rationalists. Their somewhat nebulous pantheistic views would accord well with the theological orientation that has usually grounded the rationalist stress on intelligibility.

73. E. Schrödinger, "Quantisierung als Eigenwertproblem: Dritte Mitteilung," *Ann. der Phys.* 80 (1926): 437–90.

74. WM, p. 83. *Zusammenbestehen* (simultaneous existence) is italicized in the original text, p. 465.

75. Przibram, *Letters*, pp. 55–66.

76. E. Schrödinger, "Quantisierung als Eigenwertproblem," Vierte Mitteilung. *Ann. der Phys.* 81 (1926): 109–39, trans. in *WM*, pp. 102–23.

77. WM, p. 120.

78. WM, pp. 111–12.

79. Hund, chap. 12, claims that in the spring of 1926 there were four apparently equivalent versions of quantum mechanics: the matrix formulation, the formulation based on Dirac's

distinction between c and q numbers, the preliminary version of the operator formalism developed by Born and N. Wiener, and Schrödinger's wave mechanics. Only the Schrödinger formulation, however, led to a physical interpretation different from that given by matrix mechanics.

80. M. Born, "Zur Quantenmechanik der Stossvorgänge," Zs. f. Phys. 37 (1926): 863–67. The interpretation presented here is a summary of that given in Linda Wessels's "What Was Born's Statistical Interpretation?" PSA 1980, vol. 2.

81. M. Born, "Quantenmechanik der Stossvorgänge," Zs. f. Phys. 38 (1926): 803–27.

82. W. Pauli, "Über Gasentartung und Paramagnetismus," Zs. f. Phys. 41 (1927): 81–102; reproduced in Pauli, Works, vol. 2, pp. 284–305. The probability interpretation is given in note 1, p. 83 (286).

83. Hund, who began his career as a quantum physicist during this period, gives a good summary of these developments in chaps. 12, 13, 14. See also George Birtwistle, The New Quantum Mechanics (Cambridge: Cambridge University Press, 1928), chaps. 21–32.

84. Born explained the differences between his and Schrödinger's interpretations of the Ψ function in a paper "Physical Aspects of Quantum Mechanics," read at Oxford on 10 August 1926 (after Q3 was published but before Q4). This is reprinted in Max Born, Physics in My Generation, 2d rev. ed. (New York: Springer-Verlag, 1969), pp. 6–12. An excellent account of the further development of the probability interpretation may be found in Jammer, History, pp. 281–93, 305–7, 320–21.

85. A detailed calculation of this difference is given in Wessels's dissertation, "Schrödinger's Interpretation of Wave Mechanics," pp. 255–66.

86. This letter, without the calculations, is in Przibram, Letters, pp. 67–75.

87. Wessels, "Intellectual Sources."

88. This is summarized in W. Heisenberg's reminiscences, Physics and Beyond: Encounters and Conversations, trans. A. J. Pomerans (New York: Harper, 1971), chap. 6.

89. In Physics and Beyond, p. 73, Heisenberg reminisced, "It must have been that same evening that I wrote to Niels Bohr about the unhappy outcome of this discussion." This letter is not in Bohr's Scientific Correspondence from Heisenberg.

90. SHQP, Heisenberg correspondence, microfilm 80.

91. Heisenberg, who was at Bohr's Institute during Schrödinger's visit, summarized the Bohr-Schrödinger interaction in "Quantum Theory and Its Interpretation," in S. Rozenthal, ed., Niels Bohr (Amsterdam: North-Holland, 1967), pp. 103–4, and in Physics and Beyond, pp. 73–76, citation from p. 75.

92. BSC 16, sec. 2.

93. E. Schrödinger, "Über den Comptoneffekt," Ann. der Phys. 82 (1927): 257–64, trans. in WM, pp. 124–29; and "Der Energieimpulssatz der Materiewellen," Ann. der Phys. 82 (1927): 265–72, trans. in WM, pp. 130–36.

94. See note 11.

95. W. Gordon, "Der Comptoneffekt nach der Schrodingerschen Theorie," Zs. f. Phys. 40 (1926): 117–33. The continuity equation is

$$\sum_{\alpha=1}^{4} \partial S_\alpha / \partial x_\alpha = 0,$$

where

$$S_\alpha = (1/\alpha)(\bar{\psi}\partial\psi/\partial x_\alpha - \psi\partial\bar{\psi}/\partial x_\alpha - 4\pi(ie/hc)\Phi_\alpha\psi\bar{\psi})$$

and Φ_α gives the four components of the electromagnetic potential. The first three components of S_α may be interpreted as the spatial distribution of a current density and the fourth component as a charge density.

96. WM, p. 129.

97. P. A. M. Dirac, "Relativity Quantum Mechanics with an Application to Compton Scattering," *Proc. Roy Soc. A* 111 (1926): 405–23. Dirac's gradual development of a unified quantum theory exerted an influence on Schrödinger. Among his unpublished papers is a twenty-four-page notebook entitled "Dirac" relating Dirac's developments to Schrödinger's own work. This is undated, but was probably written in the summer or fall of 1926. A microfilm copy is in the archives for the "History of Quantum Theory" (SHQP, 41, 2).

98. P. A. M. Dirac, "The Quantum Theory of the Emission and Absorption of Radiation," *Proc. Roy. Soc. A* 114 (1927): 243–65.

99. W. Heisenberg, "Mehrkörperproblem und Resonanz in der Quanten-Mechanik," *Zs. f. Phys.* 38 (1926): 411–26. His explanation of the distinction is: "In der physikalischen Interpretation der mathematischen Formalisms besteht jedoch ein Unterscheid zwischen Schrödingers und unserer Auffassung. Wegen der mathematischen Äquivlenz des Schrödingerschen Verfahrens mit den Quantenmechanik könnte allerdings die Frage nach den Gleichungen zugrunde liegenden physikalischen Geschehen eissteilen als eine unsere Anschauung betreffende Zweckmässigkeitsfrage betrachtet werden; aber nur solange wir nicht versuchen, auf Grund der einmal gewählten anschaulichen Bilder die Grundlage dieser Quantentheories zu erweitern." (p. 411).

100. W. Heisenberg, "Spektra von Atomsystemen mit zwei Elektronen," *Zs. f. phys.* 39 (1926): 499–519. This was sent from Copenhagen on 24 July 1926, before Bohr and Heisenberg developed the Copenhagen interpretation.

101. W. Heisenberg, "Quantenmechanik," *Naturwiss.* 14 (1926): 989–95.

102. W. Heisenberg, "Schwankungerscheinungen und Quantenmechanik," *Zs f. Phys.* 40 (1927): 501–6. When he sent a copy of this to Pauli, Heisenberg sent an accompanying letter expressing guilt feelings because the paper really contained no new physics, and then he explained, "The reason why I have written it is really only a pedagogical one, in opposition to the Lords of the Continuum theory." (*Pauli Briefwechsel*, p. 352). Fifty years later he expanded the details of this pedagogical motivation: "Since Schrödinger was not quite convinced it seemed to me extremely important to decide beyond any doubt whether or not quantum 'jumps' were an unavoidable consequence, if one accepted that part of the interpretation of matrix mechanics, which already at that time was *not* controversial, namely the assumption that the diagonal element of a matrix represents the time average of the corresponding physical variable in the stationary state considered. Therefore I discussed a system consisting of two atoms in resonance." This is from Heisenberg's "Remarks on the Origin of the Relations of Uncertainty," in W. C. Price and S. S. Chissock, eds., *The Uncertainty Principle and Foundations of Quantum Mechanics: A Fifty Years' Survey* (New York: Wiley, 1977), p. 3.

103. P. Jordan, "Über quantenmechanksche Darstellung van Quantensprugen," *Zs. f. Phys.* 40 (1927): 661–66.

104. Schrödinger's paper was received on 10 June 1927, Heisenberg's on 23 March 1927. However, Heisenberg's paper was published in an issue of the *Zeitschrift* marked "Abgeschlossen" 31 May, which I assume to mean 'closed by the editor and sent to the publisher'.

105. "Energieaustausch nach der Wellenmechanik," *Ann. der Phys.* 83: (1927) 956–68, trans. in WM, pp. 137–46.

106. WM, p. 142.

107. SHQP, interview 8, pp. 14–15. Heisenberg has given retrospective accounts of his discussions with Bohr in "Erinnerungen an die Zeit der Entwicklung der Quantenmechanik," in Fierz and Weisskopf, *Pauli Memorial*, pp. 40–47; in "Remarks"; and in his more or less philosophical works.

108. Pauli and Schrödinger exchanged a number of letters during this period. It is interesting to note that they use the familiar 'du', while Pauli and Heisenberg were still referring to each other as 'Sie'.

109. *Pauli Briefwechsel*, pp. 340–49, citation from pp. 346–47.

110. Heisenberg to Pauli, 28 October 1926, *Pauli Briefwechsel*, pp. 349–51.

111. Ibid., pp. 354–56.

112. Ibid., pp. 357–60.

113. P. A. M. Dirac, "The Physical Interpretation of the Quantum Dynamics," *Proc. Roy Soc. A* 113 (1927): 621–41, citation from p. 623.

114. This was reported by Léon Rosenfeld in "Men and Ideas in the History of Atomic Theory," reproduced in *Selected Papers of Léon Rosenfeld*, ed. R. S. Cohen and J. J. Stachel (Dordrecht: Reidel, 1979), pp. 266–96.

115. Letter of 23 February 1927, *Pauli Briefwechsel*, pp. 376–81.

116. W. Heisenberg, "Über den anschaulichen Inhalt der quantentheoretischen Kinematik und Mechanik," *Zs. f. Phys.* 43 (1927): 172–98.

117. "Here we discover the deep sense of the linearity of the Schrödinger equations. For this reason they can be understood only as equations for waves in phase space and for this reason we might regard as fruitless every attempt to replace these equations, e.g., in the relativistic case (with many electrons) with non-linear ones,"ibid., p. 184. In his draft letter to Pauli, Heisenberg had said: "This shows that Schrödinger's [wave packet] proposal does not work and that it seems completely hopeless so to come to a 'path'.

The solution can now, I believe, be pregnantly expressed through the proposition: *The path exists only inasmuch as we observe it.*" *Pauli Briefwechsel*, p. 379.

118. Heisenberg, "Über den anschaulichen," p. 196.

119. *Pauli Briefwechsel*, pp. 394–96, citation from p. 395.

120. Ibid., p. 397.

121. P. Ehrenfest, "Bemerkung uber die angenaherte Gultigkeit der klassischen Mechanik innerhalb der Quantenmechanik," *Zs. f. Phys.* 45 (1927): 455–57.

122. This claim is based primarily on the bibliography given in Scott, *Erwin Schrödinger*, pp. 151–61.

123. *Electrons et photons: Rapports et discussions du Cinquième Conseil de Physique tenu a Bruxelles du 24 au 29 Octobre 1927 sous les auspices de L'Institut International de Physique Solvay* (Paris: Gauthier-Villars, 1928), p. 288.

124. "Conceptual Models in Physics and Their Philosophical Value," trans. in Schrödinger's *Science and Theory and Man* (New York: Dover, 1957), pp. 148–65, esp. p. 161.

125. E. Schrödinger, *Four Lectures on Wave Mechanics* (London: Blackie, 1928).

126. Ibid., p. 6.

127. Ibid., p. 53.

128. E. Gaviola, "An Experimental Test of Schrödinger's Theory," *Nature* 122 (1928): 772.

129. Wessels, "Schrödinger's Interpretation of Wave Mechanics," pp. 313–24. There she summarizes all of his pertinent writings in English, German, and Spanish.

130. C. V. Raman and K. S. Krishnan, "A New Type of Secondary Radiation," *Nature* 121 (1928): 501–2.

131. G. Landsberg and L. Mandelstam, "Eine Neue Erscheinung bei der Lichtzerstreuung in Krystallen," *Naturwiss.* 16 (1928): 557–58.

132. A detailed treatment of this may be found in G. Herzberg, *Molecular Spectra and Molecular Structure*, I vol. 1, *Spectra of Diatomic Molecules*, 2d ed. (New York: Van Nostrand, 1950), pp. 82–90.

133. Dirac, "Physical Interpretation," esp. pp. 633–37.

134. P. A. M. Dirac, "The Quantum Theory of the Electron," *Proc. Roy. Soc. A* 117 (1928): 610–24; 118 (1928): 351–61.

135. "The Meaning of Wave Mechanics," in *Louis de Broglie: Physicien et penseur* (Paris: Albin Michel, 1953), pp. 17–32.

Chapter Eight

1. W. Pauli, "Die allgemeinen Prinzipien der Wellenmechanik," reproduced in his *Collected Scientific Papers*, ed. R. Kronig and V. F. Weisskopf (New York: Interscience, 1964), pp. 771–938, esp. pp. 771–92.

2. Jammer, *Development*, pp. 172–80.

3. Gerald Holton, "The Roots of Complementarity," in his *Thematic Origins of Scientific Thought* (Cambridge, Mass.: Harvard University Press, 1973), pp. 115–16.

4. Lewis Feuer, *Einstein and the Generation of Science* (New York: Basic Books, 1974), pp. 109–57.

5. According to Hans Bohr, Niels's son, the stories Niels read to his children included in particular those of Poul Martin Møller, "to whose ever young tale, 'A Danish Student's Adventure,' my father often returned. In the description of the licentiate's difficulties in coming to a decision he saw a perfect example of his ideas about complementary features in psychology," from S. Rozental, ed., *Niels Bohr: His Life and Work as Seen by His Friends and Colleagues* (Amsterdam: North-Holland, 1967), p. 330. L. Rosenfeld recounted the standard indoctrination procedure at Bohr's Institute: "For this dialectical orientation of his thought, which was not the result of any formal schooling, but the fruit of highly original meditation, we have an amusing testimony in the fascination which—more than any learned treatise—the charming 'Tale of a Danish Student,' this pearl of the romantic literature of Denmark, exerted on his mind. One may well say that the quandaries of the licentiate lost in the maze of self-reflection were the only practical lesson in Hegelian logic that Bohr ever received, and he certainly never forgot it. Every one of those who came into closer contact with Bohr at the Institute, as soon as he showed himself sufficiently proficient in the Danish language, was acquainted with the little book; it was part of his initiation. Bohr would point to those scenes in which the licentiate describes how he loses the count of his many egos or disserts on the impossibility of formulating a thought, and from these fanciful antinomies he would lead his interlocutor—along paths Poul Martin Møller never dreamt of—to the heart of the problem of unambiguous communication of experience, whose earnestness he thus dramatically emphasized." Ibid., p. 121.

6. J. Rud Nielsen, "Memories of Niels Bohr," *Physics Today* 16 (October 1963): 22–30, citation from p. 27. This issue was devoted to essays on Bohr.

7. Bohr, *Works*, vol. I, pp. 501, 503. Bohr was actually working on his Master's thesis rather than his dissertation at the time this letter was written.

8. N. Bohr, "Atomic Theory and Mechanics," in ATDN, pp. 25–51.

9. In addition to the work by Rozental (*Niels Bohr*) and *Physics Today* (vol. 16 [October 1963]), there is also the collection edited by W. Pauli, *Niels Bohr and the Development of Physics* (New York: Pergamon Press, 1955). Much has been written on Bohr's position. I will omit critical or argumentative expositions and list some basic historical outlines. In addition to the works of Jammer and Meyer-Abich already cited one could consult Aage Peterson, "The Philosophy of Niels Bohr," *Bulletin of the Atomic Scientists* 9 (September 1963): pp. 8–14, probably the best brief survey of Bohr's philosophy; Erhard Scheibe, *The Logical Analysis of Quantum Mechanics*, trans. J. B. Sykes (New York: Pergamon Press, 1973), chap 1; W. Heisenberg, "The Copenhagen Interpretation of Quantum Theory," in his *Physics and Philosophy: The Revolution in Modern Science* (New York: Harper, 1958); Jagdish Mehra, "The Quantum Principle: Its Interpretation and Epistemology," *Dialectica* 27 (1973): 76–157. Further references with evaluations may be found in B. S. Dewitt and R. Neill Graham, "Resource Letter IQM–1 on the Interpretation of Quantum Mechanics," *Amer. J. of Phys.* 39 (1971): 724–38.

10. Heisenberg, *Physics and Philosophy*, p. 42.

11. A more detailed reconstruction of these events may be found in Jammer, *Philosophy*, chap. 3.

12. Heisenberg's book *The Physical Principles of the Quantum Theory*, trans. C. Eckart and F. Hoyt (Chicago: University of Chicago Press, 1930), was based on the lectures he gave in the spring of 1929. Here, especially in chap. 5, he is willing to use whatever picture is most convenient in explaining crucial experiments.

13. Ruth Moore, *Niels Bohr: The Man, His Science, and the World They Changed* (New York: Knopf, 1966), p. 132.

14. SHQP, interview 3 (17 October 1962), pp. 23–29.

15. The version I am summarizing is the final version given in ATDN, pp. 52–91.

16. Paul Forman, "Weimar Culture, Causality, and Quantum Theory, 1918–1927: Adaption by German Physicists and Mathematicians to a Hostile Intellectual Environment," HSPS, vol. 3, pp. 1–115. A less technical updating and extension of this is in Forman's "The Reception of an Acausal Quantum Mechanics in Germany and Britain," in Seymour H. Mauskopf, ed., *The Reception of Unconventional Science* (Boulder, Colo.: Westview Press, 1978), which Forman kindly sent me.

17. Heisenberg, "Über den anschaulichen," p. 197.

18. Bohr, ATDN, p. 87.

19. "The Quantum of Action and the Description of Nature," ATDN, pp. 92–101.

20. These arguments are adapted from Bohr's introduction to ATDN, which was written in 1929.

21. Ibid., pp. 16–18.

22. Such extensions are the theme of almost every essay in Bohr's APHK.

23. ATDN, p. 54.

24. Textbook summaries of this standard experiment frequently fail to point out that it is an ideal rather than an actual experiment. Soon, however, it may be within the limits of technical feasibility. This is discussed in H. Pierre Noyes's "An Operational Analysis of the Double Slit Experiment," in P. Suppes, ed., *Foundations of Quantum Mechanics: The 1976 Standard Seminar* (Lansing, Mich.: Philosophy of Science Association, 1979), pp. 77–108.

25. ATDN, p. 19. Dirac had used this term and Heisenberg had criticized its usage in the 1927 Solvay Conference, p. 264.

26. ATDN, p. 18.

27. W. V. Quine, *Methods of Logic*, 3d ed. (New York: Holt, Rinehart & Winston, 1972), p. xii.

28. ATDN, p. 15.

29. M. Born, W. Heisenberg, and P. Jordan, "Zur Quantenmechanik II," *Zs. f. Phys.* 35 (1926): 557–615, trans. in van der Waerden, pp. 321–85.

30. P. A. M. Dirac, "The Quantum Theory of the Emission and Absorption of Radiation," *Proc. Roy. Soc. A* 114 (1927): 243–65. This is reproduced in Julian Schwinger, ed., *Selected Papers on Quantum Electrodynamics* (New York: Dover, 1958), pp. 1–23. Henceforth, this will be referred to as Schwinger. My summary of the development of quantum field theory is guided by Schwinger's preface, pp. vii–xvii; by G. Wentzel, "Quantum Theory of Fields (until 1947)," in Fierz and Weisskopf, *Pauli Memorial*; and by Steven Weinberg, "The Search for Unity: Notes for a History of Quantum Field Theory," *Daedalus* 106 (fall, 1977): 17–35.

31. P. A. M. Dirac, "The Quantum Theory of Dispersion," *Proc. Roy. Soc. A* 114 (1927): 710–28, citation from pp. 710–11.

32. W. Heisenberg and W. Pauli, "Zur Quantendynamik der Wellenfelder," *Zs. f. Phys.* 56 (1929): 1–61, reproduced in *Pauli Papers*, vol. 2., pp. 354–414.

33. E. Fermi, "Sopra l'Elettrodinamica Quantistica," *Rend. Lincei* 5 (1929): 881–97, reproduced in E. Segrè et al., *Enrico Fermi: Collected Papers*, vol. 1 (Chicago: University of Chicago Press, 1962), pp. 305–11. Henceforth this will be referred to as Segrè. The lectures on this topic that Fermi gave at Ann Arbor were soon published and became the basic text most

physicists used for learning quantum field theory; see E. Fermi, "Quantum Theory of Radiation," *Reviews of Modern Physics* 4 (1932): 87–132, reproduced in Segrè, vol. I, pp. 401–55.

34. In his obituary notice, "Paul Ehrenfest," *Naturwiss*. 21 (1933): 841–43, reproduced in Fierz and Weisskopf, *Pauli Memorial*, vol. 2, pp. 698–700, Pauli claimed that Ehrenfest, drawing on the analogy between classical and quantum field theory, concluded that quantum field theory must include infinite self-energies. Though he did not write on the topic, Fermi also recognized the difficulty. In his introduction to Fermi's articles on field theory E. Amaldi (in Segrè, p. 305) claims that Fermi filled several notebooks with unsuccessful attempts to solve the problem of divergences.

35. J. R. Oppenheimer, "Note on the Theory of the Interaction of Field and Matter," *Phys. Rev.* 35 (1930): 461–77.

36. SHQP, interview 3, p. 5.

37. P. A. M. Dirac, "The Quantum Theory of the Electron, I," *Proc. Roy. Soc. A* 117 (1928): 610—24; "The Quantum Theory of the Electron, II," ibid. 118 (1928): 351–61.

38. Dirac, "Quantum Theory of the Electron, I," p. 618.

39. O. Klein, "Die Reflexion von Elektronen an einem Potentialsprung nach der Relativistischen Dynamik von Dirac," *Zs. f. Phys.* 53 (1928): 157–65. A good survey of all these difficulties is contained in Joan Bromberg's "The Impact of the Neutron: Bohr and Heisenberg," HSPS, vol. 3, pp. 307–42. Related problems are discussed in her "Concept of Particle Creation before and after Quantum Mechanics," HSPS, vol. 7, pp. 161–91.

40. E. Schrödinger, "Über die krafterfreie Bewegung in der relativistischen Quantenmechanik," *Sitzber. Preuss. Akad. Wiss. Phys-Math.* 24 (1930): 418–28.

41. A brief history of the early development of this topic is given in R. Peierls, "Fermi-Dirac Statistics," in Abdus Salam and E. P. Wigner, eds., *Aspects of Quantum Theory* (Cambridge: Cambridge University Press, 1972), pp. 117–27.

42. W. Heitler and G. Herzberg, "Gehorchen die Stickstuffkerne der Boseschen Statistik," *Naturwiss*. 17 (1929): 673.

43. See C. S. Wu, "The Neutrino," in Fierz and Weisskopf, *Pauli Memorial*, pp. 249–303.

44. Otto Robert Frisch, "The Interest Is Focussing on the Atomic Nucleus," in Rozental, pp. 137–48, citation from p. 139.

45. BSC 4, sec. 4.

46. Ibid., letter of 26 November 1929.

47. Ibid., letter of 5 December 1929.

48. Ibid.

49. *Convegno di Fisica Nucleare, Ottobre 1931* (Rome: Reale Accademia D'Italia, 1932), pp. 119–32.

50. L. Rosenfeld, "Quantum Theory in 1929: Recollections from the First Copenhagen Conference," in R. S. Cohen and J. J. Stachel, eds., *Selected Papers of Léon Rosenfeld* (Dordrecht: Reidel, 1979), pp. 302–12, claims that Bohr stunned everyone, except Pauli, at the 1929 conference with his remarks concerning the nonobservability of electron spin. Bohr's argument was that such a purely quantal concept, vanishing in the classical limit, could not be directly related to such classical properties as angular momentum and magnetic moment. This rather conceptual argument was later given a more rigorous mathematical development by N. F. Mott and others.

51. In the Faraday lecture, delivered to British chemists, Bohr made the same point more emphatically: "At the present stage of atomic theory, however, we may say that we have no argument, either empirical or theoretical, for upholding the energy principle in the case of β-ray disintegration, and are even led to complications and difficulties in trying to do so." This is from N. Bohr, "Chemistry and the Quantum Theory of Atomic Constitution," *Chemical Society Journal*, 1932, pp. 369–84, citation from p. 383.

52. Bohr, in *Convegno*, p. 128.

53. An older, but still useful, history of this development is Norwood Russell Hanson's *The Concept of the Positron* (Cambridge: Cambridge University Press, 1963).

54. This will be discussed subsequently.

55. L. Landau and R. Peierls, "Erweiterung des Unbestimmtheitsprincips für die relativistische Quantentheorie," *Zs. f. Phys.* 69 (1931): 56–69. A summary of the background to this paper may be found in L. Rosenfeld's "On Quantum Electrodynamics," in Pauli's *Niels Bohr*, pp. 70–95.

56. N. Bohr and L. Rosenfeld, "Zur Frage der Messbarkeit der Elektromagnetischen Feldgrössen," *Det. Kgl. Dansk Vid. Selskab.*, vol. 31, no. 8 (1933), trans. in Rosenfeld's *Selected Papers*, pp. 357–400.

57. Rosenfeld, *Selected Papers*, p. 365.

58. *Structure et propriété des noyaux atomiques: Rapports et discussions du Septième Conseil de Physique (22–29 Oct., 1933)* (Paris: Gauthier-Villars, 1934). Both Carl D. Anderson's "The Positive Electron," *Phys. Rev.* 43 (1933): 491–94, and J. Chadwick's "The Existence of a Neutron," *Proc. Roy. Soc. A* 136 (1932): 692–708, are reproduced in Robert T. Beyer, ed., *Foundations of Nuclear Physics* (New York: Dover, 1949). Earlier, Anderson had published a short note reporting his discovery in *Science* 76 (1932): 238.

59. Solvay, 1933, p. 175.

60. Ibid., pp. 203–12.

61. Dirac to Bohr, 10 August 1933, in BSC 18, sec. 4.

62. Solvay, 1933, pp. 216–28.

63. Ibid., p. 139.

64. Ibid., p. 334.

65. E. Fermi, "Tentativo di una Thoria dell'Emissione dei Raggi 'Beta,'" *Ric. Scientifica* 4 (1933): 491–95, reprinted in Segrè, pp. 540–44. Rasetti's introduction to this series of papers, pp. 538–40, summarizes the steps in Fermi's development.

66. E. Fermi, "Tentativo di una Teoria dei Raggi," *Nuovo Cimento* 11 (1934): 1–19 (in Segrè, pp. 559–74); and "Versuch einer theorie der β-Strahlen. I," *Zs. f. Phys.* 88 (1934): 161–71 (in Segrè, pp. 575–590).

67. Robert S. Shankland, "An Apparent Failure of the Photon Theory of Scattering," *Phys. Rev.* 49 (1936): 8–13.

68. The various papers of Fermi and his co-workers are reproduced in Segrè, pp. 639–807. The crucial paper on slow neutrons is on pp. 759–60 (Italian version) and pp. 761–62 (English version).

69. From Frisch, in Rozental, p. 41.

70. N. Bohr, "Neutron Capture and Nuclear Constitution," *Nature* 137 (29 February 1936): 344–48; N. Bohr and F. Kalckar, "On the Transmutation of Atomic Nuclei by Impact of Material Particles," *Kgl. Dansk Vid. Selskab., Math-Phys. Medd,* vol. 14, no. 10 (1937). F. L. Friedman and V. F. Weisskopf in "The Compound Nucleus," in Pauli, *Niels Bohr*, pp. 134–62, said of this article, "During the eighteen years since its appearance, it has been the decisive influence on the analysis of nuclear reactions" (p. 134).

71. The basic idea is suggested in the Bohr-Kalckar article, pp. 9–10. Some six years earlier Gamow had suggested a liquid drop model of the nucleus containing protons, electrons, and alpha particles.

72. N. Bohr and A. Wheeler, "The Mechanism of Nuclear Fission," *Phys. Rev.* 56 (1939): 429–50. Here the authors concluded that nuclear fission was due to the relatively rare isotope ^{235}U. Very few technical papers in the history of physics have had the impact of this one.

Chapter Nine

1. Albert Einstein, "Autobiographical Notes," in Schilpp, p. 95. Henceforth this will be cited as "Autobiographical Notes."

2. This letter was published in J. Hadamard, *The Psychology of Invention in the Mathematical Field* (Princeton: Princeton University Press, 1945), pp. 142–43.

3. G. Holton, "Finding Favor with the Angel of the Lord: Notes toward the Psychobiographical Study of Scientific Genius," in Y. Elkana, ed., *The Interaction between Science and Philosophy* (Atlantic Highlands, N.J.: Humanities Press, 1974), pp. 349–87.

4. This aspect of Einstein's thought is discussed, though in a rather nonsystematic say, in Jeremy Bernstein, *Einstein* (New York: Viking, 1973).

5. "Autobiographical Notes," p. 89.

6. This was discussed earlier in the conclusion of chapter 3. In 1905, E. Zermelo translated Gibbs's *Elementary Principles in Statistical Mechanics* into German.

7. "Autobiographical Notes," p. 21.

8. "Autobiographical Notes," p. 32. Einstein's appraisal of the state of current science at this stage of his career is well explained and documented in M. Klein's articles "Thermodynamics in Einstein's Thought," *Science* 157 (1967): 509–16, and "Einstein and the Wave-Particle Duality," *Natural Philosopher* 3 (1964): 1–49. An excellent overall view is presented in A. Pais, "Einstein and the Quantum Theory," *Reviews of Modern Physics* 51 (1979): 863–914. Though this study appeared too late to be used as a source, I have introduced a few modifications based on Pais's study.

9. A. Einstein, "Bemerkung zu der von Hen. Paul Ehrenfest: *Translation deformierbarer Elektronen und der Flächensatz*," *Ann. der Phys.* 23: 206–08. Here Einstein said, "The relativity principle or—more exactly—the relativity principle together with the principle of the constancy of light is not a closed system, and in fact is not properly understood as a system, but solely as a heuristic principle, which considered by itself contains no information concerning rigid bodies, time, or light signals" (206).

10. A. Einstein, "Zum gegenwärtigen Stand des Strahlungsproblems," *Phys. Zeit.* 10 (1909): 185–93.

11. A. Einstein, "Über die Entwicklung unserer Anschauungen über das Wesen und die Konstitution der Strahlung," *Phys. Zeit.* 10 (1909): 817–25. The analogy is developed on pp. 824–25.

12. A. Einstein, "Theorie der Opaleszenz von homogenen Flüssigkeiten und Glüssigkeitgemischen in der Nähe des kritischen Zustandes," *Ann. der Phys.* 33 (1910): 1275–98.

13. A. Einstein, "Bemerkung zu dem Gesetz von Eötvös," *Ann. der Phys.* 34 (1911): 165–69.

14. A. Einstein, "What Is the Theory of Relativity?" reprinted in *Ideas and Opinions by Albert Einstein*, trans. Sonja Bargmann (New York: Crown Publishers, 1954), pp. 227–32. This book is essentially a translation and reordering of the articles in Einstein's collection of nontechnical articles and addresses *Mein Weltbild*, ed. Carl Seelig, and some articles from two other collections, *Out of My Later Years* and *The World as I See It*. The final revised edition of *Mein Weltbild* (Frankfurt/M: Ullstein, 1970) contains the articles in *Ideas and Opinions*, though the ordering is different. Henceforth I will cite *Ideas and Opinions*.

15. A. Einstein, "Principles of Research," a 1918 address for Max Planck's sixtieth birthday, in *Ideas and Opinions*, pp. 224–27, citation from p. 226.

16. A. Einstein and L. Hopf, "Uber einen Satz der Wahrscheinlickeitsrechnung und seine Anwendung in der Strahlungstheorie," *Ann. der Phys.* 33 (1910): 1096–1104.

17. In a footnote to this paper, on p. 1096, they define what they mean by 'probability of a coefficient'. It is essentially an adaption to Fourier coefficients of Einstein's basic idea that the probability of a state is proportional to the percentage of time the system spends in that state. Thus, probability notions presuppose an underlying determinism.

18. A. Einstein and L. Hopf, "Statistische Untersuchung der Bewegung eines Resonators in einen Strählungsfeld," *Ann. der Phys.* 33 (1910): 1105–15.

19. Ibid., p. 1105.

20. A. Einstein and O. Stern, "Einige Argumente für die Annahme einer molekularen Agitation beim absoluten Nullpunkt," *Ann. der Phys.* 40 (1913): 551–60.

21. This result is still surprising. In quantum field theory one can obtain similar results through quantization of the radiation field. In this case, at least in the older quantum field theory, one assumed fictitious oscillators with zero-point energy of $h\gamma/2$. This gives an infinite contribution which is then removed by renormalization. Hence, only the integral values of $h\gamma$ contribute, as in the Einstein-Stern paper.

22. A. Einstein, "Beiträge zur Quantentheorie," *Deutsche physikalische Gesellschaft Verhandlungen* 16 (1914): 820–28.

23. A. Einstein, "Strahlungs-Emission und Absorption nach der Quantentheorie," *Verhandlungen der Deutschen Physikalischen Gesellschaft* 18 (1916): 318–23.

24. This paper was discussed in chapter 5. See also Pais, "Einstein and the Quantum Theory," pp. 886–89.

25. Jagdish Mehra in "Albert Einstein's First Paper," *Science Today* (April 1971), has translated an essay Einstein wrote around 1895 entitled "Concerning the Investigations of the State of Aether in Magnetic Fields." Though this paper, by the sixteen-year-old Einstein, does not adumbrate the principles of special relativity, it does consider the interrelation of the electric and magnetic components of the electromagnetic field. I thank Professor Mehra for sending me this paper.

26. The books they read are discussed briefly in the introduction to *Albert Einstein, lettres à Maurice Solovine* (Paris: Gauthier-Villars, 1956). Henceforth, this will be referred to as Solovine Correspondence.

27. A general survey of the philosophy of science of this period, together with representative texts from the authors the Olympia Academy read, may be found in Joseph Kockelmans, ed., *Philosophy of Science: The Historical Background* (New York: Free Press, 1968).

28. My appraisal of the influence of philosophers on Einstein was partially influenced by a talk, "Einstein and the Philosophers," which Robert S. Cohen gave at California State University, Hayward, spring, 1979.

29. This is cited from Albert Rothenberg, "Einstein's Creative Thinking and the General Theory of Relativity: A Documented Report," *American Journal of Psychiatry* 136 (1979): 38–43, citation from p. 39. A similar account written a few years later is Einstein's "Notes on the Origin of the General Theory of Relativity," in *Ideas and Opinions*, pp. 285–90.

30. This is from Mach's "The Economical Nature of Physical Inquiry," reproduced in his *Popular Scientific Lectures*, trans. J. McCormack (Chicago: Open Court, 1895), citation from p. 199. The same idea was presented in his *The Science of Mechanics*, trans. T. J. McCormack, 6th ed. (LaSalle, Ill.: Open Court, 1960) (German original, 1883). This is a source Einstein was quite familiar with.

31. Einstein, *Ideas and Opinions*, p. 287.

32. A. Einstein, "Über dem Einfluss der Schwerkraft auf die Ausbreitung des Lichte," *Ann. der Phys.* 35 (1911): 898–908; English trans. in A. Einstein et al., *The Principle of Relativity*, trans. W. Perrett and G. B. Jeffrey (New York: Dover, 1923), pp. 99–108.

33. There is as yet no history of the development of the theory of relativity at all comparable with those on quantum theory. The histories of general relativity I found most helpful were Whittaker, *History*, vol. 2, chap. 5; J. D. North, *The Measure of the Universe: A History of Modern Cosmology* (Oxford: Clarendon Press, 1965); and Jagdish Mehra, *Einstein, Hilbert, and the Theory of Gravitation* (Dordrecht: Reidel, 1974).

34. Albert Einstein/Arnold Sommerfeld, *Briefwechsel* (Basel: Schwabe & Co., 1968), p. 26. Henceforth, this will be referred to as Sommerfeld Correspondence.

35. A. Einstein, "Die Grundlage der allgemeinen Relativitätstheorie," *Ann. der Phys.* 49 (1916): 769–822, trans. in *The Principles of Relativity*, pp. 111–64.

36. Ibid., p. 141.

37. Ibid., p. 144.

38. Mehra, *Einstein, Hilbert, and Gravitation*.

39. A. Einstein, "Kosmologische Betrachtungen zur allgemeinen Relativitätstheorie," *Sitz. Preuss. Akad. Wiss.* (1917), part 1, pp. 142–52, trans. in *The Principles of Relativity*, pp. 177–88, citation from p. 188.

40. This is from the appendix, "On the Cosmological Problem," added to the second and later editions of *The Meaning of Relativity*. The citation is from the fifth edition (Princeton: Princeton University Press, 1955), p. 127.

41. For a general survey of these theories, see W. Pauli, "Relativitätstheorie," reproduced in R. Kronig and V. F. Weisskopf, eds., *Wolfgang Pauli: Collected Scientific Papers*, vol. 1 (New York: Interscience, 1964), pp. 1–237, esp. pp. 211–37. This was translated into English by G. Field and published by Pergamon Press in 1958. See pp. 184–206.

42. A. Einstein, "Spielen Gravitationsfelder im Aufber der materiellen Elementarteilchen eine wesentliche Rolle?" *Sitz. Preuss. Akad. Wiss.* (1919), part 1, pp. 349–56, trans. in *The Principles of Relativity*, pp. 191–98.

43. See Ronald W. Clark, *Einstein: The Life and Times* (New York: Avon, 1972), p. 34. I am accepting this as the most reliable of the various biographies of Einstein.

44. See József Illy, "Albert Einstein in Prague," *Isis* 70 (1979): 76–84.

45. Max Born, ed., *The Born-Einstein Letters*, trans. Irene Born (New York: Walker & Co., 1971), p. 4. Henceforth, this will be referred to as Born Correspondence.

46. Ibid., p. 7. This letter is undated, but was probably written in August 1919.

47. Albert Reiser, *Albert Einstein: A Biographical Portrait* (New York: Albert & Charles Boni, 1930), citation from introduction, p. xiii.

48. Thus, Kant says: "Geometry is based on the pure intuition of space" (p. 30); "that everywhere space . . . has three dimensions and that space cannot in any way have more is based on the proposition that not more than three lines can intersect at right angles in one point" (p. 32), from Kant's *Prolegomena to Any Future Metaphysics*, trans. L. W. Beck (Indianapolis: Bobbs-Merrill, 1950).

49. A. Einstein, "Geometry and Experience," a 1921 address trans. in *Ideas and Opinions*, pp. 232–46, citation from p. 241.

50. In addition to those contained in *Ideas and Opinions*, see also Einstein's quasi-Nobel address, reprinted in *Nobel Lectures: Physics, 1901–1921*, pp. 482–90. This address was given some time after the prize was awarded. While the epistemological foundations of relativity are stressed, the photoelectric effect, for which the prize was awarded, is never even mentioned.

51. Thus, in his 1921 London address, "On the Theory of Relativity," in *Ideas and Opinions*, pp. 246–49, Einstein said: "It is in general one of the essential features of the theory of relativity that it is at pains to work out the relation between general concepts and empirical facts more precisely. The fundamental principle here is that the justification for a physical concept lies exclusively in its clear and unambiguous relation to facts that can be experienced" (p. 247).

52. *Ideas and Opinions*, p. 233.

53. Ibid., p. 234. In a letter to Max Born, 9 December 1919, Einstein said, "Schlick has a very good head on him; we must try to get him a professorship. He is in desperate need of it because of the devaluation of property. However, it will be difficult as he does not belong to the philosophical established Church of the Kantians," Born Correspondence, p. 18.

54. Adolf Grünbaum, *Philosophical Problems of Space and Time* (New York: Knopf, 1963), pp. 131–51.

55. *Ideas and Opinions*, citations from pp. 225, 226.

56. Max Planck, "The Unity of the Physical World Picture," trans. in S. Toulmin, ed., *Physical Reality* (New York: Harper, 1970), pp. 1–27, citation from p. 4.

57. Paul Forman, "Weimar Culture, Causality, and Quantum Theory, 1918–1927: Adap-

tion by German Physicists and Mathematicians to a Hostile Intellectual Environment," HSPS, vol. 7, pp. 1–116, esp. pp. 91–96, for the role of Planck and Einstein.

58. An odd example of this is Einstein's "The Cause of the Formation of Meanders in the Course of Rivers and of the So-called Baer's Law," a paper read to the Prussian Academy on 7 January 1926, trans. in *Ideas and Opinions*, pp. 249–53. Here Einstein was explicitly and purposefully seeking a causal account of phenomena which seemed to be adequately expressed through general laws. Similarly his 1927 paper commemorating the two hundredth anniversary of Newton's death, "The Mechanics of Newton and Their Influence on the Development of Theoretical Physics," *Ideas and Opinions*, pp. 253–261, claims that Kepler gave a phenomenological account of planetary motion: "But these rules do not satisfy the demand for causal explanation" (p. 254). Newton's great achievement centered on supplying the requisite causal account.

59. A. Einstein, "Zum Quantensatz von Sommerfeld und Epstein," *Verhandlungen der Deutschen Physikalischen Gesellschaft* 18 (1916): 318–23.

60. Sommerfeld Correspondence, p. 40.

61. Ibid., p. 44.

62. Born Correspondence, p. 10.

63. Ibid., p. 23.

64. Ibid., letter of 30 December 1921, p. 65.

65. Pierre Speziale, ed., *Albert Einstein: Michele Besso. Correspondence, 1903–1955* (Paris: Hermann, 1972), letter of 5 January 1924, pp. 197–98. Henceforth, this will be referred to as Besso Correspondence. Ordinarily, differential equations have to be supplemented by boundary conditions to yield definite solutions. The overdetermination of these equations should render boundary conditions unnecessary.

66. Ibid., p. 202.

67. N. Bohr, "Discussions with Einstein on Epistemological Problems in Atomic Physics," in Schlipp, pp. 199–241, citation from p. 206.

68. Solvay, 1911, p. 115.

69. Sommerfeld Correspondence, p. 54.

70. The relation of her work to Einstein's development is discussed by Mehra, *Einstein, Hilbert, and Gravitation*, pp. 22, 71.

71. See Pais, "Einstein and the Quantum Theory," pp. 894–97.

72. Solvay, 1927. Einstein's intervention is on pp. 253–56, his question on p. 266.

73. BSC 10.

74. Besso Correspondence, p. 202.

75. Sommerfeld Correspondence, p. 108.

76. Born Correspondence, p. 97.

77. Solvay, 1927, p. 248.

78. A brief retrospective account of the form these discussions took may be found in W. Heisenberg's *Physics and Beyond* (New York: Harper & Row, 1971), pp. 80–81.

79. Solvay, 1927, p. 248.

80. Bohr, "Discussions with Einstein," p. 212.

81. Ibid., p. 213.

82. "Autobiographical Notes," p. 81.

83. This was a reply to Philipp Frank, who vividly depicts the surprise he felt at Einstein's rejection of the Copenhagen interpretation in his *Einstein: His Life and Times*, trans. George Rosen (New York: Knopf, 1970), pp. 214–18.

84. "Autobiographical Notes," p. 89.

85. Banesh Hoffmann, who was one of Einstein's collaborators on this project, presents a nontechnical outline of Einstein's efforts in his *Albert Einstein: Creator and Rebel* (New York: Viking Press, 1972), pp. 221–31.

86. My treatment of this difficulty was stimulated by a discussion with C. W. F. Everitt and J. Heilbron.

87. The paper in question is Einstein's "On the Influence of Gravitation on the Propagation of Light," reproduced in *The Principles of Relativity*, pp. 99–108.

88. A. Einstein, R. C. Tolman, and B. Podolsky, "Knowledge of Past and Future in Quantum Mechanics," *Phys. Rev.* 37 (1931): 780–81.

90. In a 1931 address, "Maxwell's Influence on the Evolution of the Idea of Physical Reality," trans. in *Ideas and Opinions*, pp. 266–70, Einstein said, "Dirac, to whom, in my opinion, we owe the most perfect exposition, logically, of this theory, rightly points out that it would probably be difficult, for example, to give a theoretical description of a photon such as would give enough information to enable one to decide whether it will pass a polarizer placed (obliquely) in its way or not" (p. 270). This is the example Dirac repeatedly used to explain the superposition principle. Though Dirac does not utilize the ideas or methods of formal logic, his *Principles of Quantum Mechanics* has a carefully developed logical ordering. I have made this logical structure explicit in my "Ontic Commitments of Quantum Mechanics," in *Boston Studies in the Philosophy of Science*, vol. 13, pp. 225–308.

91. A. Einstein and W. Mayer, "Semi-Vektoren und Spinoren," *Preuss. Akad. Wiss., Phys-Math., Klasse, Sitzungsberichte*, 1932, pp. 522–50; "Die Dirac Gleichungen für Semi-Vektoren," *Akad. van wetenschappen* (Amsterdam), *Proceedings*, vol. 36, part 1, pp. 497–516; "Darstellung der Semi-Vektoren als gewöhnliche Vektoren von besonderem Differentiations Charakter," *Annals of Mathematics* 35 (1934): 104–10.

92. A. Einstein, "On the Method of Theoretical Physics," in *Ideas and Opinions*, pp. 270–76, citation from p. 275.

93. Ibid., p. 276.

94. Einstein and Mayer, "Darstellung," p. 109.

95. A. Einstein and N. Rosen, "The Particle Problem in the General Theory of Relativity," *Phys. Rev.* 48 (1935): 73–77.

96. The two papers are A. Einstein, B. Podolsky, and N. Rosen, "Can Quantum-Mechanical Description of Physical Reality Be Considered Complete?" *Phys. Rev.* 47 (1935): 777–80; and N. Bohr, "Can Quantum-Mechanical Description of Physical Reality Be Considered Complete?," *Phys. Rev.* 48 (1935): 696–702. Both are reproduced in Toulmin, *Physical Reality*. In summarizing these, it seems overly pedantic to give repeated page references to citations drawn from a three- or four-page article. Two surveys of this confrontation deserving of special recommendation are Jammer's historical analysis in *Philosophy*, pp. 166–251; and Clifford Hooker's "The Nature of Quantum Mechanical Reality: Einstein Versus Bohr," in Robert Colodny, ed., *Paradigms and Paradoxes* (Pittsburgh: University of Pittsburgh Press, 1972), pp. 67–302. I think that Hooker's presentation of Bohr's position is correct in its essential features. However, he clearly distorts Einstein's position by attributing to him an all-purpose classical ontology devised by Hooker himself.

97. A detailed examination of the conditions under which this experiment could be performed is given in H. Pierre Noyes, "An Operational Analysis of the Double Slit Experiment," in Patrick Suppes, ed., *Studies in the Foundations of Quantum Mechanics* (East Lansing, Mich.: Philosophy of Science Association, 1980), pp. 77–108.

98. Bohr discusses this in his reply to the EPR paper. However, the illustration is taken from his contribution to the Schlipp volume, p. 219.

99. Cited from Toulmin, *Physical Reality*, p. 136.

100. Ibid., pp. 138–39.

Chapter Ten

1. The classical sources for the interpretation of Bohr as a positivist are Mario Bunge's "Strife about Complementarity," *British Journal for the Philosophy of Science* 6 (1955): 1–12,

141–54; and Rom Harré, *Theories and Things* (London: Sheed & Ward, 1961), pp. 93–106, and subsequent writings of both men. Two recent articles which associate Bohr more with the trend of positivism than with any particular school are Bernard d'Espagnat, "The Quantum Theory and Reality," *Scientific American* 241 (November 1979): 158–81; and Stephen Brush's "The Chimerical Cat," a preprint of which he kindly sent me. For the reasons given in this chapter I believe that the differences between Bohr and positivism require stressing more than the similarities.

2. Aage Peterson, "The Philosophy of Niels Bohr," *Bulletin of the Atomic Scientists* 9 (September 1963): 8–14, citation from p. 10. Bohr's views on the primacy of ordinary language are also summarized in T. Bergstin's *Quantum Physics and Ordinary Language* (London: Macmillan, 1972). In spite of the title, no relation with ordinary language analysis is developed.

3. Bohr's views on language are discussed in W. Heisenberg's *Physics and Beyond: Encounters and Conversations* (New York: Harper & Row, 1971), chap. 11. Thus, as Heisenberg reconstructs it Bohr said, while doing the dishes in a mountain cabin, "We use dirty water and dirty dishcloths, and yet we manage to get the plates and glasses clean. In language, too, we have to work with unclear concepts and a form of logic whose scope is restricted in an unknown way, and yet we use it to bring some clarity into our understanding of nature" (p. 137). This aspect of Bohr's thought is developed in detail in John Honner's "The Transcendental Philosophy of Neils Bohr," preprint.

4. Heisenberg, SHQP, interview 3, p. 14.

5. Ludwig Wittgenstein, *Tractatus Logico-Philosophicus*, (1921), trans. D. F. Pears and B. F. McGuinnes (London: Routledge & Kegan Paul, 1961). Any further references will simply cite Wittgenstein's numbering of the sentences.

6. Heinrich Hertz, *The Principles of Mechanics: Presented in a New Form*, trans. D. E. Jones and J. T. Walley (New York: Dover, 1956), citation from p. 1.

7. This is quoted from Anthony Kenny's *Wittgenstein* (London: Allen Lane, Penguin Press, 1973), p. 106.

8. Letter from W. Heisenberg to the author, 12 July 1974.

9. P. F. Strawson, *Individuals: An Essay in Descriptive Metaphysics* (London: Methuen, 1959).

10. For a recent perceptive criticism, see Susan Haack, "Descriptive and Revisionary Metaphysics," *Philosophical Studies* 35 (1979): 361–71.

11. Niels Bohr, *Atomic Physics and Human Knowledge* (New York: Wiley, 1958), pp. 67–68.

12. One of Bohr's clearest historical summaries is given in his address "Unity of Knowledge," in ibid. His way of interpreting the history of physics was fleshed out by his disciple Léon Rosenfeld. See esp. part 1 of Robert S. Cohen and John J. Stachel, eds., *Selected Papers of Léon Rosenfeld* (Dordrecht: Reidel, 1979).

13. W. Heisenberg, *Physics and Philosophy: The Revolution in Modern Science* (New York: Harper, 1958), p. 200.

14. Niels Bohr, *Essays, 1958–1962, on Atomic Physics and Human Knowledge* (New York: Interscience, 1963). Henceforth, this will be referred to as *Essays*. The citation is from p. 2.

15. Ibid., p. 2.

16. Ibid., p. 3.

17. John von Neumann, *Mathematical Foundations of Quantum Mechanics*, trans. R. T. Beyer (Princeton: Princeton University Press, 1955), chap. 6. I have discussed the differences between von Neumann's and Bohr's views on the problem of measurement in quantum mechanics in more detail in my article-length review of Jeffrey Bub's *The Interpretation of Quantum Mechanics*, in *Philosophia* 10 (1981): 89–124.

18. This idea was introduced by H. Everett III in "Relative State Formulation of Quan-

tum Mechanics," *Reviews of Modern Physics* 29 (1957): 454–62. A collection of articles on this topic including de Witt's original article is contained in Bryce S. de Witt and R. Neill Graham, eds. *The Many-Worlds Interpretation of Quantum Mechanics* (Princeton: Princeton University Press, 1973).

19. Eugene Wigner, *Symmetries and Reflections: Scientific Essays* (Bloomington: Indiana University Press, 1967), esp. pp. 171–84.

20. Bohr, *Essays*, pp. 60–61.

21. Bohr's views on this are summarized by L. Rosenfeld in his "The Measuring Process in Quantum Mechanics" in *Selected Papers*, pp. 536–46.

22. Bohr, *Essays*, p. 60.

23. This is the central theme in almost every essay in Bohr's *Atomic Physics.*

24. N. Bohr, "Physical Science and the Problem of Life," in *Atomic Physics*, pp. 94–100. Though the address was given in 1949, the printed version was not completed until 1957. The citations are from p. 98 and 101.

25. Bohr's views on biological explanation influenced his student Max Delbrück. He in turn brought a new methodology to the study of genes. This change is discussed in Gunther Stent, *Molecular Genetics: An Introductory Narrative* (San Francisco: W. H. Freeman, 1971), pp. 23–24.

26. See note 5 in chapter 8.

27. Bohr, *Essays*, p. 14.

28. Such an argument has been developed by Henry J. Folse in "Complementarity and the Description of Experience," *International Philosophical Quarterly* 17 (1977): 378–92, and esp. in his "Kantian Aspects of Complementarity," *Kant-Studien* 69 (1978): 58–66.

29. See chapter 2, note 99.

30. Bohr, SHQP, interview 5, p. 3.

31. It does, for example, in the exchange of letters with Pauli cited by Folse, "Kantian Aspects", pp. 64–65.

32. Heisenberg discussed this in SHQP, interview 3.

33. Albert Einstein, "Religion and Science," *New York Times Magazine*, 9 Nov. 1930, pp. 1–4, reproduced in *Ideas and Opinions.*

34. From *Ideas and Opinions*, p. 266.

35. Citation from Clark's *Einstein*, p. 287.

36. A good survey of Einstein's epistemology, written by someone who worked with him, is Ilse Rosenthal-Schneider, "Presuppositions and Anticipations in Einstein's Physics," in Schlipp, pp. 129–46.

37. A. Einstein, "Physik und Realität," *Journal of the Franklin Institute* 221 (1936): 313–47, with an English trans. by J. Piccard, pp. 349–82. This is reproduced in *Ideas and Opinions*, pp. 290–323. Future references to this paper will cite the page number in *Ideas and Opinions.*

38. Ibid., p. 290.

39. A. Einstein, "The Common Language of Science," in *Ideas and Opinions*, pp. 335–37.

40. See Bertrand Russell, "Language," in E. Nagel and R. Brandt, eds., *Meaning and Knowledge: Systematic Readings in Epistemology* (New York: Harcourt, Brace & World, 1965), pp. 87–93. In his contribution "Remarks on Bertrand Russells' Theory of Knowledge," in P. A. Schlipp, ed. *The Philosophy of Bertrand Russell*, 3d ed. (New York: Harper Torchbooks, 1963 [original 1944], vol. 1, pp. 279–91), Einstein says, "I owe innumerable happy hours to the reading of Russell's works, something which I cannot say of any other contemporary scientific writer, with the exception of Thorstein Veblen" (p. 279). Einstein uses this occasion to defend his own version of scientific realism.

41. *Ideas and Opinions*, p. 292.

42. Ibid.

43. Ibid., p. 294.

44. Ibid.

45. The second half of "Physics and Reality" fits the history of physics into the interpretative schema just summarized. *Ideas and Opinions* also contains other historical sketches and addresses Einstein gave on such scientific figures as Kepler, Newton, Maxwell, Lorentz, and Planck. Though it is difficult to determine how much of *The Evolution of Physics* (3d ed. [New York: Simon & Schuster, 1963]) was due to Einstein and how much to his collaborator, Leopold Infeld, the interpretation of every major development in physics seems consistent with what Einstein has written elsewhere. In the preface they say: "Our intention was rather to sketch in broad outline the attempts of the human mind to find a connection between the world of ideas and the world of phenomena. . . . Facts and theories not reached by this road had to be omitted" (p. xv). Einstein's way of using historical examples as a kind of intuitive archeology is examined in Patrick H. Byrne's "The Significance of Einstein's Use of the History of Science," *Dialectica* 34 (1980): 263–67.

46. Einstein-Infeld, *Evolution*, p. 295.

47. Ibid., p. 9.

48. This treatment of mechanism is a summary of material in *Ideas and Opinions*, pp. 295–303.

49. *Ideas and Opinions*, p. 302.

50. *Ideas and Opinions*, p. 303–7, Einstein-Infeld, *Evolution*, pp. 125–63.

51. *Ideas and Opinions*, pp. 306–7.

52. Ibid.,pp. 315–16.

53. Ibid., pp. 322–23.

54. Einstein's stress on the need for an abandonment of quantum theory in its present form and a consequent radical redevelopment on a new foundation was particularly stressed by Pais, "Einstein and the Quantum Theory," *Reviews of Modern Physics* 51 (1979): 863–914, esp. pp. 907–8. Pais's evaluation was based on frequent discussions he had with Einstein on this point.

55. *Ideas and Opinions*, pp. 322–23.

56. Cited from Pais, "Einstein and the Quantum Theory," p. 910.

57. A. Einstein, "Considerations concerning the fundamentals of Theoretical Physics," *Science* 91 (1940): 487-92, reproduced in *Ideas and Opinions*, pp. 323–35, citation from pp. 334–35.

58. A. Einstein, "Notes on the Origin of the General Theory of Relativity," in *Ideas and Opinions*, pp. 285–90, citation from p. 289.

59. This is from his 1933 Herbert Spencer lecture, "On the Method of Theoretical Physics," in *Ideas and Opinions*, pp. 270–76, citation from p. 274.

60. *Ideas and Opinions*, p. 292.

61. "Autobiographical Notes," p. 5.

62. From Clark's *Einstein*, p. 35.

63. "Autobiographical Notes," p. 5.

64. From *Ideas and Opinions*, p. 262.

65. From *Ideas and Opinions*, p. 39.

66. A. Einstein, "Science and Religion," in *Ideas and Opinions*, pp. 41–49, citation from p. 46.

67. Solovine Correspondence, pp. 114–16.

68. Cited from Clark's *Einstein*, p. 502.

69. A detailed exposition of Spinoza's doctrine on this point may be found in Harry Wolfson, *The Philosophy of Spinoza* (New York: Meridian, 1958), vol. 2, chap. xvi.

Index